Advances in Spatial Science

The Regional Science Series

More information about this series at http://www.springer.com/series/3302

Randall Jackson • Peter Schaeffer
Editors

Regional Research Frontiers - Vol. 2

Methodological Advances, Regional Systems Modeling and Open Sciences

Springer

Editors
Randall Jackson
Regional Research Institute
West Virginia University
Morgantown
West Virginia, USA

Peter Schaeffer
Division of Resource Economics
and Management
Faculty Research Associate
Regional Research Institute
West Virginia University
Morgantown, WV, USA

ISSN 1430-9602 ISSN 2197-9375 (electronic)
Advances in Spatial Science
ISBN 978-3-319-50589-3 ISBN 978-3-319-50590-9 (eBook)
DOI 10.1007/978-3-319-50590-9

Library of Congress Control Number: 2017936672

Printed on acid-free paper

This Springer imprint is published by Springer Nature
The registered company is Springer International Publishing AG
The registered company address is: Gewerbestrasse 11, 6330 Cham, Switzerland

Preface

The idea for this book emerged as we prepared the celebration of the 50th anniversary of the Regional Research Institute (RRI) at West Virginia University in 2016. The Institute was founded in 1965, and the personalities who helped shape it include founding director William Miernyk, Andrew Isserman, Luc Anselin, Scott Loveridge, and Randall Jackson. The Institute reflected the research focus and personalities of each of these directors, flavored by the diversity of personalities and scholarship of others with RRI ties. Yet throughout its history, the primary mission remained: engaging in and promoting regional economic development research, with a special emphasis on lagging and distressed regions. RRI scholars have come from economics, geography, agricultural and resource economics, urban and regional planning, history, law, engineering, recreation and tourism studies, extension, etc. Over the half century of RRI's existence, regional research has grown and developed dramatically, with members of the Institute contributing to scholarship and leadership in the profession. Reflecting on the history of the RRI made us wonder about the next 50 years of regional research, so we decided to ask colleagues in our field to share their thoughts about issues, theories, and methods that would shape and define future regional research directions. Many responded to our call for contributions, and in the end we accepted 37 chapters, covering many aspects of regional research. Although the chapters are diverse, several share common ideas and interests, so we have grouped them into seven parts. As with most groupings, of course, there are chapters whose content would have been appropriate in more than one part.

The large number of contributions resulted in a much greater number of pages than planned, but their quality made us reluctant to cut some or to significantly shorten them. We are, therefore, grateful to Johannes Glaeser, Associate Editor for Economics and Political Science at Springer, and to the Advances of Spatial Sciences series editors, for suggesting that we prepare two volumes instead of only one, as initially proposed. We also thank Johannes Glaeser for his advice and support throughout the process of preparing the two volumes. Volume 1 carries the subtitle "Innovations, Regional Growth and Migration" and contains 20 chapters in its four parts. In addition to the topics named in the subtitle, Volume 1 also contains three chapters on disasters, resilience, and sustainability, topics that are of growing interest to scholars, policy makers, and agency and program administrators alike. The subtitle of Volume 2 is "Methodological Advances, Regional Systems Modeling and Open Sciences." Its 17 chapters are organized into the three parts named in the volume's subtitle. The two volumes are roughly equal in length.

The chapters reflect many of the reasons why research methods and questions change over time. A major reason for recent developments in regional research is the digital revolution, which made vastly increased computational capacities widely available. This made possible methodological advances, such as spatial econometrics or geographic information systems (GIS), but perhaps more importantly, it changed fundamentally the way empirical modeling is conducted. Furthermore, it has become possible to integrate different tools, such as spatial econometrics and GIS, and generate graphical displays of complex relationships that enrich our analyses and deepen our understanding of the processes that underlie empirical patterns. Overall, the impact of technological changes on regional research has been pervasive and, judging by the contributions to this volume, will likely continue to be so, and this can be seen in most book parts. In *Modeling Regional Systems*, the chapters' authors rely on recently developed methodological tools and approaches to explore what future research directions could be. In the part *Disasters and Resilience*, Yasuhide Okuyama proposes a future modeling system that would be unthinkable without modern computational tools. All contributions in the part *Spatial Analysis* depend heavily on computational spatial analytical tools, including visualization (e.g., Trevor Harris' contribution on exploratory spatial data analysis). Particularly interesting in this context is the part *Open Source and Open Science*, because it is dealing with aspects of the computational revolution and the Internet that are only now starting to become a major force in our fields, and the collective development and integration of software proposed by Jackson, Rey, and Járosi is still in its infancy.

The evolution of technologies not only drives much of societal change but also has changed how we look at economic growth. While early models of economic growth focused on the capital-labor ratio and treated technology as an exogenous variable, current research in economic growth includes technology as an endogenous variable and stresses entrepreneurship. It is, therefore, not surprising to see an entire part focused on technology, innovation, and entrepreneurship. This part confronts gender issues explicitly in the chapter by Weiler and Conroy, further reflecting changing social attitudes. Gender issues are also addressed in the *Regional Growth, Regional Forecasts, and Policy* part. As Chalmers and Schwarm note, gender is still a relatively neglected topic in regional research, but social trends and forces will likely increase the attention it receives in the future.

The digital revolution that made mobile phones ubiquitous has also had another important effect, namely the emergence relatively recently of "big data" (e.g., the chapters by Newbold and Brown, and Harris). Even more importantly, vastly improved communication technologies and faster means of transportation are changing the nature of agglomeration. Timothy Wojan reminds us that Alfred Marshall anticipated some of these changes more than a century ago, a remarkable feat of foresight. Because of improved communication technologies, the gap between geographic and social distance is likely to widen in the future, particularly among the highly skilled. Those of us working in research settings at universities or institutes are already experiencing this phenomenon, as it has become common to collaborate with distant colleagues, a sharp contrast to the case until the late twentieth century. It seems certain that the impact of digital technologies on traditional views of geographical space as separation and differentiation will raise new regional research questions. Woodward provides a complement to Wojan's chapter when he speculates about the effects of the interplay of agglomeration and automatization, which is yet another example of the pervasive influence of technology on the future of spatial organization of our societies.

Wojan is not the only one looking to the past to glance into the future. David Bieri studies neglected contributions in regional monetary economics of such foundational scholars of regional research as Lösch and Isard. His chapter presents a genealogy of regional monetary thinking and uses it to make a strong case for renewed attention over the next 50 years to this neglected branch of our intellectual family tree.

While most regional scholars are well aware of the impacts of the digital revolution, there is less awareness of the impacts of an ongoing demographic revolution. This may be because the revolution is far advanced in the economically most successful countries, mostly the members of the Organisation for Economic Co-operation and Development (OECD). But while England became the first country to be more urban than nonurban in the mid-nineteenth century, the world as a whole has reached this threshold less than 10 years ago. Indeed, urbanization in the southern hemisphere is proceeding at a very rapid pace that poses significant policy challenges in the affected nations. As part of industrialization and urbanization, the world is also experiencing a dramatic decline in effective fertility, with the number of births per female of child-bearing age declining. Since longevity is increasing, this is resulting in demographic structures unlike any in the past. This phenomenon is most advanced and dramatic in places such as Germany, Japan, and most recently China—where government policies contributed mightily to demographic restructuring—and challenges the future of public social safety programs, particularly provisions for the financial security of the elderly and their healthcare. In such cases, immigration may be seen as a way to slow the transition from a predominantly young in the past to a much older population. Franklin and Plane address issues related to this unprecedented demographic shift.

Migration, domestic and international, is also of growing importance because of the disruptions caused by industrialization in many countries. The "land flight" that once worried today's industrial powers is now occurring in the southern hemisphere. Migration is also fueled by political change in the aftermath of the end of colonialization. The new nations that emerged were often formed without regard for historic societies and traditions, and tensions that had been held in check have sometimes broken out in war between neighboring countries or civil war. As a result, the world as a whole has seen an increase in internally displaced persons as well as refugees who had to leave their home countries. In an overview of directions in migration research, Schaeffer, therefore, argues for more work on migrations that are rarely completely voluntary because traditional models have been developed primarily for voluntary migrations.

Demographic shifts are also driving reformulations and advances in *Regional Systems Models*, as evidenced by new directions in household modeling within the chapter on household heterogeneity by Hewings, Kratena, and Temurshoev, who touch on these and enumerate a comprehensive research agenda in the context of dynamic interindustry modeling, and Allen and his group identify pressing challenges and high potential areas for development within computable general equilibrium models. Varga's chapter contributes to this part's topic and to technological change, as his Geographic Macro and Regional Impact Modeling (GMR) provides explicit mechanisms for capturing the impacts of innovation and technology.

The chapters in these volumes reflect the changing world that we live in. While some new directions in regional research are coming about because new technologies allow us to ask questions, particularly empirical questions that once were beyond the reach of our capabilities, others are thrust upon us by political, economic, social, demographic, and environmental events. Sometimes several of these events combine to effect change. A primary task of a policy science is to provide guidelines for the design of measures to address problems related to change. So far, regional researchers seem to have been most successful in making progress toward completing this task in dealing with environmental disasters, addressed in the *Disasters and Resilience* part. Rose leverages decades of research in regional economic resilience to lay the foundation for this part.

These chapters will certainly fall short of anticipating all future developments in regional research, and readers far enough into the future will undoubtedly be able to identify oversights and mistaken judgements. After all, Kulkarni and Stough's chapter finds "sleeping beauties" in regional research that were not immediately recognized, but sometimes required long gestation periods before becoming recognized parts of the core knowledge in our field, and Wojan and Bieri also point to and build upon contributions that have long been neglected. If it is possible to overlook existing research, then it is even more likely that we are failing to anticipate, or to correctly anticipate, future developments. Nonetheless, it is our hope that a volume such as this will serve the profession by informing the always ongoing discussion about the important questions that should be addressed by members of our research community, by identifying regional research frontiers, and by helping to shape the research agenda for young scholars whose work will define the next 50 years of regional research.

Morgantown, WV

Randall Jackson
Peter Schaeffer

Contents

Editors and Contributors

About the Editors

Randall Jackson is professor, Department of Geology and Geography, West Virginia University (WVU), and Director of the Regional Research Institute. His primary research interests are regional industrial systems modeling; energy, environmental, and economic systems interactions; and regional economic development. He is an adjunct professor in WVU's Department of Economics and Division of Resource Management, and in Geography at the Ohio State University (OSU). Previous faculty positions were at OSU and Northern Illinois University. Dr. Jackson earned his PhD in geography and regional science from the University of Illinois at Urbana-Champaign in 1983.

Peter Schaeffer is professor, Division of Resource Economics and Management, West Virginia University (WVU). His primary research interests are regional economic policy, international labor migration, job mobility, natural resource management, and historic preservation. He is a faculty research associate in WVU's Regional Research Institute and adjunct professor in the Department of Economics. Previous faculty positions were at the Universities of Colorado–Denver, Illinois at Urbana–Champaign, and one year as visiting professor at the Swiss Federal Institute of Technology–Zurich. Dr. Schaeffer earned the Ph.D. in economics from the University of Southern California in 1981.

Contributors

Grant J. Allan Fraser of Allander Institute and Department of Economics, Strathclyde Business School, University of Strathclyde, Glasgow, UK

Yudhie Andriyana Statistics Department, Universitas Padjadjaran, Kabupaten Sumedang, Indonesia

Daniel Arribas-Bel Department of Geography and Planning, University of Liverpool, Liverpool, UK

Zhenhua Chen Austin E. Knowlton School of Architecture, The Ohio State University, Columbus, OH, USA

Graham Clarke School of Geography, University of Leeds, Leeds, UK

Gary Cornwall Department of Economics, Carl H. Lindner College of Business, University of Cincinnati, Cincinnati, OH, USA

Sandy Dall'erba Department of Agricultural and Consumer Economics and Regional Economics Applications Laboratory, University of Illinois at Urbana-Champaign, Champaign, IL, USA

Thomas de Graaff Department of Spatial Economics, Vrije Universiteit Amsterdam, Amsterdam, The Netherlands

Erik Dietzenbacher Professor of Economics, University of Groningen, The Netherlands

Fang Fang Graduate Interdisciplinary Program in Statistics and Regional Economics and Spatial Modeling laboratory, University of Arizona, Tucson, AZ, USA

Henk Folmer Faculty of Spatial Science, University of Groningen, Groningen, The Netherlands

Trevor M. Harris Department of Geology and Geography, West Virginia University, Morgantown, WV, USA

Kingsley E. Haynes Schar School of Policy and Government, George Mason University, Arlington, VA, USA

Kristinn Hermannsson School of Educaction, University of Glasgow, Glasgow, UK

Geoffrey J.D. Hewings Regional Economics Applications Laboratory, University of Illinois, Urbana, IL, USA

Randall Jackson Regional Research Institute, West Virginia University, Morgantown, WV, USA

Amir Jamali Civil Engineering Department, Montana State University, Bozeman, MT, USA

Péter Járosi West Virginia University, Morgantown, WV, USA

I Gede Nyoman Mindra Jaya Statistics Department, Universitas Padjadjaran, Kabupaten Sumedang, Indonesia

Dongwoo Kang Korea Labor Institute, Sejong, South Korea

Changjoo Kim Department of Economics, Carl H. Lindner College of Business, University of Cincinnati, Cincinnati, OH, USA

Kijin Kim Regional Economics Applications Laboratory, University of Illinois, Urbana, IL, USA

Asian Development Bank, Manila, Philippines

Tae-Jeong Kim Bank of Korea, Seoul, Korea

J. Kim Swales Fraser of Allander Institute and Department of Economics, Strathclyde Business School, University of Strathclyde, Glasgow, UK

Kara Kockelman Department of Civil, Architectural, and Environmental Engineering, University of Texas at Austin, Austin, TX, USA

Kurt Kratena Centre of Economic Scenario Analysis and Research, Department of Economics, Loyola University Andalucía, Spain

Farah Kristiani Mathematics Department, Parahyangan Catholic University, Kota Bandung, Indonesia

Donald J. Lacombe Regional Research Institute, West Virginia University, Morgantown, WV, USA

Michael L. Lahr EJB School of Planning and Public Policy, Rutgers University, New Brunswick, NJ, USA

Patrizio Lecca European Commission, DG Joint Research Centre, Seville, Spain

Peter G. McGregor Fraser of Allander Institute and Department of Economics, Strathclyde Business School, University of Strathclyde, Glasgow, UK

Stuart G. McIntyre Fraser of Allander Institute and Department of Economics, Strathclyde Business School, University of Strathclyde, Glasgow, UK

Alan T. Murray Department of Geography, University of California at Santa Barbara, Santa Barbara, CA, USA

Olivier Parent Department of Geography, University of Cincinnati, Cincinnati, OH, USA

Seryoung Park Bank of Korea, Seoul, Korea

Sergio Rey Arizona State University, Phoenix, AZ, USA

Budi Nurani Ruchjana Mathematics Department, Universitas Padjadjaran, Kabupaten Sumedang, Indonesia

Kim Swales Fraser of Allander Institute, University of Strathclyde, Glasgow, UK

Umed Temursho Centre of Economic Scenario Analysis and Research, Department of Economics, Loyola University Andalucía, Spain

Daoqin Tong School of Geography and Development, University of Arizona, Tucson, AZ, USA

Eveline van Leeuwen Vrije Universiteit Amsterdam, Amsterdam, The Netherlands

Attila Varga Faculty of Business and Economics, University of Pécs, Hungary

Yiyi Wang Civil Engineering Department, Montana State University, Bozeman, MT, USA

Sang Gyoo Yoon Bank of Korea, Seoul, Korea

Part I
Regional Systems Modeling

Chapter 1
Dynamic Econometric Input-Output Modeling: New Perspectives

Kurt Kratena and Umed Temursho

1.1 Introduction

One of the first research strategies based on input-output (IO) modelling that had as an objective a fully fledged macro-econometric IO model is the 'Cambridge Growth Project' (Cambridge DAE 1962). The focus of extending the IO model towards a full macroeconomic model was on the endogenization of parts of final demand (usually exogenous in the static IO model) and the modelling of demand components depending on (relative) prices. Another milestone of this work on the Cambridge Growth Project was the macroeconomic multisectoral model of the U.K. economy (Barker 1976; Barker and Peterson 1987). Almost at the same time, U.S. based research group known as INFORUM (Inter-industry Forecasting at the University of Maryland) developed a macroeconomic closed IO model, which is first described in Almon et al. (1974). Since then, this model family has spread worldwide and developed into an international model by linking similar national models via bilateral trade matrices (Almon 1991; Nyhus 1991). Both the Cambridge Multisectoral Dynamic Model of the British economy (MDM) as well as the INFORUM models incorporate econometric specifications that take into account economic theory but cannot be directly derived from maximization or minimization calculus of representative agents. At the regional level, different types of econometric IO models have been developed by Geoffrey Hewings and his team at the Regional Economics Applications Laboratory (REAL, University of Illinois at Urbana-Champaign) based on the Washington Projection and Simulation Model (Conway 1990). Another important example of a recently developed econometric IO model is the (fully interlinked) Global Interindustry Forecasting System (GIN-

K. Kratena (✉) • U. Temursho
Centre of Economic Scenario Analysis and Research, Department of Economics, Loyola University Andalucía, Spain
e-mail: kurt.kratena@wifo.ac.at

R. Jackson, P. Schaeffer (eds.), *Regional Research Frontiers - Vol. 2*, Advances in Spatial Science, DOI 10.1007/978-3-319-50590-9_1

FORS) model (Lutz et al. 2005), developed by Bernd Meyer and his team at the Institute of Economic Structures Research (GWS, Gesellschaft für Wirtschaftliche Strukturforschung).

The purpose of this paper is to bring to the attention of practitioners some, in our view, fruitful future directions for econometric IO modeling. Our suggestions on improving this branch of economic modeling comes from our observations that theoretical and empirical economic research of the last decades has developed completely new approaches that have not all found their representation in the econometric IO modeling strain. In this respect, we highlight the relevant developments in three subfields or schools of economics: neoclassical macroeconomics, agricultural economics, and post-Keynesian economics. Macroeconomics-related improvements have to do with an improved modeling of private consumption, production and trade, as briefly outlined below and discussed in some detail in the next two sections. Theoretical and empirical research in agricultural economics on observed data calibration seems to be a promising new addition to the econometric IO modeling. Another very important recent development in macroeconomic modeling includes the comprehensive integration of all the flows and stocks of the economy in the spirit of the post-Keynesian school of economic thought. These last two issues and their relevance for econometric IO modeling are discussed briefly in Sect. 1.4.

It is not difficult to realize that private consumption modeling should not be simplistic, because it constitutes the largest component (over 50%; close to 70% in the US) of aggregate demand (or national income) in virtually all individual economies around the world. Models based on the social accounting matrices (SAM) structure using average coefficients still dominate the modeling of the link between consumption and household income generation. That holds true for econometric IO as well as for computable general equilibrium (CGE) modeling. In both modeling families, also the concept of the representative consumer dominates and reactions of consumption of single goods to price and income changes follow simple linear approaches. In Sect. 1.2 we show how this part of an econometric IO model can be improved by introducing approaches that explicitly deal with household wealth, durables and nondurables as well as different household characteristics that have an influence at the level of consumption by commodity. The approaches presented all take into account the dynamics of structural change in society as well as in the economy.

In production theory, the important issues are imperfect competition and technical change. It is well known that both phenomena equally affect the wedge between costs and prices and, therefore, are rather difficult to disentangle. The IO model structure is fully compatible with flexible functional forms like the transcendental logarithmic (or translog) function (Jorgenson et al. 2013), which allow for a generic form of introducing different sources of technical change (i.e., total factor productivity (TFP), factor bias, embodied or induced). In Sect. 1.3 we discuss these generic forms and compare them with a more explicit treatment of technical change in an IO framework.

Another important issue, especially in the context of multi-regional modeling, is trade. As is well known, estimation of trade flows within the standard multiregional

IO framework is a challenging task mainly due to unavailability or incompleteness of the relevant data and the fact that interregional inter-sectoral flows can be quite volatile over time. Thus, in general, it is to be expected that trade flows may be one of the most important sources of uncertainty in multiregional IO modeling. It should be noted that within the traditional multiregional IO modeling, surprisingly very little attention, to the best of our knowledge, has been given to the full characterization of the IO price system. For example, multicountry IO price systems that explicitly model (changes in) exchange rates, which is a crucial factor for the analysis of open economies, seem to be largely lacking. In this respect, econometric IO modelling has gone much further, since the framework readily allows to incorporate all the real complexities of the pricing system of an economy. As an example, while prices per sector (or product) in the IO price model are identical for all intermediate and final users, in econometric IO models, prices are user-specific due to their proper account of margins, taxes and subsidies, and import shares that are all allowed to be different for each user (see e.g., Kratena et al. 2013). Trade flows of substitutes to domestic goods, as well as in terms of the country of origin and destination in most models, simply depend on the level of goods demand and relative prices. The standard workhorse in CGE modeling is still the Armington function (Armington 1969), which is calibrated to elasticity values found in two or three seminal papers. In this respect, we emphasize the necessity of new empirical work on the magnitude of Armington elasticities, and call for developing other alternatives to Armington approaches of trade modeling in IO models with clear links to the production side (for the first steps in this direction, see Kratena et al. 2013).

Section 1.4 concludes and summarizes the discussed perspectives for future econometric IO modeling.

1.2 Private Consumption, Income and Socio-economic Characteristics of Households

In this section we discuss the complex relationship between consumption and income that has been a major field of macroeconomic research during the last decades (for an overview of the debate, see e.g., Meghir and Pistaferri 2010). The SAM multiplier model as well as the standard CGE model both use a static link between income and consumption. The standard formulation of consumption in the CGE model with a static consumption function and a linear expenditure system for splitting up the consumption vector does *not* take into account the huge body of literature on macroeconomic consumption functions of the last decades. A line of development reaches from the Keynesian consumption function used in Miyazawa (1976) to the model of permanent income. As empirical research has discovered some puzzles about the dependence of consumption on income dynamics (Hall 1978) inconsistent with the predictions of the permanent income hypothesis, the 'buffer-stock model' of consumption emerged. Carroll (1997) has laid down the basis of the buffer-stock model, starting from the empirical puzzles that the permanent income hypothesis has not been able to resolve. One of the main starting points for Carroll in developing this model was the desired characteristic of a *concave* consumption function, due to a non-constant marginal propensity of consumption (MPC) along the process of income growth and wealth accumulation. This idea dates back to the work of Keynes himself, as Carroll and Kimball (1996) have shown. In general, the MPC should increase with higher income uncertainty (the main innovation of the buffer-stock model) and decrease with higher levels of wealth. Several empirical tests of the buffer-stock model have been carried out. Japelli et al. (2008) and Luengo-Prado and Sorensen (2004) are two prominent examples. The two main issues in this empirical testing were, in general, the income sensitivity of consumption and the empirical proof of a non-constant MPC. As far as the first point is concerned, the difference between permanent and transitory income shocks by the founders of the Permanent Income Hypothesis has been crucial. The MPC out of transitory income should only be significantly different from zero for households with binding liquidity constraints. This can be part of the households—in that case household heterogeneity needs to be introduced—or all households in situations of high liquidity demand, e.g., for debt deleveraging.

Whereas in the original version of the buffer-stock model income uncertainty was the main saving motive, in a new version households save for the purchase of durables, as described in Luengo-Prado (2006). Consumers maximize the present discounted value of expected utility from consumption of nondurable commodity and from the service provided by the stocks of durable commodity, subject to the budget and collateralized constraints. The consideration of the collateralized constraint is formalized in a down payment requirement parameter, which represents the fraction of durables that a household is not allowed to finance.

$$\max_{(C_t,K_t)} V = E_0 \left\{ \sum_{t=0}^{\infty} \beta^t U(C_t, K_t) \right\} \tag{1.1}$$

Specifying a constant relative risk aversion (CRRA) utility function yields:

$$U(C_t, K_t) = \frac{C_t^{1-\rho}}{1-\rho} + \varphi \frac{K_t^{1-\rho}}{1-\rho}, \tag{1.2}$$

where φ is a preference parameter and $\rho > 0$ implies risk aversion of consumers.

The budget constraint in this model without adjustment costs for the durables stock is given by the definition of assets, A_t:

$$A_t = (1 + r)(1 - t_r) A_{t-1} + YD_t - C_t - (K_t - (1 - \delta) K_{t-1}) \tag{1.3}$$

The sum of C_t and $(K_t - (1 - \delta)K_{t-1})$ represents total consumption, i.e., the sum of nondurable and durable expenditure (with depreciation rate of the durable stock, δ). The gross profit income rA_{t-1} is taxed at the rate t_r. These taxes, therefore, reduce the flow of net lending of households that accumulates to future assets. Disposable household income that excludes profit income, YD_t, is given as the balance of net wages $(1-t_S-t_Y)w_tH_t$ and net operating surplus accruing to households $(1-t_Y)\Pi_{h,t}$, plus unemployment benefits transfers with UN_t as unemployed persons and br as the benefit replacement rate, measured in terms of the after tax wage rate, plus other transfers Tr_t:

$$YD_t = (1 - t_S - t_Y)\, w_t H_t + (1 - t_Y)\, \Pi_{h,t} + brw_t\, (1 - t_S - t_Y)\, UN_t + Tr_t. \quad (1.4)$$

The following taxes are charged on household income: social security contributions with tax rate t_S, which can be further decomposed into an employee and an employer's tax rate (t_{wL} and t_L) and income taxes with tax rate t_Y. The wage rate w_t is the wage per hour and H_t are total hours demanded by firms. Wage bargaining between firms and unions takes place over the employee's gross wage, i.e.,$w_t\,(1-t_L)$.

Financial assets of households are built up by saving after durable purchasing has been financed, and the constraint for lending is:

$$A_t + (1 - \theta)\, K_t \geq 0. \quad (1.5)$$

This term represents voluntary equity holding, $Q_{t+1} = A_t + (1-\theta)K_t$, as the equivalent of the other part of the durable stock (θK_t) needs to be held as equity. The consideration of the collateralized constraint is operationalized in a down payment requirement parameter θ, which represents the fraction of durables purchases that a household is not allowed to finance. One main variable in the buffer stock-model of consumption is 'cash on hand', X_t, measuring the household's total resources:

$$X_t = (1 + r_t)\,(1 - t_r)\, A_{t-1} + (1 - \delta)\, K_{t-1} + YD_t \quad (1.6)$$

Total consumption is then defined as:

$$CP_t = C_t + K_t - (1 - \delta)\, K_{t-1} = r_t\,(1 - t_r)\, A_{t-1} + YD_t - (A_{t-1} - A_t)\,, \quad (1.7)$$

where the last term represents net lending, so total consumption is the sum of durable and nondurable consumption, or the difference between disposable income and net lending.

The model solution works via deriving the first-order conditions and yields an intra-temporal equilibrium relationship between C_t and K_t as one solution of the model, when the constraint is not binding. For all other cases, where the collateral constraint is binding, Luengo-Prado (2006) has shown that this relationship can be used to derive policy functions for C_t and K_t and formulate both as functions of the difference between cash on hand and the equity that the consumer wants to hold in the next period.

This model describes a clear *alternative* to the static model of consumption in the standard CGE model and introduces *dynamics* into the model. It allows for deriving demand for different types of durables and total non-durables as the main macroeconomic consumption functions. As an empirical application of this model, the non-linear functions for durable and nondurable consumption, depending on wealth (in this case the durable stock), cash on hand, and the down payment (θ) have been estimated for 14 EU countries[1] for which the data situation covers the main variables of the model. The non-linearity of the functions should deal with: (i) non-constant MPC (in this case with respect to cash on hand), (ii) smoothing of nondurable consumption with respect to shocks in savings requirements for the down payment. Both characteristics yield estimation results that can be, in a second step, built into an econometric IO model of the EU-27 (for details, see Kratena and Sommer 2014) that incorporates five different groups of household income (quintiles). For this purpose, the estimation results are used to calibrate the model at the level of the five quintiles of income, which are characterized by different values for the durable stocks per household. Therefore, the model contains growth rates for $C_{dur,t}$ and $C_{nondur,t}$ for each quintile (q). Once the full model is set up with the integrated consumption block, the property of 'excess sensitivity' can be tested. Excess sensitivity describes the empirical fact that the growth rate of consumption (partly) reacts to the lagged growth rate of disposable (or labour) income. This issue has been raised by Hall (1978) and confronted the Permanent Income Hypothesis with contradictory empirical findings.

The full econometric IO model (Kratena and Sommer 2014) is run until 2050, so that endogenous disposable household income is generated. Then excess sensitivity is tested by setting up the regressions that Hall (1978) proposed to test the influence of transitory income shocks on consumption. That means regressing the growth rates for $C_{dur,t}$ and $C_{nondur,t}$ for each quintile (q) on lagged disposable income growth (without profit income) for each quintile, generated by the full model. Profit income is not included, because it is endogenous and depends on equity built up, which in turn is the result of inter-temporal optimization. Luengo-Prado (2006) also carries out excess sensitivity tests with her calibrated model, based on U.S. household survey data and confronts similar results with U.S. stylized macroeconomic facts. The excess sensitivity coefficients, i.e., the MPC with respect to lagged income change, found by Luengo-Prado (2006) are 0.16 (nondurables) and 0.26 (durables). The results from the econometric IO model solution until 2050 (Table 1.1) clearly reveal that for the 5th and partly for the 4th quintile, durable and nondurable consumption do not statistically significantly depend on transitory income shocks. The MPC is higher in general for lower income households and for situations with higher liquidity constraints (higher θ). The 'low θ scenario' corresponds to a financial regime, where the relationship debt to durable stock does not significantly decrease, i.e., no major debt deleveraging by households occurs. The 'high θ scenario' corresponds to debt deleveraging so that the relationship debt to durable stock in the long-run decreases to its values before 2002, i.e., before the main expansion of household debt began.

[1]These countries include Austria, Belgium, Cyprus, Czech Republic, Denmark, Finland, France, Germany, Italy, Lithuania, Poland, Portugal, Romania, and Slovakia.

Table 1.1 Excess sensitivity of consumption with respect to lagged disposable income (without profit income), EU 14 (2005–2050)

	1st quintile	2nd quintile	3rd quintile	4th quintile	5th quintile
	Sensitivity, low θ				
dlog(C_{dur})	0.45***	0.38***	0.30**	0.21	0.14
	(0.15)	(0.16)	(0.16)	(0.16)	(0.16)
dlog(C_{nondur})	0.94***	0.76***	0.58***	0.38***	−0.03
	(0.41)	(0.20)	(0.15)	(0.12)	(0.13)
	Sensitivity, high θ				
dlog(C_{dur})	0.44***	0.40**	0.33***	0.26**	0.20
	(0.13)	(0.14)	(0.14)	(0.14)	(0.14)
dlog(C_{nondur})	1.02***	0.86***	0.69***	0.49***	0.09
	(0.37)	(0.18)	(0.14)	(0.12)	(0.09)

Note: ** and *** indicate significance at the 5%, and 1% level, respectively

This specification of the buffer-stock model that has already been built into a dynamic econometric IO model indirectly yields the following properties that make it significantly different from the standard consumption model (SAM based and linear expenditure system) applied in econometric IO and CGE modeling: (i) a non-constant MPC, (ii) a concave consumption function across household income groups, and (iii) different sensitivity of different household types in their consumption reaction on transitory income changes. This version of the buffer-stock model is data-intensive and introduces cross-section data (i.e., household heterogeneity) that are combined with time series estimation results.

A different way of ending up with a buffer-stock model that exhibits the desired properties (non-constant MPC, concave consumption function, different sensitivity of different household types), is a direct estimation of consumption functions, incorporating income, wealth and debt for different household groups. Early examples of these empirical explorations into the validity of the buffer-stock model are Japelli et al. (2008) and Luengo-Prado and Sorensen (2004). Recently, models that take into account household heterogeneity with respect to the impacts of debt deleveraging and wealth shocks have gained ground. Mian et al. (2013) show that poorer households and households with a higher debt burden react more to wealth shocks in their consumption than other households. Their specification also takes into account concavity in the consumption function with respect to the level of wealth. Eggertson and Krugman (2012) develop a theoretical model with two different household types (savers and debtors), where debt deleveraging has strong macroeconomic impacts as it reduces consumption of the debtors, which depends more on transitory income. The results presented in Table 1.1 and the findings of Mian et al. (2013), as well as of Eggertson and Krugman (2012), strongly encourage going into the direction of a model with different household groups, where the consumption of richer households is simply determined by a constant growth rate, whereas for the other groups of households, income, wealth and debt limits play a major role.

As far as the demand for nondurables at the commodity level is concerned, the alternative to the linear expenditure system could be a flexible functional form, like the widely used Almost Ideal Demand System (AIDS), starting from the cost function for $C(u, p_i)$, describing the expenditure function (for C) as a function of a given level of utility u and prices of consumer goods, p_i (see Deaton and Muellbauer 1980). The AIDS model is represented by the well-known budget share equations for the i nondurable goods in each period:

$$w_i = \alpha_i + \sum_j \gamma_{ij} \log p_j + \beta_i \log\left(\frac{C}{P}\right) \quad \text{for } i = 1 \ldots n, \tag{1.8}$$

with price index, P_t, defined by $\log P_t = \alpha_0 + \sum_i \alpha_i \log p_{it} + 0.5 \sum_i \sum_j \gamma_{ij} \log p_{it}$, $\log p_{jt}$ often approached by the Stone price index, $\log P_t{}^* = \sum_k w_{it} \log p_{it}$.

This model has been estimated by combining time series (panel data) information from 1995 to 2012 for 27 EU countries with individual data from the 2004/2006 household surveys for 6 EU countries (namely, Austria, France, Italy, Slovakia, Spain, and the UK). This cross section model introduces heterogeneity of households at the level of commodities. Several socio-economic characteristics of households can be introduced as additional variables, complementing income and prices. These variables include age group dummies for the household head, dummies if the household head is retired, unemployed, and is the owner of the house. Further, household size and population density are taken into account.

The expressions for the expenditure elasticity (η_i) and the compensated price elasticity (ε_{ij}^C) within the AIDS model for the quantity of each consumption category C_i can be written as (the details of these derivations can be found in, e.g., Green and Alston 1990)[2]:

$$\eta_i = \frac{\partial \log C_i}{\partial \log C} = \frac{\beta_i}{w_i} + 1 \tag{1.9}$$

$$\varepsilon_{ij}^C = \frac{\partial \log C_i}{\partial \log p_j} = \frac{\gamma_{ij} - \beta_i w_j}{w_i} - \delta_{ij} + \eta_i w_j, \tag{1.10}$$

where δ_{ij} is the Kronecker delta with $\delta_{ij} = 0$ for $i \neq j$ and $\delta_{ij} = 1$ for $i = j$.

[2]The derivation of the budget share w_i with respect to log (C) and log (p_j) is given by β_i and $\gamma_{ij} - \beta_i$ (log(P)), respectively. Applying Shephard's Lemma and using the Stone price approximation, the elasticity formulae can then be derived.

Table 1.2 Price and expenditure elasticity of nondurable consumption, EU 27 (1995–2012)

		Expenditure elasticity	
Nondurable consumption	Own price elasticity	Time series	Cross section
Food	−0.14	0.85	0.61
Clothing	−0.64	1.04	1.28
Furniture/equipment	−1.06	1.11	1.46
Health	−0.83	0.98	1.20
Communication	−0.89	0.96	0.68
Recreation/accommodation	−0.50	1.08	1.27
Financial Services	−0.94	1.33	1.00
Other	−0.68	1.09	1.00

As can be observed from (1.10), the parameter of the expenditure elasticity (β_i) also enters the formula for the compensated price elasticity, so that the two elasticities are tied together. Estimating both the time series and the cross section model, therefore, and combining them will also change the compensated price elasticity. This is not taken into account in the results presented in Table 1.2. These results just show the difference in expenditure elasticity values from the time series vs. the cross section model. It clearly comes out that heterogeneity in expenditure elasticity is higher in the case of the cross section model. The most important result is that introducing household heterogeneity not only introduces additional socio-economic variables that also influence behavior, besides income and prices, but that it also changes the reaction of households to income and prices and, therefore, aggregate results.

The approach presented can still be seen as sub-optimal, as a combination of time series and cross section estimation is needed, and no direct use of household group panel data has been used for estimation. This latter approach has been applied in Kim et al. (2015) and also yields considerable differences in the income and price elasticities of households, when age groups are introduced. Integrating this model into a macroeconomic IO model, Kim et al. (2015) reveal the difference for aggregate outcomes, compared to the model of the representative consumer.

1.3 Production and Technical Progress

The main workhorse in CGE modeling on the production side are nested constant elasticity of substitution (CES) functions or flexible forms like the translog function (Jorgenson et al. 2013). The translog model can be set up with inputs of capital (K), labor (L), energy (E), imported non-energy material (M^m), and domestic non-energy material (M^d), and their corresponding input prices p_K,p_L,p_E, p_{Mm} and p_{Md}.

Each industry faces a unit cost function for the price (p_Q) of output Q, with constant returns to scale:

$$\log p_Q = \alpha_0 + \sum_i \alpha_i \log(p_i) + \frac{1}{2}\sum_i \gamma_{ii}(\log(p_i))^2 + \sum_{i,j} \gamma_{ij}\log(p_i)\log(p_j) + \alpha_t t + \frac{1}{2}\alpha_{tt}t^2 + \sum_i \rho_{ti} t \log(p_i) \tag{1.11}$$

where p_i, p_j are the input prices for input quantities x_i, x_j,t is the deterministic time trend, and TFP is measured by α_t, and α_{tt}. As is well known, Shepard's Lemma yields the cost share equations in the translog case, which in this case of five inputs can be written as:

$$\begin{aligned}
v_K &= [\alpha_K + \gamma_{KK}\log(p_K/p_{Md}) + \gamma_{KL}\log(p_L/p_{Md}) + \gamma_{KE}\log(p_E/p_{Md}) \\
&\quad + \gamma_{KM}\log(p_{Mm}/p_{Md}) + \rho_{tK}t] \\
v_L &= [\alpha_L + \gamma_{LL}\log(p_L/p_{Md}) + \gamma_{KL}\log(p_K/p_{Md}) + \gamma_{LE}\log(p_E/p_{Md}) \\
&\quad + \gamma_{LM}\log(p_{Mm}/p_{Md}) + \rho_{tL}t] \\
v_E &= [\alpha_E + \gamma_{EE}\log(p_E/p_{Md}) + \gamma_{KE}\log(p_K/p_{Md}) + \gamma_{LE}\log(p_L/p_{Md}) \\
&\quad + \gamma_{EM}\log(p_{Mm}/p_{Md}) + \rho_{tE}t] \\
v_M &= [\alpha_M + \gamma_{MM}\log(p_{Mm}/p_{Md}) + \gamma_{KM}\log(p_K/p_{Md}) + \gamma_{LM}\log(p_L/p_{Md}) \\
&\quad + \gamma_{EM}\log(p_E/p_{Md}) + \rho_{tM}t]
\end{aligned} \tag{1.12}$$

The homogeneity restriction for the price parameters $\sum_i \gamma_{ij} = 0$, $\sum_j \gamma_{ij} = 0$ has already been imposed in (1.12), so that the terms for the price of domestic intermediates p_{Md} have been omitted. The immediate *ceteris paribus* reaction to price changes is given by the own and cross price elasticities. These own- and cross-price elasticities for changes in input quantity x_i can be derived directly, or via the Allen elasticities of substitution (AES), and are given as:

$$\varepsilon_{ii} = \frac{\partial \log x_i}{\partial \log p_i} = \frac{v_i^2 - v_i + \gamma_{ii}}{v_i}, \tag{1.13}$$

$$\varepsilon_{ij} = \frac{\partial \log x_i}{\partial \log p_j} = \frac{v_i v_j + \gamma_{ij}}{v_i}. \tag{1.14}$$

Here, the v_i represent the factor shares in equation (1.12), and the γ_{ij} the cross-price parameters.

The total impact of t on factor x_i is given by:

$$\frac{d\log x_i}{dt} = \frac{\rho_{ti}}{v_i} + \alpha_t + \alpha_{tt}t. \tag{1.15}$$

Table 1.3 Price elasticities of factor demand and the factor bias of technical change

Production	Own price elasticity	Cross price elasticity, E/K	Rate of factor bias
K, all industries	−0.95		0.00
L, all industries	−0.51		−0.01
E, all industries	−0.53		0.02
E, energy intensive	−0.37	0.20	0.00
All industries		0.15	
M(m)	−0.75		0.02

This expression takes into account the TFP effect on costs $(\alpha_t + \alpha_{tt}t)$, as well as the factor bias of technical change.

The systems of output price and factor demand equation by industry across the EU 27 have been estimated applying the Seemingly Unrelated Regression (SUR) estimator for the balanced panel under cross section fixed effects. This estimation was based on data from the World Input-Output Database (WIOD) that contains World Input-Output Tables (WIOTs) in current and previous years' prices, Environmental Accounts (EA), and Socioeconomic Accounts (SEA). The estimation results (Table 1.3) yield own and cross price elasticities for capital, labour, energy, and imported intermediates, respectively. The own price elasticity of labour is on average about −0.5, with relatively high values in some manufacturing industries. The own price elasticity of energy is very heterogenous across industries and slightly higher in energy intensive industries (−0.37) than for the un-weighted average of all industries (−0.53). Capital and energy are complementary in many industries, but on average are substitutes with an un-weighted cross price elasticity of 0.15. This elasticity is slightly higher for the energy intensive industries (0.2), though in two of them (paper and pulp, non-metallic minerals) energy and capital are complementary.

This simple model of production with constant returns to scale, deterministic trends for technical change and perfect competition can be extended in order to incorporate different features that have turned out to be important in the research on production and trade in the last decades.

Imperfect competition has important consequences for macroeconomic adjustment to demand shocks. If several of these components (technical progress and imperfect competition) are to be introduced into a cost/factor demand system, these components, all leading to a deviation from the perfect competition price level, have to be identified and disentangled.

The translog structure is linked to the IO system by splitting up the factor shares v_E, v_M and v_D (the residual) into the technical coefficients (in current prices) by using fixed use structure matrices $\mathbf{S}^{\mathrm{m}}_{\mathrm{NE}}$, $\mathbf{S}^{\mathrm{m}}_{\mathrm{E}}$ for imported goods and $\mathbf{S}^{\mathrm{d}}_{\mathrm{NE}}$, $\mathbf{S}^{\mathrm{d}}_{\mathrm{E}}$ for domestic goods (with E as energy and NE as non-energy goods). A single IO technical coefficient of a domestic input i in industry j(in current prices) therefore is defined as:

$$a^d_{ij} = s^d_{ij} v_D. \tag{1.16}$$

This holds for non-energy and energy inputs, where s_{ij}^{d} is the corresponding coefficient of the use structure matrix.

As far as technical change is concerned, there are two main avenues for enriching this standard model with new features. One is making technical change depend on some variable measuring innovation activity, like R&D expenditure, R&D stocks or patent stocks, instead of the deterministic trend. This approach does not deal explicitly with technical change, and still uses some 'black box' philosophy on technical change, which is seen as a mixture of technological and organizational improvement that is driven by general innovation activities. Most studies in that line still leave the deterministic trend in the estimation, and the standard result is that controlling for innovation activity still leaves a significant part of technical change explained by the deterministic trend (i.e., unexplained). The theoretical base for this endogenous explanation of technical change stems from endogenous growth theory and represents technology as a stock of knowledge (Sue Wing 2006; Gillingham et al. 2008). Technological change is then the outcome of innovative activity within the model and, therefore, endogenous. Moreover, when innovations respond to policy instruments, such as taxes, government R&D and regulations, the direction or bias of technological change itself becomes endogenous.

The other line is combining bottom-up technology information with the top-down structure of the production model, which—in the case of CGE models—mainly is a nested CES function structure. Schumacher and Sands (2007) present a CGE model, where the top-down (CES) structure of one industry (iron and steel) is split up into different technologies that are combined in the sector and in turn have a flexible input structure. One prerequisite for the application of this approach is the availability of input data, which characterize each technology. Schumacher and Sands (2007) take this information from the German Association of Steelmakers and other sources. They nest the technologies and their choice into the CES function of the steel industry. The general logic of this approach is that the unit cost function of an industry (equation (1.11)) has fixed coefficients, like in the standard IO model:

$$\log p_Q = \alpha_0 + \sum_i v_i \log(p_i) + \alpha_t t + \frac{1}{2}\alpha_{tt} t^2 + \sum_i \rho_{ti} t \log(p_i). \tag{1.17}$$

This specification *directly* uses the factor shares (v_i) and still allows for deterministic trend variables representing technical change, like TFP ($\alpha_t + \alpha_{tt} t$) and the factor bias. The main idea is that this unit cost function is the weighted sum of different (fixed) technologies, because any factor share of the industry is the weighted sum of the input coefficients of all technologies:

$$v_i = \sum_k v_{ik} \tau_k, \tag{1.18}$$

where the τ_k are the shares of the technologies in the output of the sector, i.e., the part of sector output that has been produced with the corresponding technology.

Combining (1.16) with (1.18), the IO technical coefficient of a domestic input i in industry j can be defined as the product of (fixed) technology factor shares with the coefficient of the use structure matrix:

$$a_{ij}^{d} = s_{ij}^{d} \sum_{k} v_{Dk} \tau_{k}. \tag{1.19}$$

This formulation allows for technical change via substitution of technologies only at the level of the factor shares (v_i) of the translog model. In the model presented here, this comprises the factors K, L, E, M^m and M^d. In Schumacher and Sands (2007), this includes labour, capital, different energy sources, raw material for steel production and a bundle of all other inputs. This could in principle be extended by allowing for different columns of the use structure matrix for each technology. In that case, a specific $s_{ij,k}^{d}$ for each of the k technologies exists.

Technical change in this framework can occur by shifts in the shares of technologies (τ_k) as well as by changes in the productivity that lead to changes in technology factor shares (v_{ik}). The main issue in this framework is the determining factors for shifts in the share of technologies. In the CGE framework of Schumacher and Sands (2007), this is driven by a substitution elasticity, similar to the one used in the industry CES function. As the factor shares include capital, the allocation of investment across technologies is directly determined by technical change in terms of shifts in the shares of technologies.

The approach chosen by Pan (2006) and Pan and Köhler (2007) uses an IO model as the framework and, thus, directly aims at determining the single IO coefficients as the weighted sum of technology shares (τ_k) and the fixed input coefficients of a technology ($a_{ij,k}^{d}$):

$$a_{ij}^{d} = \sum_{k} a_{ij,k}^{d} \tau_{k}. \tag{1.20}$$

Pan (2006) presents a profound critique of the standard way of including technical change in economic models, i.e., via a trend or an accumulated stock of knowledge. His concept is based on the lifecycle of technologies and describes a discontinuous process of new technologies substituting old technologies. The R&D activities and the allocation of investment across technologies are driving this substitution process in Pan (2006). It can be shown that technical coefficients exhibit considerable long-run changes through this substitution process. This approach as well as the one lined out in Schumacher and Sands (2007) present options to describe technical change as an explicit process of change, driven by prices, investment and innovation activities.

1.4 Calibration and Stock-flow Consistency

Very often the results of econometric IO models show simplistic straight lines/trends into the future, which seem quite unrealistic. Partly, this has to do with the fact that in such cases the forecasts of exogenous data are not accounted for in the model. On the other hand, it is also due to the fact that the observed data are not or, most probably, cannot be (closely or perfectly) replicated by the model at hand, especially over time whenever the model claims to be a dynamic model.

By now there is a vast amount of literature in agricultural economics on farm-level production modeling focusing solely on perfect or incomplete calibration techniques. It turns out that until the late 80s, agricultural economists for policy analyses widely used linear programming (LP) models, and as such had to introduce (many) calibration constraints in order to solve the problem of overspecialization. However, this solution is not really a reasonable solution, since "models that are tightly constrained can only produce that subset of normative results that the calibration constraints dictate" (Howitt 1995, p. 330). Therefore, a more formal approach called Positive Mathematical Programming (PMP) was developed that solved the calibration issues in agricultural policy analysis modeling. Technically, this was implemented by introducing *non-linear* terms in the objective function of a model such that its optimality conditions are satisfied at the observed levels of endogenous (or decision) variables without introducing artificial calibrating constraints. Thus, inclusion of the so-called "implicit total cost function" captures the aggregate impact of all other relevant factors that are not explicitly modeled. Applications of the PMP approach date back to Kasnakoglu and Bauer (1988), but it was first rigorously formalized and developed by Howitt (1995). The last paper, consequently, led to an immense amount of empirical applications of the PMP approach and further raised extensive theoretical discussions within the field of agricultural economics. Review papers on the theory, applications, criticisms and extensions of the PMP approach include Heckelei and Britz (2005), Henry de Frahan et al. (2007), Heckelei et al. (2012), Langrell (2013), and Mérel and Howitt (2014).

Recently, Temurshoev et al. (2015), and Temurshoev and Lantz (2016) have borrowed ideas from the PMP literature for economic modeling of the global refining industry and proposed a perfect calibration procedure for multi-regional or global refining modeling, adopting a PMP-like technique of calibration of *spatial models of trade* introduced by Paris et al. (2011). One could also adopt the Bayesian highest posterior density estimator of Jansson and Heckelei (2011) from the same literature, if there exist a time series of observed data to be closely replicated and, as such, also accounting for the impact of other variables (not necessarily economic ones) not modeled. Given the success of the numerous and diverse applications of PMP-related literature, we tend to believe that their adoption in econometric IO modeling would be equally fruitful.

The second line of research from which, in our view, econometric IO modeling would gain, is to consider seriously the issue of consistency of the real and financial flows and stocks. This issue has recently gained particular importance in what is now called the Stock-Flow Consistent (SFC) models within the post-Keynesian school of thought (see Godley and Lavoie 2007). SFC models are a type of macroeconomic model that rigorously take into account the accounting constraints, which are, for example, not fully accounted for with SAM modeling or the standard textbook macromodels. Referring to such standard economic models, Godley and Lavoie (2007, p. 6) state that "this system of concepts is seriously incomplete. Consideration of the matrix [i.e. the standard macro-framework] immediately poses the following questions. What form does personal saving take? Where does any excess of sectoral income over expenditure actually go to—for it must all go somewhere? Which sector provides the counterparty to every transaction in assets? Where does the finance for investment come from? And how are budget deficits financed?" These are apparently all legitimate questions, and equally important for a full-fledged, realistic analysis.

It is, of course, true that some stock-flow relationships are present in the existing dynamic econometric IO models, e.g., equations relating investment to capital stock, or consumption of durables to the stock of the durable goods. The consumption model described in Sect. 1.2 takes into account this type of stock-flow consistency within the household sector, by making income relevant flows (property income, debt service payments) depending on stocks as well as stocks on income and expenditure flows (gross saving and net lending). However, this is only one part of the stock-flow consistency requirement. What is important is that such consistency in accounting has to cover all stock-flow aspects of all sectors (households, firms, government, and the external sector) in the sense that 'everything comes from somewhere and everything goes somewhere,' which thus requires adequate consideration of not only real (tangible) assets, but also financial assets (cash, deposits, loans, shares, bonds, etc.). In this respect, Godley and Cripps (1983, p. 18) state that "the fact that money stocks and flows must satisfy accounting identities in individual budgets and in an economy as a whole provides a fundamental law of macroeconomics analogous to the principle of conservation of energy in physics". The important implication of being stock-flow coherent in economic modeling is that it allows for realistic restraining of the space of possible outcomes of economic agents' behavior, which would otherwise be almost surely an impossible task, especially with the medium- to large-scale economic models. In the words of Taylor (2004, p. 2), an explicit account of the stock-flow restrictions "remove[s] many degrees of freedom from possible configurations of patterns of payments at the macro level, making tractable the task of constructing theories to "close" the accounts into complete models".

Although SFC modeling is by now a rather well-established approach, its extension to multi-sectoral and/or multi-product modeling is still in the stage of its infancy. The first such contributions, to the best of our knowledge, include SFC IO model of Berg et al. (2015), and the multisectoral SFC *macro* model of Naqvi (2015); we are not aware of any work on the integration of the SFC techniques into the econometric IO modeling. Therefore, we expect that such attempts in the future would definitely benefit this modeling strain in particular, and regional research in general.

1.5 Conclusion

In this chapter we have presented our views on the prospective future research directions in the strain of econometric input-output (IO) modeling. We think that some important recent developments, both theoretical and empirical, in other fields of economics, in particular, in macroeconomics, agricultural economics, and post-Keynesian economics, have been completely ignored in this type of modeling. Given their importance and usefulness for a sound economic analysis, regional research in general would benefit in the future, if these issues were incorporated into and/or appropriately adopted to the needs of econometric IO modeling.

The issues discussed in this chapter that could very well become the forefront topics of research and empirical applications in econometric IO modeling could be briefly summarized as follows:

- Importance of modeling *consumers' heterogeneity*, which includes, among other issues, using a concave consumption function across household income groups indicating non-constant marginal propensities to consume, different sensitivity of different household types in their consumption reaction to transitory income changes, heterogeneity with respect to the impacts of debt deleveraging and wealth shocks, concavity in the consumption function with respect to the level of wealth, and heterogeneity of households at the level of commodities.
- Importance of accounting for several *socio-economic characteristics* of households as additional variables, complementing income, wealth and debt limits. These variables include age group dummies for the household head; dummies if the household head is retired, unemployed, and is the owner of the house; household size; population density; etc. Introducing household heterogeneity not only introduces additional socio-economic variables other than income and prices that also influence behavior, but it also changes the reaction of households to income and prices and, therefore, aggregate results.
- Importance of *imperfect competition* and *technical change* in production modeling. Imperfect competition has important consequences for macroeconomic adjustment to demand shocks. Two approaches of modeling technical change (one in which technical change depends on innovation activities, and second where the bottom-up technology information and the top-down structure of the production model are combined) are discussed.
- Complete or close *calibration* of the observed data implies accounting for many relevant factors that are not explicitly modeled, which is essential for (more) realistic analysis of simulation scenarios. Here adoption of the discussed approaches of positive mathematical programming and related techniques seems to be promising.
- Importance of *stock-flow consistency*, i.e., full integration of stock and flow variables, both real (tangible) and financial assets. This would also greatly contribute to the more realistic economic modeling since then the diverse budget constraints imposed on all economic agents would be respected. Here the techniques developed in stock-flow consistent models could be readily used or adopted for the purposes of econometric IO modeling.

References

Almon C (1991) The INFORUM approach to interindustry modeling. Econ Syst Res 3(1):1–7

Almon C, Buckler M, Horwitz L, Reimbold T (1974) 1985: interindustry forecasts of the American economy. D.C. Heath, Lexington, MA

Armington PS (1969) A theory of demand for products distinguished by place of production. IMF Staff Pap 16:159–178

Barker T (ed) (1976) Economic structure and policy. Chapman and Hall, London

Barker T, Peterson W (1987) The cambridge multisectoral dynamic model of the British economy. Cambridge University Press, Cambridge

Berg M, Hartley B, Richters O (2015) A stock-flow consistent input-output model with application to energy price shocks, interest rates, and heat emissions. New J Phys 17. doi:10.1088/1367-2630/17/1/015011

Cambridge, DAE (Dept. of Economic Analysis) (1962) A programme for growth. A computable model for economic growth, vol 1

Carroll CD (1997) Buffer-stock saving and the life cycle/permanent income hypothesis. Q J Econ 112:1–55

Carroll CD, Kimball MS (1996) On the concavity of the consumption function. Econometrica 64(4):981–992

Conway RS (1990) The Washington projection and simulation model: a regional interindustry econometric model. Int Reg Sci Rev 13:141–165

Deaton A, Muellbauer J (1980) An almost ideal demand system. Am Econ Rev 70(3):312–326

Eggertson G, Krugman P (2012) Debt, deleveraging, and the liquidity trap: a Fisher-Minsky-Koo approach. Q J Econ 2012:1–45

Gillingham K, Newell RG, Pizer WA (2008) Modeling endogenous technological change for climate policy analysis. Energy Econ 30:2734–2753

Godley W, Cripps F (1983) Macroeconomics. Fontana, London

Godley W, Lavoie M (2007) Monetary economics: an integrated approach to credit, money, income, production and wealth. Palgrave Macmillan, New York

Green RD, Alston JM (1990) Elasticities in AIDS models. Am J Agric Econ 72:442–445

Hall RE (1978) Stochastic implications of the life cycle-permanent-income hypothesis: theory and evidence. J Polit Econ 86:971–987

Heckelei T, Britz W (2005) Models based on positive mathematical programming: state of the art and further extensions. In: Arfini F (ed) Modelling agricultural policies: state of the art and new challenges. Proceedings of the 89th European seminar of the European association of agricultural economics, University of Parma, Parma, Italy, pp 48–73

Heckelei T, Britz W, Zhang Y (2012) Positive mathematical programming approaches—recent developments in literature and applied modelling. Bio-based Appl Econ 1:109–124

Henry de Frahan B, Buysse J, Polomé P, Fernagut B, Harmignie O, Lauwers L, van Huylenbroeck G, van Meensel J (2007) Positive mathematical programming for agricultural and environmental policy analysis: review and practice. In: Weintraub A, Romero C, Bjorndal T, Epstein R, Miranda J (eds) Handbook of operations research in natural resources, International series of operations research & management science, vol 99. Springer, New York, pp 129–154

Howitt RE (1995) Positive mathematical programming. Am J Agric Econ 77:329–342

Jansson T, Heckelei T (2011) Estimating a primal model of regional crop supply in the European Union. J Agric Econ 62:137–152

Japelli T, Pistaferri L, Padula M (2008) A direct test of the buffer-stock model of saving. J Eur Econ Assoc 6(6):1186–1210

Jorgenson DW, Goettle R, Ho M, Wilcoxen P (2013) Energy, the environment and US economic growth. In: Dixon P, Jorgenson DW (eds) Handbook of CGE Modeling, vol 1. Elsevier, Amsterdam

Kasnakoglu H, Bauer S (1988) Concept and application of an agricultural sector model for policy analysis in Turkey. In: Bauer S, Henrichsmeyer W (eds) Agricultural Sector Modelling. Proceedings of the 16th Symposium of the EAAE, Wissenschaftsverlag Vauk, Kiel, pp 71–84

Kim K, Kratena K, Hewings GJD (2015) The extended econometric input-output model with heterogenous household demand system. Econ Syst Res 27(2):257–285

Kratena K, Sommer M (2014) Policy implications of resource constraints on the European economy. WWWforEurope Policy Brief, No 6, November 2014

Kratena K, Streicher G, Temurshoev U, Amores AF, Arto I, Mongelli I, Rueda-Cantuche JM, Andreoni V (2013) FIDELIO 1: fully interregional dynamic econometric long-term input-output model for the EU 27. JRC Scientific and Policy Reports, JRC 81864, EU Commission, Joint Research Centre

Langrell S (ed) (2013) Farm level modelling of CAP: a methodological overview. JRC Scientific and Policy Report, EUR 25873 EN, Publications Office of the European Union, Luxembourg

Luengo-Prado MJ (2006) Durables, nondurables, down payments and consumption excesses. J Monet Econ 53:1509–1539

Luengo-Prado MJ, Sorensen BE (2004) The buffer-stock model and the aggregate propensity to consume: a panel-data study of the US states. CEPR Discussion Papers, No 4474, July 2004

Lutz C, Meyer B, Wolter MI, (2005) GINFORS-Model, MOSUS Workshop. IIASA Laxenburg, 14–15 April 2005

Meghir C, Pistaferri L (2010) Earnings, consumption and lifecycle choices. NBER Working Paper Series, 15914, April 2010

Mérel P, Howitt R (2014) Theory and application of positive mathematical programming in agriculture and the environment. Ann Rev Resour Econ 6:451–447

Miyazawa K (1976) Input-output analysis and the structure of income distribution. Springer, Berlin

Mian A, Rao K, Sufi A (2013) Household balance sheets, consumption, and the economic slump. Q J Econ 2013:1687–1726

Naqvi AA (2015) Modeling growth, distributions and the environment in a stock-flow consistent framework. WWWforEurope Policy Paper 18

Nyhus D (1991) The INFORUM international system. Econ Syst Res 3(1):55–64

Pan H (2006) Dynamic and endogenous change of input-output structure with specific layers of technology. Struct Chang Econ Dyn 17:200–223

Pan H, Köhler J (2007) Technological change in energy systems: learning curves, logistic curves and input-output coefficients. Ecol Econ 63:749–758

Paris Q, Drogué S, Anania G (2011) Calibrating spatial models of trade. Econ Model 28:2509–2516

Schumacher K, Sands RD (2007) Where are the industrial technologies in energy-economy models? An innovative CGE approach for steel production in Germany. Energy Econ 29:799–825

Sue Wing I (2006) Representing induced technological change in models for climate policy analysis. Energy Econ 28:539–762

Taylor L (2004) Reconstructing macroeconomics: structuralist proposals and critiques of the mainstream. Harvard University Press, Cambridge, MA

Temurshoev U, Lantz F (2016) Long-term petroleum product supply analysis through a robust modelling approach. Loyola Econ Working Paper, Loyola University, Andalusia

Temurshoev U, Mraz M, Delgado SL, Eder P (2015) EU petroleum refining fitness check: OURSE modelling and results. JRC Science for Policy Report. EUR 27269 EN. doi:10.2791/037768. Publications Office of the European Union, Luxembourg

Kurt Kratena is director of the Centre of Economic Scenario Analysis and Research, CESAR and lecturer at the Department of Economics, Loyola University Andalucía. His primary research interests are macroeconomic input-output (IO) modeling, applied to energy-environment policy and to labor market issues. Previously, he has worked for the Austrian Institute of Economic

Research, WIFO (1993–2015) where he still works as a part-time consultant. Dr. Kratena earned the Ph.D. in economics from the Vienna University of Economics and Business Administration in 1988.

Umed Temursho is associate professor, Department of Economics, Loyola University Andalucía. His primary research interests are energy-environment-economy policy modeling and input-output (IO) economics broadly defined (ranging from IO data construction to IO theory and applications, both at the national and multiregional levels). Previously, he has worked for the Joint Research Centre of the European Commission (2012–2015) and had a faculty position at the University of Groningen (2009–2012). Dr. Temurshoev earned the Ph.D. in economics from the University of Groningen in 2010.

Chapter 2
Unraveling the Household Heterogeneity in Regional Economic Models: Some Important Challenges

Geoffrey J.D. Hewings, Sang Gyoo Yoon, Seryoung Park, Tae-Jeong Kim, Kijin Kim, and Kurt Kratena

2.1 Introduction

Torsten Hägerstrand (1970), in his presidential address to the Regional Science Association, raised the question about the neglect of people in regional science. In the intervening decades, there has been a great deal of work elaborating on the role of movement of people, some significant attempts to create demographic-economic models (or in the terminology of Ledent 1977, *demometric* models) but relatively little work unraveling the heterogeneity of households in terms of their consumption behavior. This chapter documents some current and continuing research, primarily focused on the Chicago economy, exploring the role of households, tracing impacts of ageing, income distribution, consumption expenditure patterns, in- and out-migration and retirement. Thereafter, some remaining challenges will be presented since demographic influences on regional economic development are likely to assume even greater importance in the decades ahead.

As consumption by households plays a dominant role in both national and regional economies (accounting for about 70% of gross domestic product in the

G.J.D. Hewings (✉) • K. Kim • K. Kratena
Regional Economics Applications Laboratory, University of Illinois, Urbana, IL, 61801-3671, USA
e-mail: hewings@illinois.edu

S.G. Yoon • S. Park • T.-J. Kim
Bank of Korea, Seoul, Korea

K. Kim
Asian Development Bank, Manila, Philippines

K. Kratena
Centre of Economic Scenario Analysis and Research, Department of Economics, Loyola University Andalucía, Spain

R. Jackson, P. Schaeffer (eds.), *Regional Research Frontiers - Vol. 2*, Advances in Spatial Science, DOI 10.1007/978-3-319-50590-9_2

U.S.), any change in the composition of this consumption could have important direct and indirect (ripple) effects on the economy. These changes could be generated by:

- changes in the age composition of households since consumption patterns change with age;
- changes in income distribution, since there are important differences in the way income is allocated depending on the level of income;
- changes in in- and out-migration, not only in terms of volume but also in terms of composition (e.g., skills or human capital endowments);
- changes in the way and when individuals invest in human capital;
- changes in retirement patterns and especially the propensity for retirees to remain in a region;
- the changing role of non wage and salary income (wealth) over time;
- changes in social security costs and the way these are allocated across households over time;
- changes in the way households evaluate the role of savings and precautionary measures to address idiosyncratic risks and retirement.

In many cases, these changes occur at the same time, generating important synergies that complicate the outcomes. The Chicago region[1] is selected for a reference region since it has long been both a leading immigration destination and, further, it is expected to face a significant demographic change with increasing retirement out-migration as the population ages over the next two decades.

There can be little doubt that the lower level of relative (to the U.S.) economic performance of both Chicago and Illinois partly resulted from the successive recessions in the manufacturing sector starting from the early 1980s. Between 1990 and the end of 2015, the state has lost 335,000 manufacturing jobs at a rate that is almost twice as high as that for the Midwest as a whole. Slow population growth and changing structure of population in this region have also contributed. In fact, population growth (through natural increase or immigration) turns out to be one of the two main engines of economic growth (the other being technological change). The production system provides income to labor that in turn is spent on the consumption of goods and services, generating potential for change in the production structure. The labor component is further influenced by changes in supply (for example, with retirees leaving and immigrants entering the labor force). All of these dimensions have a significant spatial component since changes in goods demanded may signal production increases in one region over another. In the last two decades, there have been some dramatic changes in the spatial structure of production systems. However, by contrast, relatively modest attention has been given to the spatial structure of labor and its concomitant influence on production.

[1]The Chicago area is the MSA, comprising the counties of Cook, Will, DuPage, McHenry, Lake, and Kane.

Although international (legal and illegal) immigration is an increasingly important component of national population change, the region's demographic structure is determined by the combination of natural increase (births—deaths), and two types of migration, international and interregional. However, as regional fertility and mortality have become more uniform throughout the United States, migration has become by far the more important factor in changing regional populations. One of the most important reasons, of course, is that fertility changes may take many years to register in terms of a significant change in the labor force; in contrast, immigrants have an instantaneous impact on labor supply. Hence, part of the reason for the slower pace of population growth in Chicago might be traced to the out-migration of retirees, because Chicago is the second largest loser, next to New York, in retirement out-migration. Moreover, over the next couple of decades, retiree migration may be expected to have a dramatic impact on the Chicago economy because of the rapid transition to a status where the ageing population will comprise a larger share (20% by 2030) of total population than at the present time.

The rest of this chapter describes some of the analyses that have been conducted in the Regional Economics Applications Laboratory (REAL); the outcomes provide a mix of results that meet a priori expectations, produce some surprises and also create outcomes whose impacts depend on the time period chosen. Thus, policy formation needs to be considered carefully and while a great deal has been accomplished, the research agenda is still incomplete. In the next section, attention focuses on the changing composition of population; Sect. 2.3 explores ways of estimating consumption by households of different types. Sect. 2.4 addresses the assessment of ageing and the macro economy while Sect. 2.5 considers the impact of immigration. The impact of changing the retirement age is explored in Sect. 2.6 while Sect. 2.7 considers the role of endogenous investment in human capital. A summary of the contributions of these various components on the ageing problem is provided in Sect. 2.8. The final section presents some important challenges that arise from the work completed to date.

2.2 Population Composition and Changes Over Time

The population over 65 in both Chicago and the U.S. is expected to exceed 20% by 2030. Figure 2.1 reveals the expected aggregate consumption growth by six age groups in comparison to aggregating the effects into a single household type. The evidence suggests that it is important to pay attention to age if for no other reason than changes in the rate of growth by age are so different.

However, it is not just the rate of growth but also differences in consumption patterns; there are some important differences in the way households allocate income. For example, on average in 2003, households allocated almost 13% of their income for food, 36% for housing (including mortgage, other loans, maintenance expenditures etc.) and 17% for all forms of transportation. The food expenditure allocation varied from 12.4% (45–54 age group) to 14.5% (under 25) while the

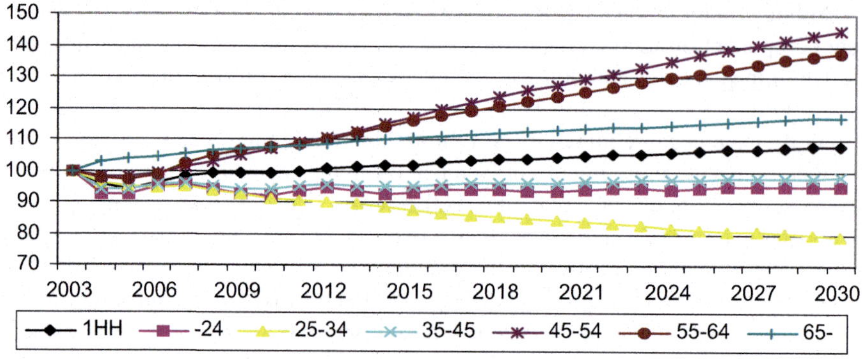

Fig. 2.1 Consumption growth by households of different ages (2000 = 100)

transportation allocations varied from 18.1 (under 25) to 14.7 (over 65). Over time, many of these expenditures are forecast to change. For example, people over 65 will spend a declining share of their income on food but an increasing share on other goods and services that include restaurants. Given the current and projected increases in obesity and eating-related disorders, this is not altogether good news!

The health care allocations generate some interesting outcomes; while all age groups will experience an increase in the share of income allocated to health care, the greatest increases occur not in the over 65 age group but in the other age groups, increasing from 3.9 to 5.9% (35–44), from 4.4 to 5.9% (44–54) and 6.2 to 8.1% (55–64). Since income usually follows a growth path that peaks in middle to pre-retirement, the implication here is that not only will a larger share of income go towards health care but the volume of expenditures on health care will increase as well. Further, as shown in Kim et al. (2015, 2016), the household disaggregation makes a significant difference in the forecasts for the region's economy.

2.3 Consumption by Households of Different Types

Different consumption patterns caused by demographic changes such as an ageing population will change the industrial production structure of the Chicago region in the future. In turn, these changes in production structure will have important implications on the profile of activities that remain competitive in the Chicago region, creating further feedback effects on the nature of local jobs and wage and salary income. The analysis was conducted using an extended econometric-input-output model of the region (see Israilevich et al. 1997); the household sector was disaggregated by income and age. The consumption behavior of these disaggregated households was modeled using an Almost Ideal Demand System (AIDS) originally proposed by Deaton and Muellbauer (1980a, b).

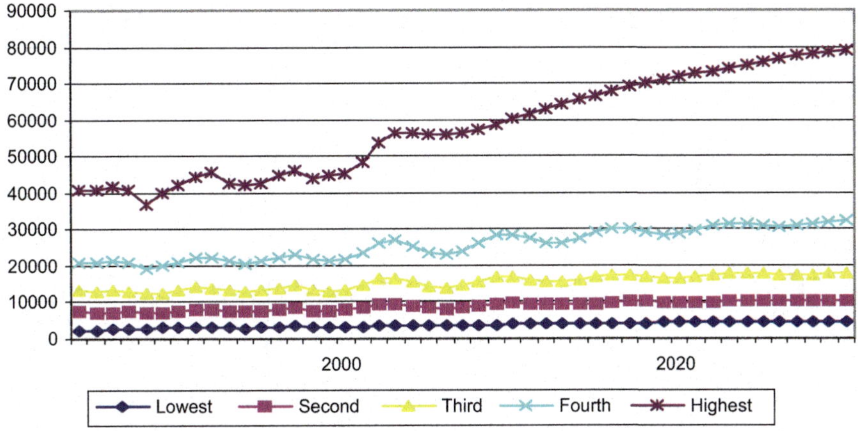

Fig. 2.2 Income growth by quintiles, 1980–2030

The AIDS model of Deaton and Muellbauer (1980a, b) gained popularity from its functional form that allows flexibility in income elasticity as well as substitutability and complementarity among goods (for details of the application, see Kim et al. 2015). A concern in this phase of the analysis was the implications for the distribution of income; in parallel to the division of consumption expenditures by age, differences due to levels of income were also explored. Over time, the changing structure of production (for example, the continued erosion of manufacturing employment that accounted for a large percentage of middle-income jobs) generates an outcome that can be presented in Fig. 2.2 (for more detail, see Yoon and Hewings 2006).

A combination of factors will see the income inequality rise in Chicago through 2030; in work that will be discussed later in this chapter, this result is modified by the effects of migration and non wage and salary income.

2.4 Ageing and the Macro Economy

Whereas the analysis presented thus far still explores a set of households that are reacting to changes in the economy rather than generating those changes, a slightly different version of our model was constructed on the same database to explore changes in household behavior on the economy. To accomplish this, behavior by households of different ages (from 21 on up) was considered through integration of an overlapping generations framework inside a computable general equilibrium model; to simplify the analysis, it was assumed that individuals were forward looking (i.e., they considered the future in making decisions about whether to spend or save) that they had some uncertainty about how long they would live and that their income consisted of wage and salary (and dividends) while they were working

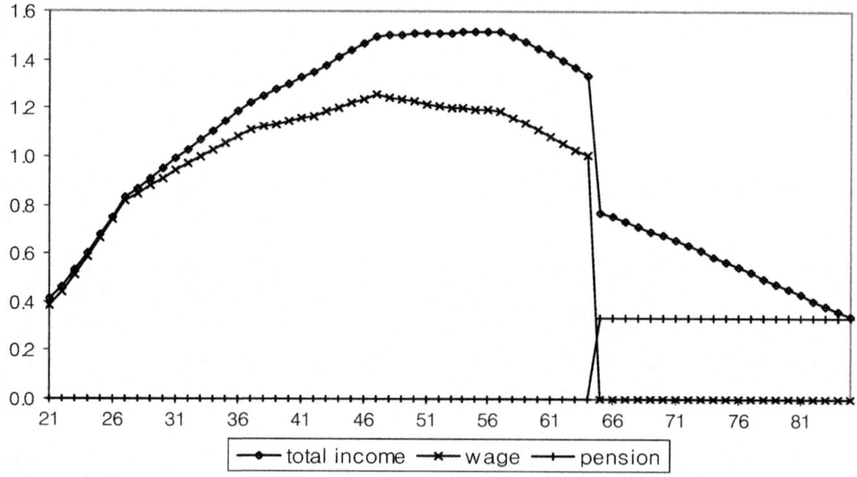

Fig. 2.3 Contributions to income over a lifetime (no change in population structure)

and only dividends and pensions in retirement. Further, it was assumed that all individuals retired at age 65 and died at 85 (for more details see Park and Hewings 2009). One additional feature of this analysis was the inclusion of non-wage and salary income since, as Fig. 2.3 suggests, this component becomes an increasingly important share of total income as an individual ages. While conceptually this accords with empirical data, capturing the full accounting (e.g., the geographical source) of this part of total income is exceedingly difficult.[2]

Each individual makes lifetime decisions about consumption and savings at the beginning of his/her adult life, leaving no voluntary bequests and receiving no inheritances. Since each agent is represented as forward looking and having perfect foresight, the evolution of consumption and savings depend on all future interest rates and after tax wages. Representative agents of each age cohorts maximize a time-separable expected lifetime utility function that depends on streams of aggregate consumption goods. Once these optimal conditions governing the aggregate consumption levels at each period are established, the consumption choice is made between goods produced in Chicago and the Rest of the U.S.; an Armington elasticity of substitution assumes that goods produced in these two regions are imperfectly competitive.

Figure 2.3 shows the various components of income over a typical household's lifetime; since we assume individuals die at 85 (or unexpectedly earlier), their

[2]Consider for an example, an individual with shares in a diversified mutual fund that invests in a range of domestic and international companies. While the fund manager might send a dividend check each quarter from one location to the owner of the shares, the source of that income would be difficult to trace since a single company might have operations in a variety of locations.

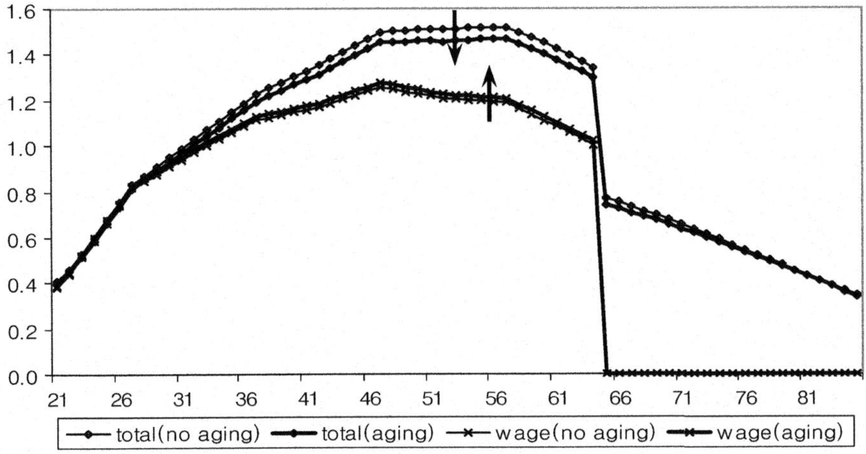

Fig. 2.4 Contributions to income over a lifetime (ageing population)

consumption patterns reflect a finite expectation for the calculation of expenditures from income (drawing down their non pension assets over the period from 65 to 85).

Figure 2.4 presents the outcomes under an ageing population scenario. Not surprisingly, untaxed wages increase under an ageing population, reflecting the relative scarcity of labor. Nonetheless, total income decreases over almost all age cohorts. For working age cohorts, this happens because the sharp increase in social security tax under an ageing population reduces the net wage income from labor supply. For early retirees, the fall in the interest rate caused by relatively abundant capital contributes to reducing the capital income from savings. With these different changes in income by age, the effect of an ageing population on savings is also sensitive to the age cohorts. That is, before the retirement, the difference in saving is not large enough to generate major interest. The possible reason is that even an ageing population will motivate precautionary saving for the working age cohorts but they cannot afford to sufficiently increase savings due to the fall in total income. As a result, consumption under an ageing population drops significantly, except for the oldest cohorts, reflecting a decline in total income and strong precautionary saving motives.

Figure 2.5 shows the transitional path of Gross Regional Product (GRP). The fall in aggregate savings accompanied by the smaller labor force eventually leads to the fall in the GRP compared to the before-ageing population. However, in the initial periods, the transition to an ageing population helps to increase the absolute level of effective labor and capital stock because baby boomers are still at work enjoying higher productivity and accumulating a larger amount of assets preparing for ageing. Both the increases in labor and capital necessarily drive the regional output above the level of GRP before the ageing population. However, in the subsequent period, GRP starts to decrease up to the 2040s, and then converges at the level that is lower by approximately 9% compared to the base year (2005). This happens because after

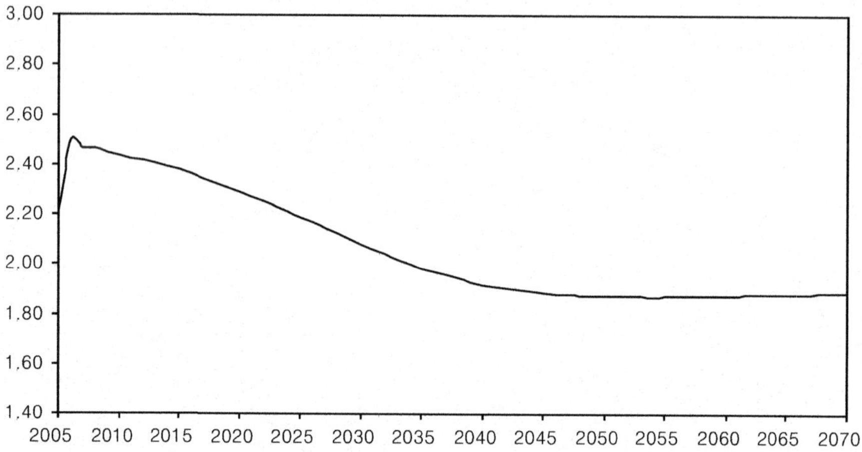

Fig. 2.5 Gross regional product (ageing population)

an initial overshoot, the capital stock starts to decrease, gradually reflecting the fall in aggregate savings; thus, two negative impacts, smaller capital stock and labor force, fuel the decline in GRP. The decreasing GRP leads, in turn, to a fall in the per capita GRP.

In contrast to the earlier finding, when an ageing population is considered in this more behavioral manner, the income inequality declines rather than increases. A major reason for this outcome may be traced to changes in social security payments by wealthier workers, increased returns from assets and, with more forward-looking behavior, retirees will have more assets from which to draw income in retirement. The earlier analysis failed to include the effect of assets (non wage and salary income) and, increasingly, these will form a major part of the income base for retirees.

2.5 Immigration, Ageing and the Regional Economy

This part of the analysis explores changes in the impacts of immigration policies; it is assumed that the immigration policies between local and federal government are differentiated. This differentiation is not in terms of issues such as quotas, visa requirements, or guest worker programs but more in terms of a region's ability to compete more effectively for the pool of in-migrants. Hence, it is assumed that the local governments in the Chicago region implement a more favorable set of incentives to attract more immigrants than the federal government (as a share of total population). These might include housing subsidies, enhanced social and health care programs, pro-active recruiting policies (through public-private partnerships) and general enhancement of the current process of channelization of immigrants flows (regions with high existing levels of immigrants have a higher probability to compete more effectively for new immigrants using family and community (Chicago)-to-community (source of immigrants in their home country) ties.

International immigration has become one of the most debated topics because it has both positive and negative impacts on the host economy. One of the biggest costs that immigration may create is through "crowding out;" increased immigration could reduce wages and exhaust employment opportunities for native workers, especially for those who are young and have low skills. Also, high income disparities could be generated due to the large decline in the income of low-skilled workers. On the other hand, however, immigration fundamentally changes the age structure, and may very helpful in contributing a solution to the demographic imbalance caused by an ageing population. In addition, one of the most common arguments in favor of immigration is that it will significantly alleviate the solvency problem of the social security program because immigrants pay social security tax, and usually have no parents who are currently drawing on the system. Of course this assumes that the immigrants participate in the formal economy (whether they are legal or not) and thus contribute through direct and indirect taxes.

Among U.S. states, Illinois has long been a major immigrant settlement place as the fifth leading immigrant-receiving state. It has admitted the nearly 0.4 million legal immigrants in the last decade, an average of 40,000 immigrants per year. The cumulative total of legal immigrants in Illinois between 1965 and 2002 was estimated to be 1.3 million. In addition, according to the Immigration and Naturalization Service (INS), over 0.4 million illegal immigrants reside in Illinois, and most of them are concentrated in the Chicago region. Among these immigrants, more than three-fifths (64.7%) of all immigrants since 1993 came from Mexico, Poland, India, Philippines, former Soviet Union, and China. Mexico alone accounted for nearly one-quarter of all new immigrants (24.8%). This continuing influx of new immigrants will account for a much more significant share of Chicago's population; now, the Latino population of Chicago slightly exceeds that of the African-American population and is growing more rapidly as a result of higher rates of natural increase as well as through in-migration (including both interregional and international contributions).

Simulations were conducted for the following three scenarios, which are differentiated by the size of immigrants for both regions, Chicago and rest of the U.S. Scenario 1 assumes that each region admits new immigrants amounting to 0.6% of the regional population every year, which is equivalent to the historical average of immigrants admitted into the Chicago region between 1993 through 2002. Scenario 2 assumes that only the Chicago region admits more immigrants, while rest of the U.S. fixes the share of immigrants at 0.6%. That is, in Scenario 2, the proportion of newly admitted immigrants into the Chicago region is adjusted to 1.2% of the population, or about 0.1 million per year while Scenario 3 assumes that the number of annual immigrants admitted to the Chicago region increases to 1.5% of its population, or about 0.12 million. According to these scenarios, the dependency ratio [the percentage of the dependent old age populations (those $\geq$65) to the population in the working age groups (between 15 and 64)] in the Chicago region is expected to be substantially reduced over the next several decades. Without immigration, the model projects a significant increase in the dependency ratio from 19% to 32% over the next 30 years, whereas new immigrants admitted following Scenario 3 contribute to dropping the dependency ratio in the 2030s to 19%, which is the same level (in 2005) as before the impacts of an ageing population. Taking into account the characteristics of immigrants, who are usually younger and lower-skilled than the resident population, newly admitted immigrants are assumed to be equally distributed between the ages of 21 and 35, and their average productivity is about 60% of the peak at 47 years of age. The baseline scenario, whose results are compared with Scenarios 1 through 3, assumes an ageing population with no immigration. This is the scenario that was introduced in the previous section.

Figure 2.6 examines the impacts on wages. The inflow of young immigrants, initially, lowers the capital/labor ratio, which, in turn, contributes to a decrease in wages. However, after the initial period, the fall in the capital/labor ratio corresponding to accumulating immigrants decreases and ceases its downward trend around 2040, about 5 years earlier than the case of baseline (no ageing or immigration). After 2040, the wages under favorable immigration remain higher than the baseline case. This result is somewhat counter intuitive because large immigration should be expected to exert a strong downward impact on wages. One possible reason for this result is that the first immigrants start to retire in the early 2040s, resulting in an increase in the capital/labor ratio. However, there are two more important factors at work for this result to happen. The first factor is that the more immigrants that are admitted, the more native workers can save since immigrants will significantly reduce the social security tax burden (by increasing the after-tax income of native workers). Further, at the time of immigration, it is assumed that the capital does not flow into the host country with immigration, but once immigrants start to work and acquire the higher levels of productivity, they can accumulate more savings, thereby increasing aggregate capital stock. This is a critical assumption;

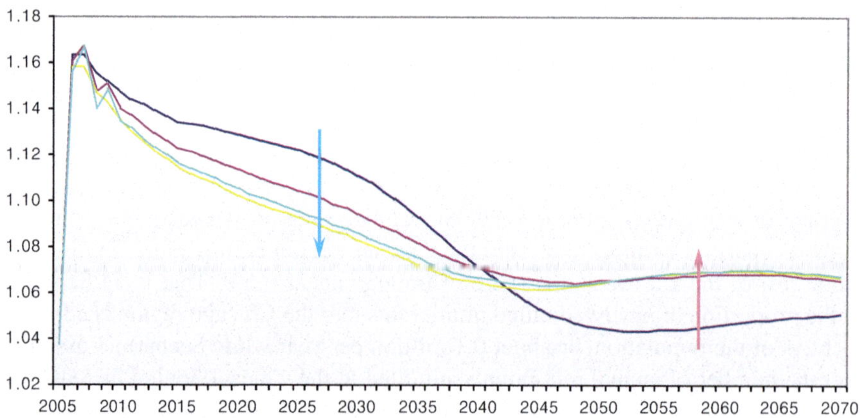

Fig. 2.6 Impacts of immigration on wages

there is likely to be some return migration and the empirical evidence has revealed significant transfers of income back to families in the countries from which the immigrants originated.

Figure 2.7 shows how the regional output would be changed by immigration streams over time. For example, in the case of the maximum contribution by the most favorable policy (Scenario 3), the Chicago region appears to grow annually by 0.9% between 2005 and 2070, while without immigration it will face negative growth (−0.2% per year) over the same period due to an ageing population. This result can be fully expected because immigration provides a positive labor supply shock to the local economy.

However, the transitional profile of per capita GRP (Fig. 2.8) is not similar to that of aggregate GRP as shown in Fig. 2.7. During the initial period, relatively larger immigration, as in Scenarios 2 and 3, keeps the per capita GRP remaining

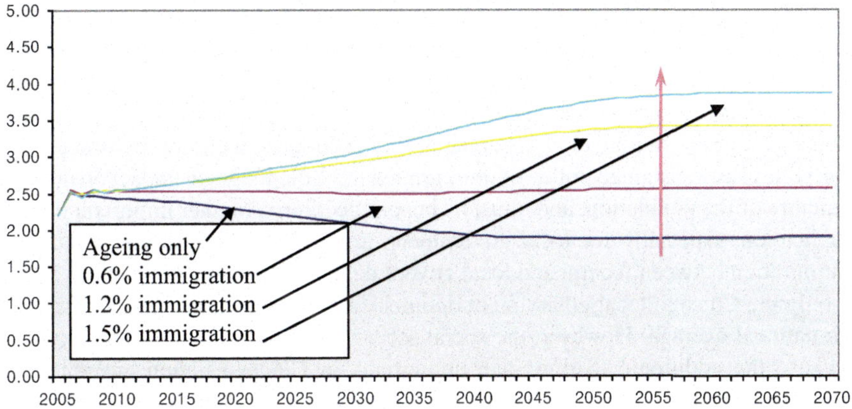

Fig. 2.7 Impact of immigration on Chicago gross regional product

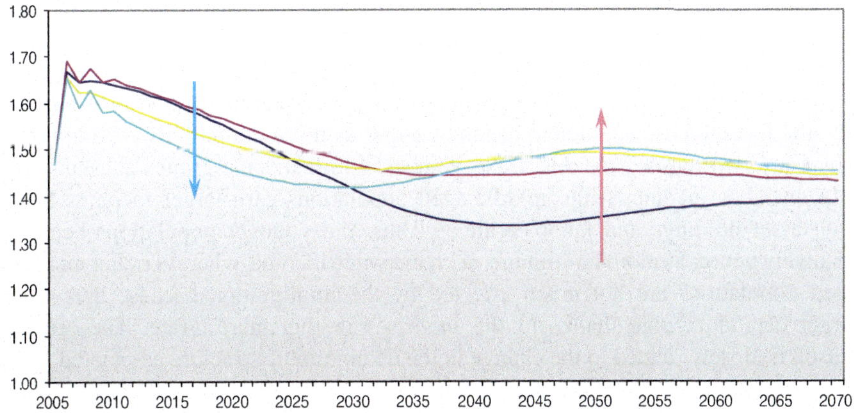

Fig. 2.8 Per capita Chicago gross regional product

at a lower level than that of the baseline case because the immigration increases (by assumption) only the supply of low skilled workers. However, after the 2030s, when the first immigrants really begin to acquire higher levels of productivity, per capita GRP reveals an upward trend and grows faster than the baseline case. This positive trend also contributes substantially to reducing the decline of per capita GRP under an ageing population. For example, between 2005 and 2070, the negative 5.5% per capita GRP growth under an ageing population is reduced, ranging from negative 2.6% in scenario 1 to negative 1.9 and negative 1.2% in scenario 2 and 3, respectively. The national GRP share of the Chicago region noticeably increases from 3.0% to around 3.5–4.0% in Scenario 2 and 3 because both scenarios assume relatively higher share of immigrants are admitted only in the Chicago region.

Not surprisingly, a larger number of working-age immigrants appear to have a significant downward impact on the social security tax rate. Thanks to this downward pressure, in 2050, the social security tax rate is projected to return to the level established before the impacts of an ageing population. This is one of the most significant benefits generated from immigration. However, the benefit for the social security system is reversed when the immigrants start to retire. After 2050, the social security tax rate starts to increase and eventually converges to around 9% that is higher than the rate expected under no immigration. This result reveals that in the longer run, immigration could generate a different impact; as immigrants age, like everyone else, a sustained policy of immigration has little long-run impact on the age structure of the population, and, thus, its benefit declines. Another important policy implication, especially for local government, arises from the different stance on immigration between federal and local governments. In the cases of Scenario 2 and 3, only the Chicago local government optimistically attracts more immigrants than the national average. However, the social security tax rate changes insignificantly because the additional working-age immigrants in Chicago region are not of a significant size to decrease the tax rate that is influenced by changes in the national population. Therefore, locally increased immigration may only hurt the local labor market without generating additional tax benefits. This is an important point; local autonomy in the case of a small region has limited impact of national policy that, in turn, could affect the outcome in Chicago (Fig. 2.9).

Figures present the effects of immigration on income distribution; immigration turns out to have a negative impact on equality in terms of income distribution, i.e., the income Gini coefficient becomes larger as more immigrants are admitted. There are two reasons for this. First, younger, lower income groups substantially rely on labor income, while middle-aged populations earn larger incomes from both asset holdings and labor earnings. Thus, the younger populations become relatively poorer as more immigrants decrease wage income, whereas richer middle-aged populations are not much affected by the immigration because they earn larger capital income thanks to the increases in the interest rate. The second reason is closely related to the change in the demographic structure associated with immigration. Before the first immigrants start to retire around the 2040s, the share of the population with larger income increases relatively faster than the younger and older poor populations because more immigrants acquire higher skills and become

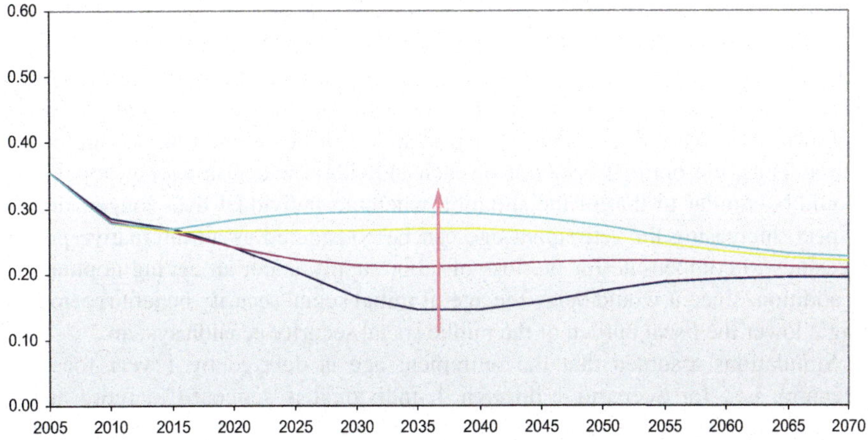

Fig. 2.9 Immigration impacts on income distribution: Gini coefficients for Chicago

richer. This structural change in population increases the aggregate income gap between the middle-aged richer population and the poor young and old populations. However, after the 2040s, since wages start to increase and immigrants start to retire, the Gini coefficients in all immigration scenarios starts to fall.

The welfare effects of the immigration were also examined.[3] The current young populations appear to be big gainers of the favorable immigration policy. The rationale for this is that even with the wage declines in the initial period, the prospect of higher disposable income for the rest of their lives obtained by both increased interest rates and reduced social security taxes outweighs the negative effect from reduced wages. This is good news for current young generations. However, unlike the assumption of this model, if more immigrants fail to adapt to conditions in the host region's labor market and, thus, remain lower skilled workers, then immigration cannot make a sufficient contribution to increasing tax contributions.

2.6 Does a Change in Retirement Age Affect a Regional Economy?

The final part of the analysis considers the impact of changes in the retirement age. Recall that it was fixed at 65 but the flexibility afforded by an absence of requirements to retire at this age is generating longer attachments to the labor force. Does this have much of an impact on a regional economy?

[3]The welfare benefit is measured by a consumption equivalent variation (EV), which computes the consumption change required to keep the expected utility in the initial condition equal to that achieved in the new condition under immigration policies.

If the worker learns that he/she will live longer than previously expected, he/she would consume less or work longer before retirement to finance the additional consumption expenditure during their extended lifetime. In this model, even though the maximum lifetime is limited to the age of 85, the average *expected lifetime* is assumed to increase due to the lower probability of death under an ageing population. Thus, the optimal behavior of each individual under an ageing population should be similar to that of the situation where an individual lives longer. In this respect, increasing the retirement age can be considered as an alternative policy measures to compensate for the loss of labor supply under an ageing population. In addition, since it would delay the age of initial social security benefit receipts, it might lower the fiscal burden of the public social security pension system.

Simulations assumed that the retirement age is delayed by 1 year for each Scenario, i.e., for Scenario 1 through 4, individual is supposed to retire at 66, 67, 68, and 69, respectively. Once again, the baseline scenario is one in which the population ages as before. Increasing the retirement age generates a smaller capital/labor ratio compared to the Baseline Scenario since the labor force increases as much as the working age is expanded. The lower capital/labor ratio leads to a fall in wage as shown in Fig. 2.10. According to the simulation results, if the retirement age is delayed by 4 years, i.e., retirement at the age 69, then wages fall by 7–8% until the 2030s compared to the baseline. Figure 2.11 shows that the rise in the retirement age contributes to an increase in the output, and, thus, the per capita GRP also increases since there is no change in the size of population. In particular, if individuals could continue working beyond the age 65 for at least 2 or 3 years longer, then the per capita GRP around the 2050s starts to rise above the level before the ageing population occurs. However, the additional gain in per capita GRP

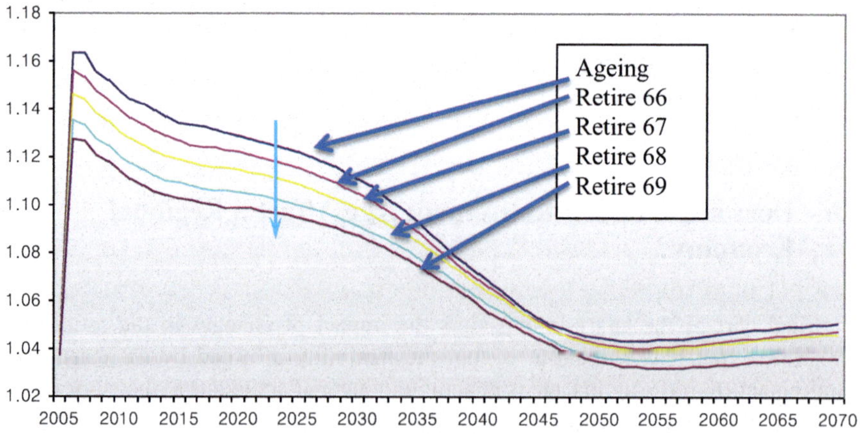

Fig. 2.10 Extending retirement age and the impact on wages

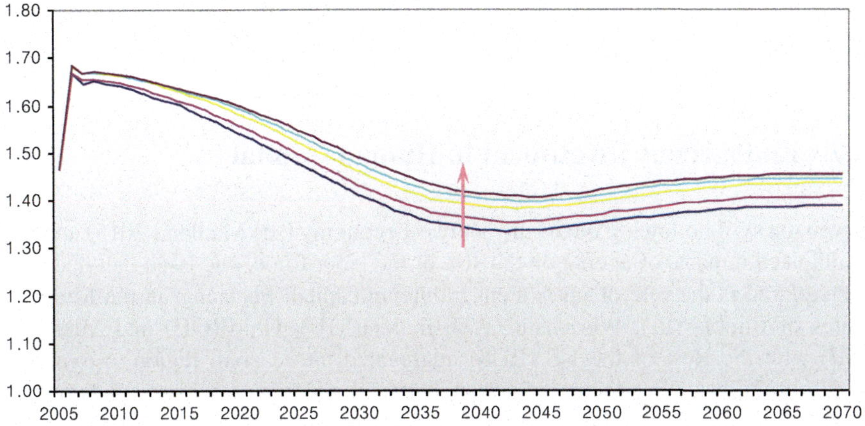

Fig. 2.11 Retirement age and Chicago per capita gross regional product

corresponding to a 1-year increase in retirement age becomes smaller, reflecting the fact that the productivity of population decreases dramatically from age 65.[4]

Not surprisingly, there is a marked decline in the social security tax rate over the transition period. For example, the maximum tax rate around the 2030s decreases from 11% in the Baseline Scenario to below 6% in Scenario 4, which is even lower than before the ageing population. The significant fall in tax rate becomes possible thanks to both increases in pension contribution by increased working-age populations and the delay in the payment of pension benefits.

By affecting the social security tax rate, the increasing retirement age influences the allocation of consumption over the lifetime, and this reallocation may cause either an increase or decrease in welfare. The welfare benefits change depending on an increase in retirement ages. All individuals over the whole age cohorts appear to favor the increasing retirement age. Furthermore, younger generations gain more than older generations who have already retired. For younger generations, they benefit from the longer payrolls with smaller taxes until far into the future, whereas for the older generations welfare gains are limited since all the benefits are generated from increasing capital income arising from the increases in the interest rate. What happens when immigration is also considered? According to the simulation, the optimal immigration occurs at the share of immigrants in the neighborhood of 0.6%. However, beyond this point, like pension reforms, an increase in immigrants generates welfare cost. The policy implications become complicated when immigration, pension reform, changing retirement age and skill acquisitions of the immigrant children are considered—as well as the effects of differential in-migration rates for Chicago and the Rest of the U.S. This is an area of

[4]As the economy shifts increasingly to non-physical labor, this assumption may not longer be valid.

research that needs far more attention—especially for the development of optimal policies.

2.7 Endogenous Investment in Human Capital

Given the skill hollowing out of the Midwest economy (see Madland 2015) and the anticipated impacts of ageing on the size of the labor force, one issue that needs to be explored is the role of investment in human capital. Focusing on the Midwest states of Illinois (IL), Wisconsin (WI), Indiana (IN), Ohio (OH) and Michigan (MI) with the Rest of the US (RUS) aggregated into a sixth region, a dynamic general equilibrium model incorporating inter-regional transactions and endogenous growth mechanisms within an overlapping generations (OLG) framework was used in conjunction with two different age-cohort population structures corresponding to years 2007 and 2030. The growth rate of the per-capita output is projected to be heterogeneous across the regions: regions with high-skilled workers hold the potential threat that population ageing could yield more negative impacts on the economy due to the relatively sluggish growth of the human capital stock

Human/occupational capital has increasingly been identified as a critical factor in attraction and retention of industry—ageing may reduce a region's "stock" of capital absent significant investment. The issue is further complicated by the presence of significant heterogeneity—both ethnic and income based. Educational investment in developing workers' human capital might improve the overall productivity in the corresponding economy and, thus, significantly attenuates the negative impacts generated by a shrinking labor force. The work of Sadahiro and Shimasawa (2002) and Ludwig et al. (2012) has been influential in motivating this exploration. The economy is closed to the rest of the world; no foreign imports or exports are considered in the model. There are two types of economic agents in each region: a representative firm and households. Each year, there are 65 overlapped generations (age 21–85) in the household sector and the federal government operates a social security system in each region. The economy produces physical goods as well as human capital; physical goods are tradable across regions and the firms can purchase intermediate goods from each region. Consumers and investors purchase goods from all the regions for consumption and investment purposes respectively.

Households now have three decisions:

- Allocation between consumption and saving (inter-temporal)
- Allocation between goods produced in any region (inter-regional)
- Allocation between education and working (human capital)

Drawing on ideas of Sadahiro and Shimasawa (2002), the model estimates parameters for the accumulation efficiency of human capital, the portion of physical capital stock used for producing the human capital stock, the depreciation rate of human capital stock and, most critically, the parameter of human capital transmission factor. This latter parameter can be interpreted as the degree of quality or efficiency to pass the available stock of knowledge from generation to generation in the workplace. If a society can provide the individual with a successful educational environment (either formally or informally) in childhood and youth so that the individual accumulates the cognitive ability and creativeness in these periods, this parameter value should be high since the human ability acquired early will make post-secondary learning easier (for more details, see Kim and Hewings 2015).

The steady state simulation results were based on the age-cohort population structure from the Census Bureau's estimation for the year 2007. Table 2.1 reveals that OH has the highest dependency ratio while IL has the lowest. For the steady state analysis, this age-cohort population structure is assumed to be maintained in the long-term; further, it is assumed that there will be no change in output, consumption and investment prices as well as factor prices such as the rental return and the wage rate. These assumptions will not be maintained in the dynamic simulation.

There exists a noteworthy gap in per-capita output across the regions according to the simulation results (Table 2.2). Simulation and actual statistics point out that the state with the lowest per-capita output among the five Midwest states is MI; and the state with the highest per-capita output is IL. It should be noted that one of the reasons for the discrepancy between the simulation result and the actual data could be attributed to ignoring the differences of the technology level across the regions in the simulation model.

The gaps of investment in physical capital and human capital play a key role in achieving different levels of per-capita output in the simulation model. The ROUS and IL invest 17.1% and 16.2% of their output while IN, WI, MI and OH allocate only 12.2%, 13.1%, 13.6% and 14.2% of their output in physical investment. This difference in investment tendencies is related to the rate of rental return; household agents would be more inclined to consume goods rather than save and invest them when the rental return becomes relatively low (or is expected to become low in the dynamic model.)

In addition, educational attainment could be a major factor in determining the difference of economic performance (here, per-capita output) since the educational investment is directly linked to the improvement of the human capital stock or

Table 2.1 Dependency ratio of each region in 2007

IL	IN	MI	OH	WI	ROUS
18.04%	18.54%	18.33%	20.11%	19.39%	18.70%

Table 2.2 Per-capita output

	IL	IN	MI	OH	WI	ROUS
Simulation	0.9704	0.8036	0.7286	0.7990	0.7996	1.0000
Actual data	1.0729	0.8885	0.8442	0.8835	0.9197	1.0000

Note: Numbers for ROUS are normalized to unity. Actual data are calculated GSP (Gross State Product) excluding public sectors ÷ population estimation in 2007
Source: BEA (www.bea.gov) for GSP; and Census Bureau for population estimation

Table 2.3 Steady-state results-time share of educational investment and average human capital stock

	IL	IN	MI	OH	WI	ROUS
Time share in education (%)	13.18	10.42	10.55	11.42	10.97	13.55
Avg. human capital stock	2.27	1.77	1.78	1.94	1.85	2.39
Gross State Product/Annual Employment: 1998 thru 2007[1]	80.52	67.77	74.88	68.94	65.06	78.66

[1]Unit: thousand dollars chained with 2000 price level
Source: Bureau of Labor Statistics and Bureau of Economic Analysis

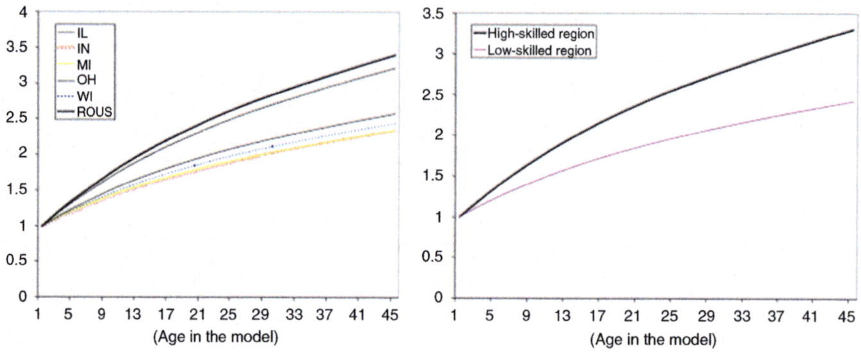

Fig. 2.12 Age profile of human capital stock

productivity in the model; the regions with higher per-capita output tend to combine inputs such as physical capital and labor force with a higher level of productivity. Table 2.3 shows the average time share spent in educational investment across the regions: IN, MI, WI and OH spend apparently less time in education than ROUS and IL. Accordingly, there should be subsequent gaps in human capital stock across the regions: Fig. 2.12 shows the discrepancies of the age-productivity profile (or human capital stock).

There is a notable gap between two groups: high skilled (IL and ROUS) and less skilled (IN, MI, OH and WI) regions. For example, the average worker at retirement age in the high skilled region is 36.8% more productive than the worker at the same age in the less skilled region. This simulation result is consistent with the statistics of labor productivity between the regions: the labor statistics show that IL and ROUS is the leading region in terms of labor productivity (the last row in Table 2.3). Again, these gaps in productivity are attributed mainly to the differences in time spent on educational investment (Table 2.3) and also the level of physical capital stock in the six regional economies according to the model specifications.

Finally, Table 2.4 presents the regional prices such as output, consumption and investment price as well as production factors. The gaps of goods prices between the regions are larger than the actual CPI; however, the order of prices matches well with the actual CPI level except MI: the simulation results underestimate the

Table 2.4 Steady state results-prices

		IL	IN	MI	OH	WI	ROUS
Goods price	Production	0.9783	0.7619	0.7611	0.8816	0.8316	1.0000
	Consumption	0.9720	0.8085	0.8057	0.9010	0.8608	0.9963
	Investment	0.9701	0.7841	0.8011	0.8892	0.8457	0.9968
Rental return (physical capital)		0.0857	0.0648	0.0662	0.0723	0.0690	0.0888
Wage rate		1.5363	0.9494	0.9717	1.2228	1.1041	1.6090

Table 2.5 Steady state result-per-capita output under the alternative age-cohort structures

		IL	IN	MI	OH	WI	ROUS
Per-capita output	2007: A	7.9932	6.6194	6.0017	6.5813	6.5866	8.2374
	2030: B	7.3336	6.1256	5.6248	6.4631	6.1252	6.6928
	B/A	0.9175	0.9254	0.9372	0.9820	0.9299	0.8125

consumption price in MI. Also, the simulation results imply that renting physical capital and hiring one unit of labor cost the most in the ROUS; on the contrary, the least expensive region is IN.

Another steady state result can be generated with the different age-cohort structure in order to obtain the insight of impact of population ageing on the economy. According to Census Bureau projections, the number of people between 15 and 64 will decline in the Midwest from 2007 to 2030. In contrast, the number of people 65 and above will grow at a significant rate. In particular, in the ROUS, the number of people of age 65+ will almost double from 2007 to 2030. Without any change of model specification, the steady state simulation was implemented with the projected age-cohort structure for the year 2030. The steady-state results in this section reflect the changes of human capital level only between the generations, but do not consider the changes of human capital stock along the time dimension.

Table 2.5 shows the comparison of per-capita output under the two different age-cohort structures. The results are quite intuitive: due to population ageing, per-capita output under the age-cohort structure in 2030 is less than the per-capita output under the age-cohort structure in 2007 in every region. It should be noted that the per-capita output in OH under the demographic scenario of 2030 does not decline so much from the level under the scenario of 2007. The number of people belonging to the working age (15–64) in OH declines faster than the other region from 2007 through 2030; subsequently, the total population size (15+) grows at only 1.4%. In contrast, it grows at 24.6% in the ROUS and 10.6% in the WI. The relative faster growth of the external demand mitigates the negative impact of population ageing to some extent. This positive effect from the external economy is reflected by the relative price changes: the demand growth from the growing population in the other regions and the limited supply of the goods produced in OH (owing to the decline in the size of the labor force) generates an improvement in the terms of trade for OH, assuming that the goods produced in each region are imperfect substitutes for each other. The growth of the relative output price of OH from 2007 through 2030 is the highest among the five Midwest states, reflecting the improving terms of trade for OH.

Kim and Hewings (2013a) also provide some dynamic simulation results. Unlike the results presented here, the dynamic simulation demonstrates that the per-capita output will grow positively even though there will be a fast growing population ageing phenomenon. Kim and Hewings (2013b, 2015) revealed that this outcome could be attributed to the individual's endogenous choice in educational investment that mitigates the negative effects of population ageing to some extent by improving the overall productivity in the corresponding economy during the transition.

2.8 Summary

Even with the caveats noted at the conclusion of the last section, several important conclusions can be drawn from the analysis conducted to date:

- Household consumption varies by age and income level; as the composition (age structure or income structure) of households change, there are likely to be important changes in the type of goods and services demanded
- Ageing in the absence of immigration will have important consequences for social security funding and the allocation of expenditures on health care by pre-retirement age cohorts
- An ageing population in the absence of immigration and with continued out-migration of retirees will likely have a longer term (next 20–30 years) impact on the Chicago economy
- Immigration at the current level (0.6% of the base population) is likely to generate positive impacts on the economy
- Expansion of the labor force and potential depressing of wage levels is more than compensated by the stimulus to demand and contributions to social security by the immigrants
- Without sustained investment in skill acquisition in the children of immigrants, the effects of immigration could turn potentially negative when the immigrants who entered in the 1980s and 1990s start to retire
- The combination of ageing and immigration is likely to change, in significant ways, what is purchased in Chicago generating an endogenous stimulus to structural change in the economy; this, in turn, could generate a positive or a negative effect on what is produced in the region to meet local consumer demand
- The synergies among ageing, immigration, retirement year and social security funding generate complex interactions that provide different effects on the Chicago economy over time

2.9 Future Research Agenda

While a great deal has been learned about the Chicago economy and the role of ageing and immigration, more research is warranted. Some of the more important issues are presented below.

2.9.1 Additional Household Disaggregation

With increased longevity, it makes no sense to continue to aggregate all households >65 years old into one category; greater disaggregation is required to explore possible changes in consumption behavior and migration. Kim and Hewings (2015b) found that between 2001–2013, unemployment rates, wage rates and labor force participation increased for persons >65 of age; but was this effect concentrated in the 65–70 age group or did it extend into the 70s?

2.9.2 Migration Dynamics

Is the return migration to city-regions like Chicago of cohorts >70 who out-migrated in their 60s a real phenomenon or an anecdote? Partridge et al. (2012) found the migration rates had declined starting in 2000, with a slight uptick in the last few years. Is this decrease in mobility spread across all age cohorts and those with different levels of human capital? The changing dynamics at the regional level are further complicated by different patterns within metropolitan regions with central cities in the U.S. once again attracting both younger working age and the older (>55) age groups in significant numbers. How are international migration patterns likely to change over the next two to three decades, especially in response to continued strife in some parts of the world? Further, there is a need to examine in more depth the pension-ageing-immigration interfaces, examining not just the short-run impacts but giving more careful consideration to the longer term. For example, a key factor centers on the role of skill acquisition of immigrant children and the potential impact of non-acquisition needs to be explored in more depth not just from an economic perspective but also with additional considerations of the impact on social cohesion.

2.9.3 Changing Demographics and Changing Regional Competitiveness

Will changes in consumption associated with changing demographics (ageing and immigration) provide firms in a region with greater opportunities to meet these new demands or will they be eclipsed by providers located outside the region? The current interest in smart specialization[5] as a policy initiative needs to be harnessed to the changing demographics of regional economies. For example, given, the findings presented in this chapter, what are some of the pro-active policies (focusing on education, skills training and re-training, immigrant attraction, affordable housing, etc.) that regions can adopt to enhance the possibilities of growth and development in the future? In this regard, the role of human capital investment may prove to be critical. What has not been addressed are issues such as: (1) how the individual will pay for this investment (assuming it is not provided by a firm); (2) when and for how long should the investment be made and (3) how many times over a lifetime in the labor force should an individual anticipate having to make this choice? Further, there are some potentially critical dynamics emerging, especially in the U.S.; many retirees have not planned for an extensive lifespan and, accordingly, many are reattaching themselves to the labor force to help fund this extended lifespan. How will they do this—returning to the one of the locations in which they last worked, seeking jobs in the same sector or will they explore other options? For example, Kim and Hewings (2015a) found that an increasing share of individuals >65 were self-employed (compared to those of prime working age). In addition, there are concerns about increasing incidence of poverty among retirees, particularly prevalent in female-headed households.

2.9.4 *Enhancing the Modeling of Consumption and the Role of Wealth*

As Kratena and Temurshoev (2016) discuss in their chapter, there is now a richer literature upon which to draw in modeling consumption. The work of Carroll (1997) in suggesting the buffer-stock savings' idea as an alternative to the traditional life cycle permanent income hypothesis to handle uncertainty and precautionary savings motives offers a richer theoretical platform on which to model consumption. In Carroll's view,

"..buffer-stock savers have a target wealth-to-permanent-income ration such that, if wealth is below the target, the precautionary saving motive will dominate impatience and the consumer will save, while if wealth is above the target, impatience will dominate prudence and the consumer will disserve."

Such a formulation could potentially enrich the intertemporal consumption function presented earlier. Note also that consumption is assumed to be a function of wealth (assets) not just wage and salary income. While conceptually appealing, assembling the necessary data for the non wage and salary components will present a challenge. There is usually a reasonably high probability of wage and salary

[5]See http://www.oecd.org/sti/inno/smartspecialisation.htm for a description and reference to more detailed analysis.

income being spent in the region in which it is earned; the same may not be true for non wage and salary income as dividends from shares may originate in companies widely scattered throughout a country or even the world. However, there are some important limitations: many investment decisions imply perfect rationality and foresight, not only in terms of the consumption decision-making but also in the context of investment in human capital. In reality, some incentives may be needed to encourage workers to invest (and continue to invest) in their human capital. From the individual perspective, the choice centers on how many times to invest in human capital and when to invest in human capital. From the firm's perspective, investment in their labor force is usually concentrated in the early age groups, but employee mobility is very high, thereby generating positive externalities on society. Munnell and Sass (2009) have been arguing for more investment in older workers who probably have a higher probability of remaining with a firm, thereby generating externalities that are internalized in the firm. Policy makers might argue that there is a need for the provision of incentives. From the government perspective, how much intervention/incentives should be considered and should incentives be provided to individuals and/or firms?

The challenges are rich ones, offering opportunities to generate new modeling systems (e.g., more extensive use of microsimulation and the application of micro-to-macro multi-level models) to address these challenges. The standard toolbox of models needs some reinvestment to be able to capture the dimensions explored in this chapter.

References

Carroll CD (1997) Buffer-stock saving and the life cycle/permanent income hypothesis. Q J Econ 92:1–55

Deaton AS, Muellbauer J (1980a) An almost ideal demand system. Am Econ Rev 70:312–326

Deaton AS, Muellbauer J (1980b) Economics and consumer behavior. University Press, Cambridge

Hägerstrand T (1970) What about people in regional science. Pap Reg Sci Assoc 24:6–21

Israilevich PR, Hewings GJD, Sonis M, Schindler GR (1997) Forecasting structural change with a regional econometric input-output model. J Reg Sci 37:565–590

Kim T-J, Hewings GJD (2013a) Inter-regional endogenous growth under the impacts of demographic changes. Appl Econ 45:3431–3449

Kim T-J, Hewings GJD (2013b) Endogenous growth in an ageing economy: evidence and policy measures. Ann Reg Sci 50:705–730

Kim K, Hewings GJD (2015) Bayesian estimation of labor demand by age: theoretical consistency and an application to an input-output model. Discussion Paper, 15-T-4 Regional Economics Applications Laboratory, University of Illinois, http://www.real.illinois.edu/d-paper/15/15-T-4.pdf

Kim T-J, Hewings GJD (2015) Ageing population in a regional economy: addressing household heterogeneity with a focus on migration status and investment in human capital. Int Reg Sci Rev 38:264–291

Kim K, Kratena K, Hewings GJD (2015) The extended econometric input-output model with heterogeneous household demand system. Econ Sys Res 27:257–285

Kim K, Kratena K, Hewings GJD (2016) Household disaggregation and forecasting in a regional economy within the framework of a regional econometric input-output model. Lett Spat Resour Sci 9:79–91

Kratena K, Temurshoev U (2016) Dynamic econometric IO modeling: new perspectives. In: Jackson R, Schaeffer P (eds) Regional research frontiers—vol. 2: methodological advances, regional systems modeling and open sciences. Springer, Heidelberg

Ledent J (1977) Regional multiplier analysis: a demometric approach. Environ Plan A 10:537–560

Ludwig A, Schelkle T, Vogel E (2012) Demographic change, human capital and welfare. Rev Econ Dyn 15:94–107

Madland D (2015) Hollowed out: why the economy doesn't work without a strong middle class. University of California Press, Berkeley

Munnell AA, Sass SA (2009) Working longer: the solution to the retirement income challenge. Brookings Institution, Washington, DC

Park S, Hewings GJD (2009) Immigration, ageing and the regional economy. Cityscape 11:59–80

Partridge MD, Rickman DS, Rose Olfert M, Ali K (2012) Dwindling U.S. internal migration: evidence of a spatial equilibrium or structural shifts in local labor markets? Reg Sci Urban Econ 42:375–388

Sadahiro A, Shimasawa M (2002) The computable overlapping generations model with an endogenous growth mechanism. Econ Model 20:1–24

Yoon SG, Hewings GJD (2006) Impacts of demographic changes in the Chicago region. Discussion Paper 06-T-7, Regional Economics Applications Laboratory, University of Illinois, http://www.real.illinois.edu/d-paper/06/06-t-7.pdf

Geoffrey J.D. Hewings is Director of the Regional economics Applications Laboratory and emeritus professor of economics, geography, agricultural and consumer economics and urban and regional planning at the University of Illinois. His primary interests are in regional economic modeling at different spatial scales with a particular focus on demographic-economic interactions. Previous appointments were at the University of Kent, Canterbury (UK) and University of Toronto. He obtained his Ph.D. from the University of Washington.

Sang Gyoo Yoon is Deputy director general of the Monetary policy department of the Bank of Korea. His primary interests are in international economics and applied econometrics including regional economic modeling. After joining the Bank of Korea in 1989, he has worked for various areas such as Research department, Statistics department, International department. He obtained his Ph.D. from the University of Illinois at Urbana-Champaign.

Seryoung Park is a senior director of the Research Department in the Bank of Korea, His primary interests are in estimating the impacts of social and population changes on regional economies using CGE models with particular attention to the role of immigration, aging and retirement. He obtained his Ph.D. from the University of Illinois at Urbana-Champaign.

Tae-Jeong Kim is currently working at the Permanent Delegation of Korea to the OECD on secondment from the Bank of Korea. His primary interests are in economic growth and macroeconomic modelling with a special attention to structural issues such as demographic changes, worsening income inequality and stagnant business investment. Formerly, he was a head of macroeconomic studies team of the Research Institute in the Bank of Korea. He obtained his Ph.D in economics from the University of Illinois at Urbana-Champaign.

Kijin Kim is economist, Economic Research and Regional Cooperation Department (ERCD), Asian Development Bank (ADB). His primary interests are in regional economics centered on the role of labor markets, impacts of environmental legislation and measurement of the effects of heterogeneity in consumption spending and labor market participation on regional economies. Dr.

Kim earned his PhD in economics from the University of Illinois at Urbana-Champaign in 2016. Prior to his PhD, he worked for the Bank of Korea (the central bank).

Kurt Kratena is director of the Centre of Economic Scenario Analysis and Research, CESAR and lecturer at the Department of Economics, Loyola University Andalucía. His primary research interests are macroeconomic input-output (IO) modeling, applied to energy-environment policy and to labor market issues. Previously, he has worked for the Austrian Institute of Economic Research, WIFO (1993–2015) where he still works as a part-time consultant. Dr. Kratena earned the Ph.D. in economics from the Vienna University of Economics and Business Administration in 1988.

Chapter 3
Geographical Macro and Regional Impact Modeling

Attila Varga

3.1 Introduction

After a long-experienced neglect of spatial issues in the mainstream of economic research (Krugman 1991b), economics becomes increasingly geographical. The appearance and success of the new economic geography (NEG) plays a key role in this development. Static NEG models extend a non-spatial macroeconomic general equilibrium framework toward a multi-regional system via the integration of agglomeration effects, transport costs and migration (Krugman 1991a; Fujita et al. 1999). Dynamic new economic geography growth models incorporate agglomeration effects in the framework of a-spatial endogenous growth theories in order to study the complex interrelationship between agglomeration and aggregate economic growth (Baldwin and Martin 2004). The extensions of non-spatial economic models in static and dynamic NEG theories underline that geography plays a substantial role in generating macro (national or supranational) economic outcomes.

The key role of geography in national economic development has also been brought into the forefront of recent policy debates (World Bank 2009; OECD 2009; Barca 2009). Advocates favoring either the place-based or the spatially blind approaches—despite the different weights the two approaches apply on various aspects of geography—agree that agglomeration, regional capabilities, or interactions at the regional and interregional scales remarkably determine national level results of development policy measures. Thus, the geography of interventions is understood as a key factor in the success of development policies.

A remarkable recent observation points to another dimension of macro-regional interactions suggesting that macro (national) level policies could significantly

A. Varga
Faculty of Business and Economics, University of Pécs, Hungary
e-mail: vargaa@ktk.pte.hu

R. Jackson, P. Schaeffer (eds.), *Regional Research Frontiers - Vol. 2*,
Advances in Spatial Science, DOI 10.1007/978-3-319-50590-9_3

influence the effectiveness of a particular geography of interventions targeting economic development (D'Costa et al. 2013). Governments' monetary and fiscal policies interact with regionally deployed development policies: macroeconomic policies could support but could also distract regionally targeted interventions.

Economic theory and policy discussions, thus, both underline that the macro (national or supranational) and regional (sub-national) spatial levels are mutually interconnected in development: the geography of interventions influence macro level policy results and the effectiveness of regionally implemented interventions is related to several macroeconomic policy conditions. Despite that theory and policy discussions emphasize the importance of both spatial levels in generating development policy impacts, the majority of economic models applied in policy evaluation consider these layers separately: models either follow the tradition of macroeconomic (national) or regional (sub-national) level of analysis.[1]

In the past decade the emergence of 'new generation impact models' (Varga 2015) has been experienced. These models undertake the initial attempts in the direction of integrating geography in traditional modeling frameworks. This chapter sheds some light on key technical challenges that geographic policy impact models currently face and illustrates the response to these challenges by outlining one of the earliest attempts in this direction, the Geographic Macro and Regional (GMR) model system. The second section briefly reviews current policy debates on the role of geography in development policy, followed by the account of some key modeling challenges. The section outlining the GMR approach follows and the chapter ends with an epilogue.

3.2 Geography in Modern Development Policy Approaches

The literature of regional development reports limited success of policies in reducing territorial disparities. For instance, the contribution of EU Cohesion policy to regional convergence in the EU appears only weakly positive (Hagen and Mohl 2009). Disappointment in traditional policy approaches has stimulated policy thinking to reconsider the old instruments in order to suggest the kinds of interventions that are expected to enhance economic development more successfully. Two streams of modern policy thinking emerged recently. The first stream, in general, does not trust regionally targeted interventions but favors space-neutral policies with universal coverage in every territory, while the second stream would continue supporting region-specific interventions and argues that properly designed place-based policies are appropriate means of economic development.

[1]The HERMIN model (ESRI 2002), the ECOMOD model (Bayar 2007) or the QUEST III model (Ratto et al. 2009) are good examples of macroeconomic modeling while the REMI model (Treyz et al. 1992) is a well-known representative of regional modeling.

In both approaches, the focus has moved towards policies that strengthen aggregate economic growth. Despite extensive debates, there is some complementarity between the two modern approaches to development policy (Farole et al. 2011; Varga 2015): the space-neutral focus does not disclose the validity of place-based policies under specific circumstances, like in the case of regional innovation policy (e.g., World Bank 2009); and the place-based approach claims that policies targeting large agglomerations could occasionally perform as better alternatives (e.g., Barca 2009; OECD 2009).

Spatially blind policies advise strengthening the self-reinforcing cycle of agglomeration and growth. The proponents suggest encouraging economic integration of lagging places with core economic areas (World Bank 2009). Economic integration is being reached when no major differences exist among territories in institutional development (e.g., provision of education, health care, security or regulations of land and labor) and when lagging regions are sufficiently interconnected with the agglomerated economic core by transportation linkages. Interventions, thus, should aim at fueling agglomeration effects in the economic core and, as such, should be designed in a space-neutral way to the greatest extent possible.

Though the proponents of place-based development do not question the relevance of spatially blind policies or the importance of agglomeration in economic growth, their main emphasis is positioned on the role of region-specific policies. It is suggested that growth potential exists in many regions outside the major agglomerations (Barca 2009; OECD 2009). Advocates of place-based development are in favor of territory-specific innovation policies as effective tools of growth promotion (McCann and Ortega-Argilés 2015).[2]

The key feature of modern policy approaches is, thus, their emphasis on geography as a significant factor in aggregate (macro-level) economic development. Agglomeration is one particular aspect of geography but local specificities, such as industrial structure, the strength of research, the size of human capital or accessibility, are at least as important geographic features as interregional linkages, such as trade flows, labor and capital migration or knowledge transfers.

Because geography is considered a key element in economic development, the spatial structure of interventions influences the outcomes of development policies: the same development policy budget may affect national level economic growth differently depending on alternative distributions of resources across different regions. Impact models incorporating geography would, thus, act as suitable tools for assessing the likely outcomes of different spatial distributions of the same aggregate policy budget.

[2]An additional reason for place-based (or region-specific) policies is politics, particularly in an ethnically and/or culturally diverse economy. The EU meets this criterion. It is also the reason why Canada has very specific region-based objectives, as does Switzerland.

3.3 Geographical Extension of Traditional Development Policy Impact Models: Critical Challenges

In the following I detail some key economic modeling challenges of incorporating geography in development policy impact models. These include modeling the effect of policies on technological progress, formulating the transmission of innovation impacts to economic variables, modeling spatiotemporal dynamics of growth and incorporating the macro dimension.

The first question in model design is related to the way the impacts of policy instruments on innovation are represented in an economic model. A rich empirical literature has mapped several geographical aspects of innovation and, as such, collected important information for model builders (Varga and Horváth 2015). The observed positive association of innovation with research, human capital, physical proximity, agglomeration, entrepreneurship and knowledge networks at different spatial scales suggests that integrated policies proposed by modern development approaches aiming at stimulating R&D, education, entrepreneurial culture, transportation infrastructure investments and collaborations in research are indeed realistically expected to positively influence innovation. The question still remains however as to how these elements of innovation are integrated into a coherent empirical modeling framework. Possibilities in this respect might range from the application of geographic knowledge production function and regional computable general equilibrium approaches to dynamic evolutionary modeling techniques.

The choice of how to model empirically the transmission of policy impacts on innovation to changes in economic variables such as output, employment or inflation is the second challenge. Innovation may contribute to aggregate growth in two (not necessarily independent) ways. Technological progress either increases the production of already existing goods (a productivity impact) or results in the introduction of new or improved quality products (a variety impact). Modeling the productivity and variety effects in an integrated framework is a real challenge. Nevertheless, it is a common experience that their translation to empirical models becomes indeed difficult because of the appearance of several technical issues. Among them, data availability is a really serious problem, especially at sub-national regional levels.

The technical challenge of incorporating spatiotemporal dynamics addresses the problem of modeling policy-induced expansion of indigenous resources and their migration between regions simultaneously. Consistency with the neoclassical growth framework requires deriving saving and investment behavior from intertemporal optimization of households and firms in all locations. Development of models in this direction is slow and solutions are rare due to substantial analytical and computational difficulties involved. Alternatives include the introduction of ad-hoc investment and saving behavior in regional models, or separately modeling intertemporal optimization of investment and saving behavior at the macro level and migration and dynamic agglomeration effects at the regional level in an integrated model system.

The macroeconomic framework, including the exchange rate of the national currency, government deficit and debt, the monetary policy regime or the interest rate, could be important factors behind the impact of development policies. In a carefully designed macroeconomic policy, economic development targets would indeed be aligned with other macro framework conditions. Because the derivation of these conditions from the regional level is not theoretically clear (and most probably regional to macro aggregation is not even possible in this respect), integration of the macro and regional dimensions seems to be a desirable solution. This is an open area of research and examples are rare in the literature (Varga 2015).

3.4 The Emergence of a New Generation of Development Policy Impact Models: The Case of the GMR-Approach

Increasing activity of different research groups to develop a new generation of economic impact models indicates that the problem of incorporating geography has already been realized and the search for suitable model constructions is ongoing. These research directions include, for example, the MASST ("MAcroeconomic, Sectoral, Social, Territorial") model (Capello 2007) and the GMR ("Geographic Macro and Regional") policy impact modeling approach. The GMR-approach is followed in EcoRet (Varga and Schalk 2004), in GMR-Hungary (Varga 2007), in GMR-Europe (Varga et al. 2015; Varga 2015), in GMR-Turkey (Varga et al. 2013; Varga and Baypinar 2016) and in the European Commission's RHOMOLO ("Regional HOlistic MOdeL") model (Brandsma and Kancs 2015). Though GMR, MASST and RHOMOLO are different in many respects in their internal structures (e.g., MASST is a partial equilibrium econometric model, RHOMOLO is a general equilibrium SCGE (Spatial Computable General Equilibrium)[3] model on six industries, the GMR model is an integrated econometric-SCGE-DSGE[4] model), they share the common interest of incorporating geographical effects into their model structures. Below we outline how GMR policy impact models reflect the challenges indicated in the previous section.

[3]SCGE models extend the more conventional CGE (Computable General Equilibrium) approach with geographic effects such as agglomeration, interregional migration and transport costs. An SCGE model is formulated as a set of (sub-national) regions where regions are not independent but connected by linkages like transportation and migration. The short run equilibrium of the model is reached when supply and demand equals in each market in each of the regions. However this does not necessarily mean that this equilibrium is stable because differences in factor prices might induce interregional migration. Equilibrium becomes stable in the long run when no motivation for further factor migration is present.

[4]DSGE stands for Dynamic Stochastic General Equilibrium modeling. These models are dynamic because they explicitly take into account intertemporal decisions of economic actors; they are stochastic as the structural relationship and variables of the model can be hit by different shocks driving the economy away from the equilibrium path; they are general equilibrium as they assume market clearing (even if markets are not perfect).

The GMR approach is an economic development policy impact modeling framework. In comparison with traditional approaches the novel feature of the GMR-approach is that it incorporates geographic effects (e.g., agglomeration, interregional trade, migration) while both macro and regional impacts of policies are simulated. GMR models provide ex-ante and ex-post evaluation of development policies, such as promotion of R&D activities, human capital advancement or improved physical accessibility. The models simulate macro- and regional economic impacts while taking into account geography effects, such as regional innovation system features, agglomeration, migration and costs of transportation. The intention of the GMR research program is to develop efficient and relatively simple model structures, which fits in with the generally weak quality of regional data.

The GMR-framework is rooted in different traditions of economics (Varga 2006). Romerian endogenous growth theory shapes the GMR approach to modeling knowledge generation (Romer 1990) while the spatial patterns of knowledge flows and the role of agglomeration in knowledge transfers are formulated with insights and methodologies learned from the geography of innovation field. Interregional trade and migration linkages and dynamic agglomeration effects are formed with an empirical general equilibrium model in the tradition of the new economic geography (Krugman 1991a). Specific macroeconomic theories provide the foundations for modeling macro level impacts.

The GMR approach reflects the modeling challenges outlined in the previous section by structuring its system around the mutual interactions of three sub-models such as the TFP (Total Factor Productivity), SCGE and MACRO (macroeconomic) sub-models.

3.4.1 Modeling Policy Impact on Technological Progress

Policy impact on innovation is formulated in the TFP sub-model. Following Romer (1990), development of ideas for new technologies is explained by the amount of research inputs and the stock of accumulated scientific-technological knowledge. The assumption behind this formulation is that even the same research inputs can result in a number of new technologies depending on the level of knowledge already accumulated over time. In GMR models, the impact of research expenditures on new technological ideas is influenced by the concentration of technology intensive industries in the region on the one hand and interregional research cooperation on the other.

3.4.2 Modeling the Transmission of the Technology Impact to Economic Variables

Many of the new technological ideas become introduced in production but many of them remain unexploited. The development of concrete technologies on the basis of technological ideas is formulated in the TFP equation. Therefore, innovation policy impacts on economic variables are transmitted through an increase in TFP. Policy induced change in TFP may increase output even if capital and labor stays the same. Increased output might result from new varieties and/or from growing productivity.

3.4.3 Modeling Spatiotemporal Dynamics of Economic Growth and Macro Impact Integration

A higher level of TFP resulting from innovation policy interventions may effect production partly via increased regional employment and investment and partly via labor and capital migration from other regions. Increased concentration of economic activities might strengthen dynamic agglomeration economies that could initiate a cumulative process towards further concentration. Therefore, increased capital and labor on the one hand and additional expansion in TFP sparked by agglomeration on the other hand drive policy-induced regional growth.

In modeling spatiotemporal dynamics and macro impact integration, this complex process is separated into three steps, which at the end result in a coherent macro-regional impact via mutual alignments. The first two steps reflect spatial dynamics. In their design, the solution frequently applied in many of the new economic geography models is followed. In the first step, the short run impact of a change in TFP on the values of economic variables (e.g., output, capital and labor demand, prices, wages) for each region is calculated under the assumption that aggregate supply of capital and labor and their regional distribution remain constant. In the second step, utility differences across regions motivate labor migration, which is followed by the migration of capital. The first and second steps are formulated in the SCGE model block. So far aggregate labor and capital supply have been assumed constant. Their dynamics is modeled then in the third step with the MACRO model block.

The mutually connected three model-block system is depicted in Fig. 3.1 below. Without interventions TFP follows a steady state growth rate in each region. The impacts of interventions run through the system according to the following steps.

1. Resulting from interventions related to R&D, human capital, interregional knowledge networks and entrepreneurship regional TFP increases.
2. Changing TFP induces changes in quantities and prices of output and production factors in the short run while in the long run the impact on in-migration of production factors implies further changes in TFP not only in the region where

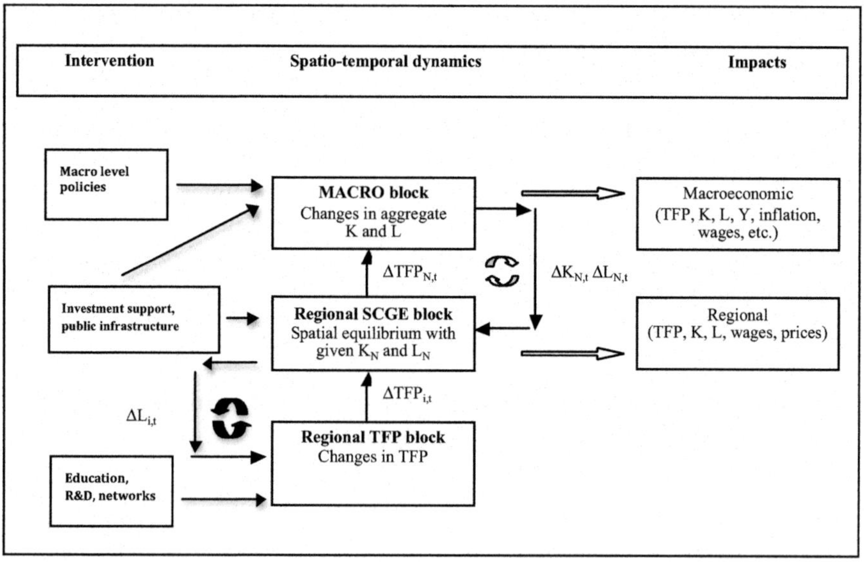

Fig. 3.1 Regional and macroeconomic impacts of the main policy variables in GMR-models

the interventions happen but also in regions that are connected by trade and factor migration linkages.

3. Increased private investments expand regional private capital, which implies further changes in regional variables (output, prices, wages, prices, TFP, etc.) in the SCGE model block. The impact of private investment support affects the macro model as well via increased private capital.
4. For each year, changes in regional TFPs are aggregated to the national level. These changes in TFP enter the macro model as time specific shocks. The macroeconomic model calculates the changes in all affected variables at the national level.
5. Changes in employment and investment calculated in the MACRO block are distributed over the regions following the spatial pattern of TFP impacts.
6. The SCGE model runs again with the new employment and capital values to calculate short run and long run equilibrium values of the affected variables.
7. The process described in steps 5 and 6 run until aggregate values of regional variables calculated in the SCGE model converge to their corresponding values calculated in the MACRO model.

3.5 Epilogue

Policy discussions highlight the key role of geography in the performance of development policies. Recently emerged policy impact models undertake the initial attempts in the direction of integrating geography in traditional modeling frameworks. To illustrate how model structures might reflect the challenges of integrating geography, I briefly outlined the GMR policy impact modeling approach in this chapter.

What can one expect in the coming 50 years in development policy impact analysis? I believe that (following the trend of "more geography in economics") models will become increasingly spatial. The sub-national (regional, local, city) and macro (national, supranational) levels will be integrated in a systematic manner incorporating findings accumulated by theoretical, empirical and policy research.

Therefore, substantial efforts and careful, professional and enduring work could lead to the development of geographic policy impact models with increasing precision. Economic theory on the one hand and empirical techniques on the other have already reached the critical intellectual mass to support this endeavor. Because technical components to address the four challenges are already accessible, the inventiveness of modelers will determine the particular characteristics of individual solutions. However, availability of detailed information on industrial sectors and innovation activities at the regional level may significantly determine the effectiveness of the models. This underlines that progress in the collection of accurate regional data will be crucial for the future success of geographic policy impact models.

References

Baldwin R, Martin P (2004) Agglomeration and regional growth. In: Handbook of regional and urban economics, vol 4. Elsevier, Amsterdam, pp 2671–2711

Barca F (2009) An agenda for a reformed cohesion policy: a place-based approach to meeting European Union challenges and expectations. Independent report prepared at the request of the European Commissioner for Regional Policy, Danuta Hübner, European Commission, Brussels

Bayar A (2007) Simulation of R&D investment scenarios and calibration of the impact on a set of multi-country models. European Commission DG JRC. Institute for Prospective Technological Studies (IPTS)

Brandsma A, Kancs A (2015) RHOMOLO: a dynamic general equilibrium modelling approach to the evaluation of the European Union's R&D policies. Reg Stud 49:1340–1359

Capello R (2007) A forecasting territorial model of regional growth: the MASST model. Ann Reg Sci 41:753–787

D'Costa S, Garcilazo E, Oliveira Martins J (2013) The impact of structural and macroeconomic factors on regional growth. OECD regional development working papers 2013/11

ESRI (2002) An examination of the ex-post macroeconomic impacts of CSF 1994–1999 on objective 1 countries and regions. http://ec.europa.eu/regional_policy/sources/docgener/evaluation/doc/obj1/macro_modelling.pdf

Farole T, Rodriguez-Pose A, Storper M (2011) Cohesion policy in the European Union: growth, geography, institutions. J Common Mark Stud 49:1089–1111
Fujita M, Krugman P, Venables A (1999) The spatial economy. MIT Press, Cambridge, MA
Hagen T, Mohl P (2009) Econometric evaluation of EU cohesion policy—a survey. Discussion paper no 09-052. Center for European Economic Research, Mannheim
Krugman P (1991a) Increasing returns and economic geography. J Polit Econ 99:483–499
Krugman P (1991b) Geography and trade. MIT Press, Cambridge, MA
McCann P, Ortega-Argilés R (2015) Smart specialisation, regional growth and applications to European Union cohesion policy. Reg Stud 49:1291–1302
OECD (2009) How regions grow. Organisation for Economic Growth and Development, Paris
Ratto M, Roeger W, In't Veld J (2009) QUEST III: an estimated open-economy DSGE model of the euro area with fiscal and monetary policy. Econ Model 26:222–233
Romer P (1990) Endogenous technological change. J Polit Econ 98:S71–S102
Treyz G, Rickman D, Shao G (1992) The REMI economic-demographic forecasting and simulation model. Int Reg Sci Rev 14:221–253
Varga A (2006) The spatial dimension of innovation and growth: empirical research methodology and policy analysis. Eur Plan Stud 9:1171–1186
Varga A (2007) GMR-Hungary: a complex macro-regional model for the analysis of development policy impacts on the Hungarian economy. Hungarian National Development Office
Varga A (2015) Place-based, spatially blind, or both? Challenges in estimating the impacts of modern development policies: the case of the GMR policy impact modeling approach. Int Reg Sci Rev. doi:10.1177/0160017615571587
Varga A, Baypinar M (2016) Economic impact assessment of alternative European Neighborhood Policy (ENP) options with the application of the GMR-Turkey model. Ann Reg Sci 56:153–176
Varga A, Horváth M (2015) Regional knowledge production function analysis. In: Karlsson C, Anderson M, Norman T (eds) Handbook of research methods and applications in economic geography. Edward Elgar, Cheltenham, pp 513–537
Varga A, Járosi P, Sebestyén T, Baypinar M (2013) Deliverable 6.2: detailed policy impact model. Sharing KnowledgE Assets: InteRregionally Cohesive NeigHborhoods (SEARCH) EU FP7 Project. http://www.ub.edu/searchproject/wp-content/uploads/2013/12/SEARCH-Deliverable-6.2.pdf
Varga A, Járosi P, Sebestyén T, Szerb L (2015) Extension and application of the GMR-Eurozone model towards the CEE regions for impact assessment of smart specialisation policies. GRINCOH FP 7 project deliverable
Varga A, Schalk HJ (2004) Knowledge spillovers, agglomeration and macroeconomic growth. An empirical approach. Reg Stud 38:977–989
World Bank (2009) World development report 2009: reshaping economic geography. World Bank, Washington, DC

Attila Varga is professor, Department of Economics and Econometrics, and Director of the Regional Innovation and Entrepreneurship Research Center (RIERC) at the Faculty of Business and Economics of the University of Pécs, Hungary. His primary research interests are regional innovation system modeling; and regional economic development. Previous faculty positions were at West Virginia University, the Austrian Academy of Sciences and the Vienna University of Economics, Vienna, Austria. Dr. Varga earned the Ph.D. in economics from WVU in 1997.

Chapter 4
Computable General Equilibrium Modelling in Regional Science

Grant J. Allan, Patrizio Lecca, Peter G. McGregor, Stuart G. McIntyre, and J. Kim Swales

4.1 Introduction

Why should we be interested in building regional CGEs? In general terms, because they provide a coherent framework in which to analyze the impact of any disturbances emanating from overseas, the nation or the region (or the sub-region/city) itself. The potential for assisting ex-ante and ex-post policy analysis and evaluation is clear. They can be used to explore the regional impacts of regional and national policies. We see no limit to the range of potential applications for regional CGE modelling, wherever system-wide ramifications of policy or other changes are anticipated.

It seems sensible to begin this chapter by identifying what is—and what is likely to remain—distinctive about *regional* CGE models. First and foremost, there is typically a significantly greater degree of spatial integration of factor, goods and financial markets at regional than national levels. In part, this is fostered by the history of a permanently fixed exchange rate and a common currency, subject to a national monetary policy. In goods markets, for example, this integration is reflected in the extent of trade flows and their sensitivity to relative price and other changes. Factor mobility is higher and requires explicit modelling of migration flows, in particular.

G.J. Allan (✉) • P.G. McGregor • S.G. McIntyre • J. Kim Swales
Fraser of Allander Institute and Department of Economics, Strathclyde Business School, University of Strathclyde, 199 Cathedral Street, Glasgow, G4 0QU, UK
e-mail: grant.j.allan@strath.ac.uk; p.mcgregor@strath.ac.uk; s.mcintyre@strath.ac.uk; j.k.swales@strath.ac.uk

P. Lecca
European Commission, DG Joint Research Centre, Calle Inca Garcilaso, 3, 41092, Seville, Spain
e-mail: Patrizio.LECCA@ec.europa.eu

R. Jackson, P. Schaeffer (eds.), *Regional Research Frontiers - Vol. 2*, Advances in Spatial Science, DOI 10.1007/978-3-319-50590-9_4

For regional financial markets, the default assumption is often that they are perfectly integrated with the national economy, but some dispute this and argue for a degree of spatial segmentation, even in this case (e.g. in loan markets). Regional CGEs also need to capture regional/national (and potentially local/city) government funding relationships, which can imply very different macroeconomic closures from equivalent models of the national economy. Furthermore, the models reflect available evidence of the behavior of regional markets—perhaps especially of regional labour markets, which can be crucial to overall model behavior. Typically, regional CGE modellers also have to cope with data challenges that are more severe than those for national models.

There is now a large range of CGE models, including single region and multi-regional models, that, in the latter case, may embody a bottom up model of the national economy. Spatial (regional) CGE models is a term typically only applied to those models whose specification reflects New Economic Geography (NEG) approach and incorporate transport and agglomeration economies. Early regional models often had more in common with the kinds of CGEs that were applied to developing economies than those being used in North America to explore fiscal issues. In particular, the former reflected a pragmatic concern with the impact of key market imperfections (notably in the labour market) for the efficacy of policy changes. National models often assumed entirely fixed aggregate labour supply, surely never an accurate characterization, but hopelessly inappropriate in a regional context.

Partridge and Rickman (1998, 2010) and Giesecke and Madden (2013) provide very extensive reviews, analysis and a fairly comprehensive bibliography of regional CGE modelling, Moreover, the general developments in CGE modelling are comprehensively discussed in contributions to the Dixon and Jorgenson (2013) edited volume. The interested reader is directed to these works for a literature review treatment of existing CGE models. Our purpose in this chapter, consistent with the theme of the present volume, is to provide a forward looking perspective and to anticipate major future developments in regional CGE modelling.

There are two major interrelated areas for future developments of regional CGEs. The first set of developments relate to improving and augmenting the methods currently employed by the typical regional CGE model. We want to improve the ability of these models to capture more accurately the key features of regional economies and the behavior of relevant economic actors. This involves improving the specification of CGEs, the methods used to parameterize them, and the quality of regional data. With such improvements, we expect to see regional CGEs being increasingly adopted as the standard workhorses for regional economic analysis.

The second set of developments relate to the application of the future vintages of regional CGE models. In fact, the range of regional CGE model applications is already huge (as reflected in, for example, the papers cited in Giesecke and Madden 2013); this is a manifestation of one of the great strengths of these models, namely their flexibility, which allows their adaption to address new or emerging policy (and non-policy) issues. Fundamental, and policy-relevant, concerns seem unlikely to alter, although events may lead to shifting priorities among them. These fundamental concerns are with regional (sub-regional/ local/ city) issues: economic development, environmentally sustainable growth, demographics, income-distribution, the spatial distribution of economic activity and regional finances.

Regional CGEs can, in principle at least, incorporate all of these concerns simultaneously, and indeed need to do so if they are to inform policy-makers of all the policy trade-offs they are likely to face. So we anticipate continual growth in the application of CGEs to emerging regional issues, including new policy initiatives at international, national, regional and sub-regional levels that are likely to have spatial impacts. Furthermore, as the parameterization and statistical basis of CGE models are improved, there are likely to be further applications of CGEs to (medium term) regional forecasting and historical simulation.

In Sect. 4.2 we consider likely developments in the specification, computation and parameterization of regional CGE models. Inevitably, these overlap with general developments in CGE modelling, although there are some regional specific aspects to this. Sect. 4.3 looks at one of these promising developments, namely the integration of regional CGEs with energy systems models. Sects. 4.4 and 4.5 consider two illustrative areas in which we anticipate significant further innovations in regional CGE modelling: urban modelling and regional fiscal issues. In Sect. 4.6 we provide brief conclusions.

4.2 Model Computation, Specification and Methodology

Regional CGE models are currently widely used for policy evaluation exercises. They are particularly suited to assess policies that are intrinsically supply side in nature. Additionally, regional CGE models have been valuable to analyse regional economic adjustment mechanisms (e.g. McGregor et al. 1996) in particular the nature of regional adjustment processes, and to identify the implication of alternative assumptions about regional macroeconomic processes against a national or a global economy (Deepak et al. 2001). Regional CGE models have been used to demonstrate the important role played by regional wage setting in shaping the impact of regional policies (including R&D policies) as well as alternative assumptions about the formation of agents' expectations. It is also worth mentioning efforts to identify specific regional features of financial balances and the regional macroeconomic implications of imposing balance of payment constraints (Lecca et al. 2013).

However, for the future, the focus of interest of regional CGE models should expand particularly in the area of regional macroeconomic dynamics, attempting to explain the key regularities (and irregularities) of regional business cycles. This field of research has been comparatively neglected by regional economists in general and regional modellers in particular. Hence, regional CGE models could be used to investigate issues such as the properties and the drivers of the regional business cycle, the impact of financial constraints and the role of banking systems in regional economies. However, to study these important and challenging issues, the modelling approach needs to become more sophisticated, necessitating improvements in the current, widely-used approach. What we propose here is intended to encourage an experimental approach to help to broaden the vision of regional scientists engaged in CGE modelling.

The main improvements we believe are necessary to deal with the problem defined above involve working towards abandoning the deterministic approach and systematically adopting a stochastic modelling framework to provide a statistical underpinning for regional CGE models. If, for example, we wish to use CGE models to determine the drivers of regional business cycles, the model should ideally be validated with statistical techniques; that is to say the variability obtained from the results of the model should in principle reflect that in the data we observe. Increasingly, we think regional CGE models will follow the example of Dynamic Stochastic General Equilibrium (DSGE) models and traditional Real Business Cycle (RBC) models.

The incorporation of inference in regional CGE models could result not only in a better modelling framework, but also in a more complete and potentially much-improved setup compared to conventional DSGE and RBC models. CGE models are multisectoral and often multiregional, and allow for greater flexibility in defining economic and financial closures. Imperfect markets (labour and commodity markets) as well as different treatments of the saving-investment nexus, together with flexibility in choosing financial closures are all elements that make CGE models more suitable than other general equilibrium frameworks in determining what factors govern the macroeconomic adjustments in regional economies. The regional spillover effects, agglomeration and dispersion effects are easier to capture in CGE models than in DSGE and RBC models. Furthermore, the impact of vertical linkages and asymmetric cycles could be better understood in a flexible modelling framework such as stochastic CGE models.

In order to incorporate statistical inference in CGE models we could follow the approach of Canova (1994, 1995). This is extensively applied in DSGE models, but could be adapted for CGE models as well. The way to solve the model is very similar to RBC and DSGE models. By repeatedly solving the model for the empirical distribution of parameters and structural variables it should be possible to evaluate the capacity of the model to reproduce the variance of the actual data. If the variance incorporated into the model is able to explain, say, 75% of the variance of the actual data (generally represented by the GDP growth per capita or other economic variables over a predetermined time period), the defined CGE model would be ready to be used for policy evaluation. The empirically based simulation analysis can then be augmented with bootstrapping for sensitivity analysis. The bootstrapping approach is generally performed using the same principle of the Monte Carlo simulation as in the inference of calibrated models.

The operationalization of statistical inference in regional CGE models should improve their credibility not only as tools for policy analysis, but also as an instruments capable of explaining the main economic adjustments in operation in a region. The CGE model should —in principle— be capable of replicating the most important stylized facts concerning the macroeconomic dynamics of the region. GDP, employment and consumption in the target region should move procyclically, the volatility of GDP, investment and consumption should reflect the characteristics of the region under examination. Furthermore, we would expect log-normal distributions for firm size, while firm growth distributions are expected to follow a tent-shape distribution characterized by tails fatter than the Gaussian (see Botazzi and Secchi 2003).

However, we would not expect that typical stylized facts inferred for a national economy would necessarily hold for a regional economy, given that the region is subject to a common currency area and fiscal policy constraints. What we would expect is some (significant) degree of deviation from mainstream theory as applied to national economies. As commonly found in RBC models, fluctuations in the regional economy are not necessarily determined by supply-side shocks. It is likely that demand shocks will prove to be part of the cause of regional economic fluctuations. Such a finding would not be puzzling; in general, and particularly for regional economies, technological shocks are unlikely to be the only driving force behind cyclical fluctuations.

To allow CGE models to better capture economic fluctuations in the regional economic system, the modelling setup is expected to be closer to new developments in the micro and macro-economic literature than in the past. It is useful to draw from the behavioural theories of the firm by, for example, assuming boundedly rational expectation formation rather than profit maximizing behaviours. It is also important to explore several other possible treatments of dynamic agent behaviour. Efforts to investigate alternatives to perfectly myopic households and fully perfectly foresighted consumers should be on the agenda. Indeed, dynamic choice can also be specified using a hyperbolic Euler relation as in Laibson (1998), where consumers' preferences are dynamically inconsistent since the discount rate should decline as the time horizon increases.

Given that regional economies do not have control over monetary policy, the interaction between financial intermediations and the business cycle may be important for regional economies. Some regions might be more likely to suffer from financial constraints. Under some regional financial systems, severe regional payment problems could arise if there is a continuous decline in bank reserves (for example, due to negative competitiveness effects). In this case, banks might not have sufficient generalized claims to meet the loss and loans would need to be reduced, producing a multiplicative contraction in the regional money supply. The adjustment would then require reductions in income and change in prices with a continuous drain of bank deposits that in turn could generate further income effects. Therefore, there is the potential to extend the focus of CGE models to accommodate regional financial market segmentation as appropriate, by incorporating the banking system and therefore the credit/deposit adjustment mechanisms in the model. These would all be fruitful areas for future research.

4.2.1 Statistical Inference in CGE Models

A CGE model can be formalized with the relation $X_t = f(Z_t, \beta, \pi)$ where the endogenous variables of the model X_t, which typically would include variables such as value added, labour demand, capital stock, consumption, investment, are a function of a set of behavioural parameters β(elasticities of substitution in trade, production and consumption, intertemporal elasticities), structural parameters π (for instance, the depreciation rate and the interest rate) and state variables, Z_t (any other exogenous variables such as policy variables). The time paths of the endogenous variables X_t are obtained by repeatedly solving the model for random (Z_t, β). Therefore one could draw with replacement *i.i.d.* β and Z_t and generate a simulated distribution of $\mu(X_t)$ where $\mu(X_t) \xrightarrow{d} E[\mu(X_t)]$ given a larger number of replications N.

The empirical distribution of β is constructed by assuming a specific distribution based on an a-priori interval. A more straightforward approach suggests assuming the density for β as the product of univariate densities thereby imposing no correlation between the parameters of the model. See Canova and Marrinan 1998, for the case of correlation among parameters. In a typical CGE model, the main parameters β are subject to perturbations that can be identified in the elasticities of substitution in trade σ^T, in production σ^P, the wage curve elasticity ε, the intertemporal elasticity of substitution, $\boldsymbol{v}$ and the elasticity in the migration equation φ.

Typically, the dimension of CGE models is bigger than that of their counterpart models so the major point of concern is the possibility of obtaining information about the distribution of all these parameters. We can in principle estimate their values and in turn obtain information about the statistical distributions of the elasticities. However, this is a time consuming process and researchers would have to commit significant time and effort considering the number of parameters involved. A less rigorous, but more effective path would involve a meta-analysis of the elasticities. For example, from a collection of 15 papers we observe that estimates related to the wage elasticities for the UK economy are in the range 0.03 to 0.15 with a mean in the neighbourhood of 0.1 and standard deviation of around 0.045. According to this information, the resulting empirical distribution for this parameter is normal with mean and standard deviation that equates those of our sample of estimates. Things could be more difficult for production and trade elasticities where the range of variation is generally wider and the point estimates are obtained using quite disparate modelling techniques and data. Alternatively, in the absence of data and lack of previous estimates in the literature, the parameters are generally drawn from a uniform distribution choosing a subjective but sensible range.

The CGE model f as described above is a set of linear and non-linear equations representing Euler equations, first order conditions, steady state conditions and identities. This model in the majority of cases is assumed to be correct, although in some cases, f, is unknown (see for example Canova 1994, 1995). In the case in which we assume no approximation error, the computational problem consists of finding a local or global solution using the joint density of simulated data and parameters drawn from the information set available from the literature as described above. Approximation error is only considered in very few RBC models. Typically, the approximation error is due to the transformation of a non-linear-system into a linear system (the Johansen approach). This should also contain the error of function misspecification (e.g., LES instead of CD) that is already captured by the errors in the parameters.

The next step is to choose the exogenous random shocks **Z**. The standard procedure is to simulate a random productivity shock such that the exogenous stochastic variable follows an AR(1) process. In logs:

$$\ln A_t = (1 - \gamma) \ln \overline{A} + \gamma \ln A_t + \varepsilon_t$$

Where $\overline{A} > 0$ is the steady state level of TFP, $0 < \gamma < 1$ is the first order autoregressive persistent parameter and the error term $\varepsilon_t \sim N(0, \sigma^2)$[1].

As we said above, the exogenous random shock would not necessarily have to be represented by a TFP shock as conventional RBC and DSGE models assume. Indeed, there is no reason to believe the main drivers of the business cycle are influenced solely by supply-side shocks. The issues discussed in this section provide another fruitful avenue of travel for future researchers. Having outlined some areas of focus in model development and specification, we now turn to consider potential areas for development in the *use* of regional CGE models.

4.3 Model Integration with Other Systems and Models

This section is specifically concerned with the interactions between CGE and energy system models. Energy is a vital input to economic activity, while many regions across the world have ambitious targets for reducing the environmental impacts of economic activity, of which many focus on type and scale of energy requirements. Additionally, many regions in the coming decades are likely to experience environmental and natural resource changes that will impact directly upon their economy. In this context it is becoming increasingly important to regional policy makers that they are aware of the links between the delivery of regional and national environmental targets, the changing scale and shape of energy system in a rapidly decarbonising world, and economic activity.

An easy criticism of the use of CGE models in specific cases for policy analysis is where they neglect an important aspect of the "real world". In that case, the CGE model results reflect a set of incorrect treatment(s) of the nature of, or interaction within, the economic system. CGE models have significant advantages as an ex ante modelling system. However other model types —such as "bottom-up" energy

[1]The unconditional mean of the process states that $E(\varepsilon_t) = 0$ and $E(\varepsilon_t^2) = \sigma^2 \ln A$.

systems models of MARKAL or TIMES— can have different foci and strengths of their own. A developing literature brings the features of both model types closer, and so permits more useful advice to policy. Such attempts can include "interacting," "linking," or (in the extreme) "coupling" models together.

Attempts in these directions are to be welcomed. The insights from appropriately connected models can be particularly useful for policymakers seeking to understand the trade-offs in energy and economic policy at the regional level (a recent paper doing so is Santos et al. 2013). This section continues as follows. First, top-down (CGE) and bottom-up energy systems model forms are reviewed. Second, examples from recent efforts to connect the features of CGE models to models of the energy system are examined. Third, the challenges and questions that regional integrated modelling systems might be used to address over the coming decades are explored.

4.3.1 Model Strengths and the Benefits of Interaction

Regional CGE models have obvious strengths for analysing the impacts of policy and non-policy disturbances on an economy. Such models —as discussed elsewhere in this chapter— have been widely applied to a range of issues. CGE models have been termed "top-down" in that they explicitly capture the whole economic system, being based on national or regional economic accounts. Bottom-up energy models on the other hand are characterised by their focus on specific technologies, costs, and resource availability.

What model features are important to appropriately capture the most salient issues? Clearly, these will differ based on the specifics of the job at hand, and the specific characteristics of the region (or area) being modelled. For instance, is the region an energy exporter, an exporter of energy intensive products, or both? Such issues will need to be appropriately and explicitly captured by the chosen model. As Glynn et al. (p. 385, 2015) note, characteristics such as energy intensity, trade, competitiveness and the level of development will be critical for the economic consequences of energy and environmental targets and constraints.

In this context, the strength of CGE models is particularly clear: the models are constructed within, and so constrained to, economic accounts. The dataset at the heart of the analysis reconciles the specific characteristics of the region in question. These will be multisectoral in nature, permitting a detailed assessment of production and consumption within the region. CGE models, however, do not typically capture potential or new technologies, while also tending not to address specific natural resource constraints (either within or external to the region of interest).

Bottom-up energy systems models however, such as MARKAL and TIMES, capture in fine granularity the technological constraints on energy systems, including energy resources, transformation technologies and energy use (including by location). The MARKAL model, for instance, is an optimisation model that minimises the system cost of delivering energy demands where such resources, technologies, uses and demands are exogenously imposed and known with perfect foresight. The model simulations can impose environmental constraints, such as those imposed by stated environmental policies of carbon reduction targets. In the UK, for instance, Kannan and Strachan (2008) reported the energy system requirements compatible with a stated 60% CO2 reduction target.

Early energy systems models were primarily focused on examining the shape and scale of the energy system. The term "hybrid" was developed for models that combined energy systems with other model frameworks, where the purpose was to reconcile "technological explicitness, microeconomic realism and macroeconomic completeness" (Glynn et al. 2015, p. 362). For an early paper on the potential for reconciling bottom-up and top-down systems, see Bohringer (1998). For a view on the rise and usefulness of "hybrid" models for energy issues, see the special issue of the Energy journal, introduced in Hourcade et al. (2006).

More recent versions of hybrid models have acknowledged that CGE models can provide a useful companion modelling system. Different connections have been made between CGE and energy systems models (e.g., see Glynn et al. 2015). We term these "soft-linking" and "full-linking", although, as noted, many such models at the national and global level currently do not adopt full-linking systems. Softlinked models, on the other hand, typically either adopt a single sector connection, or "interact" in more incremental ways.

For instance, Glynn et al. (2015) note how even within this field of study —and with an acceptance of the usefulness of these hybrid approaches— different forms of interaction between CGE and energy systems models occur. These might include soft-linking, e.g., using the CGE system to forecast energy service demands, which are then inputs to the TIMES system, where prices and technologies are reconciled in the energy system (e.g. Fortes et al. 2014). Similar softlinking practices are described in Glynn et al. (2015) for national models of Portugal, Sweden and South Africa. There are clear advantages to soft-linking models: "by soft-linking energy system models and CGE models the energy and climate policy analysis becomes more transparent" (Glynn et al. 2015, p. 385). Inputs and shared "connection points" between the CGE and energy system modelling make the process clear, and the assumptions at each stage obvious. Alternative approaches can be used and compared to earlier interactions between these systems, for instance. Other more formal connections, including full-form linking or coupling appear likely but are not part of national models to date.

A "whole economy-energy" system model would offer a technically pleasing solution to reconciling energy and economy models; however, it must be done with care. One clear advantage of CGE systems is the (in principle) traceability of the results, and regional scientists would do themselves no favours by combining two complex models without regard for the clear communication —and validation— of model results. Interactions between energy and economy are likely to become more important in the future, and so frameworks that permit these to be jointly considered —such as hybrid CGE-energy systems models— may become a central part of regional scientists' toolkits. As energy prices evolve, new technologies develop and old technologies become more niche in their applications, notions of regional economic competitiveness may increasingly be affected by a region's ability to meet its energy requirements. In a decarbonised future, access to energy resources will increasingly drive the timing, spatial pattern and level of economic activity.

Ambitions in energy security are driving many regions towards increasing interconnectivity through enhanced and extended networks and connections with future political, as well as economic, implications. At the other extreme, many regions are focusing on energy independence through the increased use of region specific resources, particularly renewables. As regions face the future, decisions taken over their energy policy will have profound implications for their economic directions.

4.4 Urban Modelling

In recent years there has been increased interest in developing and extending CGE models to better capture more localized economies. In this short section we consider the future use of CGE models at the urban level. In doing so, a necessarily brief outline of existing work in the area will be given. In using the term 'urban' in this context, we have in mind an economy characterized by a central city and surrounded by a number of other jurisdictions as part of the broader urban hinterland. However, within-city analysis can also be incorporated in our discussion, although this would limit the scope for the exercise of many economic policy levers which CGE models are used to analyse.

One of the main reasons for the increase in interest in developing CGE models at a more localized level is the interest in the operation of fiscal powers at such a level. Historically, in the UK case, this has been as a result of the devolution of economic powers to different regions of the UK, and this has spurred the development of more elaborate and detailed CGE models at the regional level (e.g. Ferguson et al. (2007), Harrigan et al. (1991), Harrigan et al. (1992), Lisenkova et al. (2010) and Lecca et al. (2014)). This has now extended further to an interest in better capturing city economies as a result of the growth of so-called 'city deals' for a number of UK cities which includes potential for greater fiscal autonomy. In the U.S. case, such fiscal devolution, or "fiscal federalism" as it is known., has been embedded in the fabric of governance structures much longer, and it is this literature which provides our point of departure in this section.

There is a series of papers seeking to reflect the fiscal federalism structure in the U.S. within the CGE modelling framework. Notable contributions in this area are the works of Nechyba (see for instance Nechyba (1996a, b, 1997)). In these papers, the aim is to capture the complex interactions which exist when examining local tax policy and its impacts on economic agents and the aggregate economy. Nechyba (1996a), for example, presents a model with heterogeneous agents who are endowed with property and income, who can move freely between jurisdictions, and who can vote to determine local and national/state tax policy. Using this model Nechyba (1996a) shows the impact of different models of revenue sharing between local authorities. Importantly, and unlike most regional or national CGE models, these multi-level urban CGE models capture local taxation through labour ***and*** property taxation.

Another series of papers have used similar modelling approaches to study the impact of changes from one tax base to another at a local and state (regional) level (England (2003); England and Zhao (2005)).

Specifically, England (2003) looked at a move from a uniform property tax to a land value tax, keeping revenue constant. England and Zhao (2005) took this a stage further and considered the distributional impacts of such a change. This fiscal literature has largely focused on better capturing the nature and operation of local taxation in CGE models, however there is more to the urban economy than simply the operation of fiscal powers (even if these may be the most obvious policy issue to focus on).

Kilkenny (1999) made an early effort to extend CGE models to capture more explicitly the interconnectedness of the urban economy with its wider spatial economy. This paper raised a number of issues; some which have subsequently been neglected in the broader applied literature, while some (e.g., issues around transportation costs) have been formally embedded within (particularly spatial) CGE models. The embedding of ideas of urban agglomeration within the CGE model framework is key to understanding the operation of the urban economy. Just as capturing the economic and social openness and interconnectedness of a region within a country is critical to meaningful policy evaluation at the regional level, so too is capturing the interconnectedness of our cities with their broader hinterland in economic and social terms in urban policy evaluation.

The obvious starting point here is the operation of the labour market. Regional CGE models implicitly assume that region of residence and work are the same, and that workers migrate between regions, or from outside the region more generally, in response to economic incentives (usually wage differences). Yet in the urban setting, workers may operate in either the urban labour market or in the labour market of the broader hinterland with some ease, while not moving their place of residence and main place of consumption. Better capturing and embedding the operation of the urban labour market within CGE models will be key to understanding the incentives faced by different forms of local taxation. For instance, an income based local taxation will likely induce movement of firms and workers, just as one would expect in an interregional setting, but is less likely to affect the location of residence of households directly. Meanwhile, the operation of a property tax in an urban setting is likely to impact on the location decision of households and firms, and thus indirectly on the location of work.

Relatedly, a better understanding of the operation of the housing market at a local level is likely to be critical to modelling the location decisions of households in terms of access to work and residence. Similarly, understanding the commercial property market in terms of firm location decisions will be important. While regional CGE models capture migration into the region from outside, and in an interregional setting one could endogenise migration between regions, this does not address the assumed link between working and living in the same area in these CGE models.

One example that brings these issues to the fore is some recent work at the regional level, discussed in the next section on fiscal issues. This has sought to consider the valuation of public spending more explicitly in the utility function of workers; something which is simplified by the implicit assumption that workers both work and reside — and thus are taxed and consume— within the region in question. This implicit assumption is infeasible at the urban scale. Capturing the interconnectedness between place of residence, place of work, and the places of consumption is going to be crucial for meaningful analyses at the urban level. Economic agents do not have spatially limited consumption behaviour, and with the provision of locally paid for public goods, the utility function of the consumer needs to embody this interdependence.

While spatial CGE models are not the subject of this section, or indeed this chapter, it is worth noting that while the main focus of these models has been on transportation analysis and modelling, significant strides have been made in incorporating some of the household and firm location decisions discussed above. In many cases, the uses of spatial CGE models has been to understand travel and location decisions (e.g., the studies summarized in Anas 2013). While this is unlikely to be the focus of any of the types of analyses discussed here, it is nevertheless the case that better specifying firm and household location decisions will enhance the empirical results derived from CGE modelling at the urban scale. In developing urban CGE models further, there is much to learn from existing spatial CGE modelling work (Choi and Sjoquist (2015) is a useful recent paper in this area).

Finally, we should note that one consequence of more elaborately capturing the operation of the urban economy within CGE models is that it will necessitate the specification of more key parameters. While in many cases at a more aggregate level, key elasticities and other parameters are assumed or sourced from other studies, given the localized focus of urban models, a number of these estimates will ideally be produced for the urban area in question, or at least produced at the urban level. While this is an econometric challenge, it is not something beyond the abilities of existing methods, even if the existing data are sometimes lacking.

There is, of course, much more in this literature than has been sketched out here, but the purpose of this section has been to give some flavour for the types of issues that have already been investigated using CGE models at the more localized level, and to identify areas where further work and development would be welcomed. The key areas we identified here relate to better modelling of urban migration, urban land and property markets, and more generally urban amenity and agglomeration. In policy terms, local fiscal analysis is likely to be one of the key beneficiaries of improvements in this modelling framework, and the next section of this chapter considers regional fiscal issues in more depth.

4.5 Regional Fiscal Issues

A major attraction of regional CGE models is their ability to capture alternative fiscal frameworks and to allow analysis of issues relating to fiscal federalism. There has already been considerable work on regional fiscal issues, aspects of which are discussed in Partridge and Rickman (1998, 2010) and Giesecke and Madden (2013). We focus here on possible future applications and extensions. The fact that alternative inter-governmental financing arrangements can be captured within a regional CGE offers considerable potential for the comparative analysis of alternative regional fiscal arrangements that has yet to be fully explored. This could, for example, be used to analyse the properties of alternative real-world fiscal systems, but on a common database (so that differences in structure are controlled for).[2] Equally, any proposed changes in regional fiscal systems can be subjected to rigorous analysis.

Given our location and interests, this section will use examples from recent changes in the devolution settlement in the UK to illustrate broader issues of importance in regional fiscal CGE modelling. For example, the Scottish Government has recently experienced a substantial increase in its revenue-raising powers and is set to enjoy further enhancement of these powers to the point where, in terms of its control over both revenues and public spending, it will become one of the most devolved regions in the world. One impact of this is to enhance the Scottish economy's sensitivity to policy and other shocks: government expenditure will now have to adjust to track changes in tax revenues (given limited borrowing powers), whereas previously public spending was unaffected by changes in Scottish tax revenues.

The endogeneity of public spending may increase the incentives both for the Scottish Government to adopt growth enhancing policies and for the electorate to vote for such policies. However, this also implies that negative asymmetric shocks will exert a more significant contractionary impact on the Scottish economy and that there is some scope for borrowing to mitigate the scale of such impacts. Furthermore, the dynamic response of the fiscal system to shocks may inhibit adoption of policies that take a long time to stimulate economic activity. Supply side policies, including balanced budget changes in corporation tax, can take a long time to generate positive effects—longer than the typical lifetime of a government (Lecca et al. 2016). Most regional CGE analyses of fiscal issues have not explored the implications of alternative fiscal systems, but rather have focused on particular changes in taxes or public spending, with the emphasis typically on the former. One regional fiscal issue that is due more systematic investigation than it has had to date, concerns the treatment of public expenditures within regional CGEs.

[2]We do not mean to imply that there has been no work of this kind (e.g., Nechyba (1996a, b), whose work is discussed above, and Ferguson et al. (2007)) rather that there is considerable potential for further analysis.

One aspect of this issue, linked to the fiscal federalism literature, is the extent to which potential migrants value the amenity provided by current public spending and take that into account in their location decisions. In the presence of imperfectly competitive labour markets, it is also possible that such amenity effects would be taken into account in the wage bargaining process (Lecca et al. 2014). The idea here is that workers may bargain over a "social wage" that attributes positive value to public spending as well as to private consumption. This can matter a great deal for the macroeconomic impact of balanced budget fiscal changes, since bargaining over the social wage eliminates the adverse supply-side effect that would otherwise predominate as workers seek to restore their post-tax real wage. However, there is — as yet— little evidence from opinion polls that the Scottish people would be willing to sacrifice —at least to any significant degree— public for private consumption. Internationally, however, Scandinavian systems appear to be based on these kinds of considerations, where a centralized bargaining system takes account of the provision of public services as well as the level of post-tax wages.

Furthermore, there is evidence that the public (and migrants) value some elements of public spending more than others: health and education tend to be highly valued and welfare spending much less so. This opens the possibility of heterogeneous system responses to balanced budget fiscal changes depending on the composition of public spending, a possibility that may well be worth further exploration.

One aspect of the composition of public spending that has not been neglected is the distinction between government capital and current expenditure. The common assumption that current government expenditure has no supply effects cannot legitimately be extended to the case of public capital expenditure in general, and infrastructure spending in particular (e.g., Giesecke 2008). Again, the potential long delay until beneficial supply side effects predominate can raise particular issues for regional governments that are subject to a balanced budget constraint. (Lecca et al. 2016). Furthermore, what is classified as current government spending in fact represents investment in human capital and, as such, would be expected to have potentially important supply-side effects, as well as the expenditure effects which are the focus of conventional "impact studies". Clearly, education is an example, and while there has been some regional CGE analysis much remains to be done (for example, in terms of the system-wide impacts of early years interventions and work-based learning).[3]

A further under-researched example is health, where there is compelling evidence both of the impact of the economy on health and vice versa (e.g. reduced days lost through sick leave and enhanced working life durations), but as yet there appears to have been no attempts to provide a fully system-wide regional analysis.[4] Of course, recognition of the supply-side impacts of public spending in these areas again works

[3]See e.g.Giesecke and Madden (2006); Hermannsson et al. (2014); Kim et al. (2016).

[4]However, Mayeres and Van Regemorter (2008) provide an analysis of this kind for the economies of the EU.

to mitigate or offset any adverse supply effects of wage claims to restore workers' take home pay following a balanced-budget fiscal expansion.

In multi-regional models interdependence among regions becomes central. The presence of spillovers and feedback among regions has potentially important implications for policy, as regional economists have long recognized. The proposals for enhanced fiscal devolution to Scotland made by the Smith Commission proposed a "no detriment" principle, according to which fiscal decisions by the Scottish government or the UK government that impacted adversely on the other should be compensated. In fact, in the new Fiscal Framework automatic compensation is restricted to "direct" effects only, but, at least in principle, an explicitly interregional CGE analysis can quantify any system-wide spillover and feedback effects.[5]

Longer term we would anticipate increasing efforts to assess the significance of spillovers to facilitate improved regional fiscal policy and, of course, not simply in the UK context, but also in the EU and North America. The presence of significant spillovers creates a potential for improved outcomes through the coordination of policies at the regional (sub-regional/ city) levels, and this could merit exploration using similar approaches to those adopted in the macroeconomics literature (e.g., McKibbin and Sachs 1989). There is also considerable scope for further development of a political economy approach in a regional CGE context (Groenewald et al. 2003).

The treatment of expectations and dynamics in regional CGE analyses of fiscal issues has typically been fairly crude. Indeed, comparative static models are still the most common form of regional CGE used for analysing regional tax changes. To the extent that models have been dynamic at all, they have tended to be recursively dynamic, with movement through time being generated by stock updating processes (for population and capital stocks). However, forward looking models have been developed and hold the potential for a more nuanced analysis of the dynamics of regional fiscal policy (and the impacts of national fiscal policies) (e.g., Lecca et al. 2014). So it becomes possible, for example, to distinguish between the impact of anticipated and unanticipated regional fiscal policy changes. Of course, while the limiting cases of (universal) perfect foresight and (universal) myopia offer useful benchmarks for likely adjustment paths, we would expect future exploration of hybrid models, which allow a degree of heterogeneity among transactors.

Even in the context of analyzing tax changes, there may also be some benefits to exploring the significance of alternative "behavioural" specifications for transactors— particularly for households. For example, there is some survey evidence in the UK to suggest that households are much more likely to respond to increases in taxation through variations in tax rates than they are to changes (or the absence of changes) in tax thresholds, though even if true, there would be concern about the legitimacy of policies that seek to exploit such "irrational" behavior.

[5]Lecca et al. (2015) attempt to do this. However, they identify a major concern here, namely absence of official measures of interregional trade flows.

Demographics are crucial in governing longer-term pressures on regional fiscal finances, and there is considerable scope for further exploration of this issue. There are a number of possible approaches. One of the simplest is to use an augmented demographic module linked to a CGE framework (e.g. Lisenkova et al. 2010). However, a comparatively recent development, at least in a regional context, uses an overlapping generations (OLG) framework to track the ageing of cohorts in a regional setting (Lisenkova et al. 2015). In the Scottish context, for example, the new fiscal framework implies that the Scottish Government budget will be under pressure as its population is projected to be ageing (on average) more rapidly than that of the rest of the UK (RUK), with implications for the composition (and likely total) of public spending. Longer-term, slower population growth is likely to add further pressure to the region's public finances.[6] Such frameworks offer the potential to capture long-term pressures on the public finances through changes in the levels and age structure more accurately. Of course, it is interesting in such contexts to explore the extent to which indigenous demographic pressures might be offset through interregional migration.

While the discussion of this section has been mainly in terms of ***regional*** public finances, a similar approach to sub-regional and local/city public finances would be appropriate, particularly where the relevant authorities are responsible for the setting of some taxes, or even where some tax revenues are assigned (e.g. replacement of council taxes with a local income tax). The issue of the timing of fiscal effects has come up in a number of contexts, and the systematic exploration of alternative means of financing fiscal changes is an area where further research would be welcome. Even where current arrangements preclude deficit financing, it would be useful to explore the potential role of borrowing in smoothing the time paths of adjustment, of course, in the context of dynamic models with an appropriate treatment of expectations formation.

4.6 Conclusions

We anticipate two major types of development in regional CGE models: improved specification, parameterization and solution methods; further applications to policy relevant issues. In the case of methods, we expect a whole range of developments, some of which have already been initiated, but none of which have yet been fully developed and applied. Some of these methods are likely to be mutually exclusive, and all are, at least in part, made possible by enhanced computing power. The development we have particularly emphasized in Sect. 4.2 is enhancing the statistical basis of CGE models, which will close the gap between CGEs and DSGEs.

[6] In Scotland this may well happen when the new Fiscal Framework comes up for review in 2020.

We have throughout referred to a number of other possible developments, including: improved modelling of expectations; the development of behavioural CGEs (incorporating e.g., risk aversion, inertia, hyperbolic discounting); endogenous technical change, learning and innovation; improved treatment of space; the further development of regional OLG models; more sophisticated treatments of imperfect competition; allowing for endogenous policy formation, for example, in the context of political economy models, and the incorporation of special modules with much more detail on the behavior of key tansactor groups (e.g., micro-simulation models of individuals' behaviour) or sectors (e.g., energy and transport systems models). Such developments will enhance the capacity of regional CGEs in forecasting and historical simulation.

However, the process should not be about increased sophistication for its own sake. As models grow in complexity, the more difficult it becomes to retain an intuitive grasp of key transmission mechanisms, and the more challenging it becomes to explain the results—to ourselves as well as to policy makers.[7] Models should be selected so as to be appropriate to the key questions of interest, and augmented in a transparent and informative way that carefully builds upon earlier models and thereby avoids the "black box" criticism.

While we have provided some examples here, including more systematic analyses of the effects of public expenditure (e.g. on health and education), it is virtually impossible to anticipate the whole range of potential applications of future generations of regional CGEs. The flexibility of the modelling approach, and past experience, tells us that we can say little about future applications, beyond acknowledging that they will be even more wide ranging and extensive. The potential seems limitless, and relevance to policy will be assured if the model specifications incorporate the range of policy objectives and the transmission mechanisms of policy instruments, so that key trade-offs and "double dividends" can be identified and quantified.

The applications will reflect emerging regional issues (which will in turn be associated with international, national, regional and local disturbances) and policy concerns. Increasing awareness of global warming, for example, stimulated numerous regional energy-economy-environment CGEs and their application to, for example, renewables, carbon taxes and emission trading schemes. However, fundamental issues will continue to reflect the long-standing concerns of regional scientists: economic development and employment; environmentally sustainable growth; skills; equity; the spatial distribution of economic activity, and regional finances. Ultimately, the value of future generations of regional CGE models will be assessed in terms of their ability to contribute to our understanding of regions

[7] Giesecke and Madden 2013, suggest a "back of the envelope" approach to enhancing understanding of model results. The approach we have adopted is to use simplified analytical models as appropriate, while using a very flexible modelling framework that allows us to track the source of any model "surprises" (e.g. Lecca et al. 2014).

and facilitate improved government policies that enhance the well-being of their inhabitants.

Acknowledgements The authors acknowledge funding from the ESRC under *The Constitutional Future of Scotland and the United Kingdom pre and post Referendum* Initiative (grant ES/L003325/1) and from ClimateXChange, the Scottish Government-funded Centre of Expertise in Climate Change. The authors are solely responsible for the content of the paper. The views expressed are purely those of the authors and may not in any circumstances be regarded as stating an official position of the European Commission or Scottish Government. The authors are grateful to the editors for comments on an earlier draft.

References

Anas A (2013) A summary of the applications to date of RELUTRAN, a microeconomic urban computable general equilibrium model. Environ Plann B Plann Des 40(6):959–970

Bohringer C (1998) The synthesis of bottom-up and top-down in energy policy modelling. Energy Econ 20(3):233–248

Bottazi G, Secchi A (2003) Why are distributions of firm growth rates tent-shaped. Econ Lett 80(3):415–420

Canova F (1994) Statistical inference in calibrated models. J Appl Econ 9:123–144

Canova F (1995) Sensitivity analysis and model evaluation in simulated dynamic general equilibrium economies. Int Econ Rev 36(2):477–501

Canova F, Marrinan J (1998) Sources and propagation of international output cycles: common shocks or transmission? J Int Econ 46(1):133–166

Choi KW, Sjoquist DL (2015) Economic and spatial effects of land value taxation in an urban area: an urban computable general equilibrium approach. Land Econ 91(3):536–555

Deepak MS, West CT, Spreen TH (2001) Local government portfolios and regional growth: some combined dynamic CGE/optimal control results. J Reg Sci 41(2):219–254

Dixon PB, Jorgenson D (2013) Handbook of computable general equilibrium modelling. North-Holland, New York

England RW (2003) State and local impacts of a revenue-neutral shift from a uniform property to a land value tax: results of a simulation study. Land Econ 79(1):38–43

England RW, Zhao MQ (2005) Assessing the distributive impact of a revenue neutral shift from a uniform property tax to a two-rate property tax with a uniform credit. Natl Tax J 58:247–260

Ferguson L, Learmonth D, McGregor PG, Swales JK, Turner K (2007) The impact of the Barnett formula on the Scottish economy: endogenous population and variable formula proportions. Environ Plan A 39(12):3008–3027

Fortes P, Pereira R, Pereira A, Seixas J (2014) Integrated technological-economic modelling platform for energy and climate policy analysis. Energy 73:716–730

Giesecke JA (2008) A top-down framework for regional historical analysis. Spat Econ Anal 3:45–87

Giesecke JA, Madden J (2006) A CGE evaluation of a university's effects on a regional economy: an integrated assessment of expenditure and knowledge impacts. Rev Urb Reg Dev Stud 18:229–251

Giesecke JA, Madden J (2013) Regional computable general equilibrium modeling, chap. 7. In: Dixon PB, Jorgenson DW (eds) Handbook of computable general equilibrium modelling. Elsevier, Amsterdam, pp 379–475

Glynn J et al (2015) Economic impacts of future changes in the energy system—National perspectives. pp. 359–387. In: Giannakidis et al (ed) Informing energy and climate policies using energy systems models. Springer

Groenewold N, Madden JR, Hagger AJ (2003) The effects of interregional transfers: a political economy CGE approach. Pap Reg Sci 82:535–554

Harrigan F, McGregor PG, Dourmashkin N, Perman R, Swales K, Yin YP (1991) AMOS: a macro-micro model of Scotland. Econ Model 8(4):424–479

Harrigan F, McGregor PG, Swales JK, Dourmashkin N (1992) Imperfect competition in regional labour markets: a computable general equilibrium analysis. Environ Plan A 24(10):1463–1481

Hermannsson K, Lecca P, Lisenkova K, McGregor PG, Swales JK (2014) The regional economic impact of more graduates in the labour market: a "micro-to-macro" analysis for Scotland. Environ Plan A 42(2):471–487

Hourcade J-C, Jaccard M, Bataille C, Ghersi F (2006) Hybrid modelling: new answers to old challenges. Hybrid Model Energy Environ Policies, Special issue of Energy J 27:1–11

Kannan R, Strachan N (2008) Hybrid modelling of longterm carbon reduction scenarios for the UK. Energy Econ 30:2947–2963

Kilkenny M (1999) Explicitly spatial rural-urban computable general equilibrium. Am J Agric Econ 81(3):647–652

Kim E, Hewings GJ, Lee C (2016) Impact of educational investments on economic losses from population ageing using an interregional CGE-population model. Econ Model 54:126–138

Laibson D (1998) Life-cycle consumption and hyperbolic discount functions. Eur Econ Rev 42(3–5):861–871

Lecca P, McGregor PG, Swales JK (2013) Forward-looking versus myopic regional CGEs: how significant is the difference? Econ Model 31:160–176

Lecca P, McGregor PG, Swales JK, Yin YP (2014) Balanced budget multipliers for small open regions within a federal system: evidence from the Scottish variable rate of income tax. J Reg Sci 54(3):402–421

Lecca P, McGregor PG, Swales JK (2015) Scotland-no detriment, no danger: the interregional impact of a balanced budget fiscal expansion, IPPI Policy Paper. http://strath-prints.strath.ac.uk/55430/

Lecca, P, McGregor PG. and Swales JK (2016) Taxes and spending, Chap. 2. In: Keating M (ed) A wealthier, fairer Scotland: the political economy of constitutional change. Oxford, Oxford University Press

Lisenkova K, McGregor PG, Pappas N, Swales JK, Turner K, Wright RE (2010) Scotland the grey: a linked demographic—computable general equilibrium (CGE) analysis of the impact of population ageing and decline. Reg Stud 44(10):1351–1368

Lisenkova K, Sanchez-Martinez M, Sefton J (2015) The sustainability of Scottish public finances: a generational accounting approach. NIESR Discussion Paper No 456

McGregor P, Swales JK, Yin YP (1996) A long-run interpretation of regional input-output analyses. Journal of Regional Science 36(3):479–501

McKibbin WJ, Sachs JD (1989) The McKibbin-Sachs global model: theory and specification. NBER paper 3100

Meyeres I, Van Regemorter D (2008) Modelling the health related benefits of environmental policies and their feedback effects: a CGE analysis for the EU countries using GEM-E3. Energy J 29(1):135–150

Nechyba T (1996a) A computable general equilibrium model of intergovernmental aid. J Public Econ 62(3):363–397

Nechyba T (1996b) Fiscal federalism and local public finance: a computable general equilibrium (CGE) framework. Int Tax Public Financ 3(2):215–231

Nechyba TJ (1997) Existence of equilibrium and stratification in local and hierarchical Tiebout economies with property taxes and voting. Econ Theory 10(2):277–304

Partridge M, Rickman D (1998) Regional computable general equilibriuim modelling: a survey and critical appraisal. Int Reg Sci Rev 21:205–248

Partridge M, Rickman D (2010a) Computable general equilibrium (CGE) modelling for regional economic development analysis. Reg Stud 44(10):1311–1328

Santos GF, Haddad EA, Hewings GJ (2013) Energy policy and regional inequalities in the Brazilian economy. Energy Econ 36:241–255

Grant J. Allan is lecturer, Department of Economics, University of Strathclyde. His research interests are primarily in regional economic analysis and modelling and energy economics. He is additionally Deputy Director of the Fraser of Allander Institute at the University of Strathclyde where he has led and been involved in a large number of research programmes funded by external organisations including the Scottish Government, national (UK) research councils and international organisations. Dr Allan earned his MSc in Economics through the Scottish Graduate Programme in Economics and his PhD in economics from the University of Strathclyde in 2016.

Patrizio Lecca is a Scientific Officer of the European Commission, DG Joint Research Centre, JRC. His primary research interests are on regional economics, environmental and energy economics. He was a research fellow at the Fraser of Allander Institute, Department of Economics, University of Strathclyde. Dr. Lecca earned the Ph.D. in Economics from the University of Strathclyde in 2011.

Peter G. McGregor is Professor (Emeritus) in the Department of Economics, University of Strathclyde and Director of the the Fraser of Allander Institute. He is a previous Director of the International Public Policy Institute and Head of Department of Economics, University of Strathclyde. He has led many externally funded research projects and published extensively in international journals. His current research interests include the impact of fiscal autonomy on the Scottish and UK economies, and modelling energy-economy-environment interdependence, including the economic and environmental impact of renewables technologies at the regional and national levels. He has held visiting academic posts in Germany, Japan, Sweden and the US.

Stuart G. McIntyre is a lecturer, Department of Economics, University of Strathclyde. He is also affiliated with the Fraser of Allander Institute at the University of Strathclyde and the Regional Research Institute at West Virginia University. His research interests are primarily in regional, spatial and energy economics and applied spatial econometrics. Dr McIntyre earned his MSc in Economics through the Scottish Graduate Programme in Economics at the University of Edinburgh, and his PhD in economics from the University of Strathclyde in 2013.

J. Kim Swales is a Professor (Emeritus) at the Fraser of Allander Institute, Department of Economics, University of Strathclyde. His research interests are primarily in Multi-sectoral economic modelling, regional economic analysis and policy and in energy modelling. He has previously been Director and Research Director of the Fraser of Allander Institute and Head of the Economics Department at the University of Strathclyde.

Chapter 5
Measuring the Impact of Infrastructure Systems Using Computable General Equilibrium Models

Zhenhua Chen and Kingsley E. Haynes

5.1 Introduction

Regional impact assessment of infrastructure systems is an important public policy concern given its relevance to economic development and homeland security. A valid understanding of the linkages between various infrastructure systems and growth of national and regional economies is vital for the development of sound policies targeting investment and system protection. The relationship between public expenditure and aggregate productivity has been explored for over for two decades following the path-breaking study conducted of Aschauer (1989), who argued that critical infrastructures, such as streets, highways, airports, mass transit, sewers, and water systems, play significant roles in promoting economic growth and productivity improvement. The positive impact of infrastructure systems on the economy has been widely confirmed. However, the extent and magnitude of its contributions are still not well understood. The marginal economic contributions of infrastructures were found to vary substantially across different studies. This is not surprising given that previous studies conducted impact assessments with focuses on different infrastructure systems, geographic locations and time periods. The different scales of analysis and data being adopted were also found to lead to different research findings.

Methodology is another key factor that affects a valid understanding of an infrastructure system's contribution to the economy. A plethora of pioneering studies evaluated the economic contribution of infrastructure to economic growth

Z. Chen (✉)
City and Regional Planning, The Ohio State University, Columbus, OH, USA
e-mail: chen.7172@osu.edu

K.E. Haynes
Schar School of Policy and Government, George Mason University, Arlington, VA, USA

R. Jackson, P. Schaeffer (eds.), *Regional Research Frontiers - Vol. 2*,
Advances in Spatial Science, DOI 10.1007/978-3-319-50590-9_5

and productivity improvement following a neo-classical approach by measuring the economic output elasticity of infrastructure through some forms of aggregated production function using a regression format (Duffy-Deno and Eberts 1991; Gramlich 1994, 2001; Harmatuck 1996; Nadiri and Mamuneas 1996; Fernald 1999; Boarnet 1997; Boarnet and Haughwout 2000; Mattoon 2002; Bhatta and Drennan 2003). The estimated output elasticities are found to vary substantially with a range between −0.15 and 0.56, due to the differences in the data and specific modeling forms (Melo et al. 2013).

Such evaluations of infrastructures' contribution to economic growth and productivity through an econometric analysis can only be considered a partial equilibrium assessment. This is because the relationship between economic growth and/or productivity improvement and infrastructure is only evaluated from the supply side, in other words, only a part of the market to attain equilibrium. This is due to the implicit assumption of a constant demand as a response to infrastructure change during the period of investigation. The indirect impact on the economy as a result of demand change cannot be adequately captured in a regression-based supply-side model. For instance, influence as a result of transportation infrastructure investment on the price change of final commodities and ultimately on change of disposable income of households cannot be captured in a partial equilibrium assessment. Hence, a general equilibrium assessment is indispensable to achieve a comprehensive economic impact of infrastructure with considerations from both the supply and the demand side.

With the advancement of computational technology and applied modeling platforms, Computable General Equilibrium (CGE) models have been widely adopted to assess the economic impact of infrastructure systems. Originally developed by Johansen (1960), CGE is an applied microeconomic modeling system that uses actual economic data to estimate interactions between the economy and changes in policy, technology or other external factors. Unlike partial equilibrium assessment, CGE captures the interactions among various markets and between both demand and supply through a simultaneous equations system that can involve thousands of equations and variables. The analysis is built on the Walras-Arrow-Debreu theory of general equilibrium (Arrow and Debreu 1954), with modern modifications and extensions allowing for imperfect markets. Because it provides clear linkages between the microeconomic structure and the macroeconomic environment, CGE can be used to simulate the interrelationships among multiple industrial sectors and markets.

A typical CGE modeling mechanism is illustrated in Fig. 5.1. The model consists of a series of simultaneous equations that are calibrated using two data sources: a social accounting matrix (SAM) which measures the initial economic activities under equilibrium, and a set of parameters including different types of elasticities of substitution. After calibration, the model is rerun to calculate a new equilibrium based on prerequisite policy shock conditions and closure rules. The output of CGE analysis normally contains indicators of welfare, the macro and the micro effects, which can be used to evaluate the magnitude of impacts as well as for policy analysis.

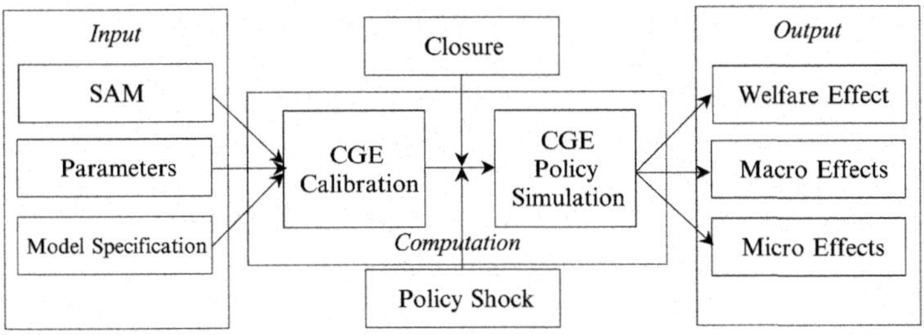

Fig. 5.1 CGE modeling mechanism

A standard CGE modeling framework consists of four sets of institutions: producer, consumer, government, and foreign trade. Each institution interacts with others while maximizing its utility or profit under relevant constraints. The production structure is often measured through either a Cobb-Douglas production function form or a constant elasticity of substitution (CES) form for aggregate factors of production, whereas fixed coefficient relationships are used for intermediate inputs. Value added from primary factors, together with intermediate inputs, generate the final output. The model specifies goods produced in different countries or regions as imperfect substitutes. Sectoral output is modeled through a constant elasticity of transformation (CET) aggregation of total supply to all export markets and supply to the domestic market by following the approach of Lewis et al. (2003). The allocation of goods between exports and domestic markets is set to maximize revenue from total sales. Government plays dual roles as both a policy maker in terms of providing exogenous shocks to the economy and a consumer in terms of allocating public funds collected from taxes and tariffs to various fields, such as public affairs, intergovernmental transfers and subsidies.

General equilibrium analysis is usually conducted at the national level due to data availability, but in recent years, the analysis has also been more widely applied for assessments of regional economic impact and related policy issues. Partridge and Rickman (1998) conducted a comprehensive appraisal of regional CGE models based on 36 empirical studies conducted between 1983 and 1997. A summary of their appraisal suggests that future research in regional CGE modeling should be focused on the following directions:

- Restrictions on functional forms of production activities should be relaxed;
- Inherent uncertainty in predictions of regional CGE should be examined systematically;
- Sensitivity should be examined for the conditioning assumptions against available data and empirical evidence;
- Attention should be paid to the dynamics, or time-paths of relationships in a regional economy.

Indeed, regional CGE modeling has been greatly enhanced in terms of computational algorithms and modeling frameworks since Partridge and Rickman's appraisal in 1998. Current CGE models are equipped with a flexible nesting structure, which allows for a detailed representation of any particular type of production activities. The bottom-up framework of many regional CGE models has generally improved with capacities to capture heterogeneous regional economic structures and diverse interregional trade flows. In addition, many recent CGE models have been upgraded with dynamic functions that allow for dynamic recursive simulations of regional economic activities in response to policy shocks.

Despite these advances, issues such as sensitivity related to parameterization, reliability of simulation and theoretical underpinnings of shock mechanisms remain critical concerns in regional CGE models and are still not well understood. This essay discusses these issues with a focus on regional impact assessments of transportation infrastructure by following the path of Partridge and Rickman (1998). The objective is to stimulate scholars and practitioners to rethink the fundamentals of regional CGE modeling and its applications in infrastructure appraisal and suggest future research directions. The rest of this chapter is organized as follows. Section 5.2 summarizes the various CGE frameworks being developed. Section 5.3 discusses the major challenges in CGE modeling, with a focus on the application of CGE for impact assessment of transportation infrastructure. Section 5.4 outlines other future research directions.

5.2 Different Frameworks of CGE Modeling

CGE inherits the advantage of Input–Output analysis in terms of capturing economic transactions among various economic sectors and entities in the form of a Social Accounting Matrix (SAM). Because the sectoral scheme of a SAM can be aggregated or disaggregated depending on any specific research purpose, the modeling framework of CGE is flexible and can be applied to impact assessments of different sectors. In addition, CGE can also be modified for impact assessment at various regional scales and it can be upgraded with dynamic functions for long-term impact assessments and forecasting. The following section summarizes the key features and status of development of four CGE modeling frameworks: a static single-region CGE, multi-regional CGE, dynamic CGE and dynamic-recursive multi-regional CGE. Understanding the various CGE modeling frameworks is important as it helps to recognize the gap between the current CGE modeling and future research needs.

5.2.1 Static Single-region CGE

Single-region static CGE is the standard modeling framework of CGE analysis and has been widely adopted for impact assessment with a focus on a single region, which is often applied to a national level assessment. The model is static because only year-one impacts of a policy shock are considered. ORANI is one of the early single-region CGE models developed by a team at the Centre of Policy Studies (CoPS) in Australia (Dixon et al. 1982). A single and static CGE model has a wide range of applications in impact assessments of transportation infrastructures. Most single-region models were originated from the tradition of Dervis-DeMelo-Robinson (Dervis et al. 1982). For instance, the International Food Policy Research Institute (*IFPRI*) model, developed by Lofgren et al. (2002), is one of the examples of the standard single region static CGE model. The role of transportation services is modeled through the related sectors and transaction cost in the SAM and the model. Conrad (1997) developed a theoretical modeling framework for a static single-region CGE model to investigate the role of transportation services on congestion and air pollution, which was primarily achieved through the development of detailed cost functions with multi-level nesting structures. McDonald (2005) also developed a single-region CGE model in the tradition of the Dervis-DeMelo-Robinson model. The model was applied by Chen and Haynes (2013) to evaluate the national economic and welfare impacts of six modes of transportation infrastructures in the U.S., including truck, rail, air, transit, water and pipeline. The impacts of transportation infrastructure investments were measured through policy shocks from investments in capital stocks of related transportation sectors. Their study found a positive but small stimulus effect of transportation infrastructure investment on growth of the national GDP and welfare in the U.S.

Economic Consequence Analysis of critical infrastructure system from unexpected events, such as natural disasters, terror attacks and technological failure, is another major application area of CGE modeling. The USCGE model has been adopted for economic consequence analysis for both natural hazards and terrorism events (Rose et al. 2009; Chen et al. 2015a). Developed by Rose and Oladosu (2002), the model consists of 58 economic sectors, along with multiple institutions including nine household income groups, three government actors (two federal and one state and local), and external agents (i.e., foreign producers). Production activities are represented in six-level nested constant elasticity of substitution (CES) function and international trade is represented through an Armington substitution function between imports and domestic production. Chen and Rose (2015) evaluated the influences of economic resilience to transportation infrastructure system failure using the USCGE model. Vulnerability and economic resilience of the different modes of transportation infrastructure, including air, road, rail, water and, local transit were assessed and compared within a modified CGE structure to allow for modal substitution.

Although the single-region and static CGE model has been widely utilized for various empirical assessments, it still has several limitations. First, the model assumes the economy is in equilibrium, though disequilibrium can be incorporated in the labor market (unemployment equilibrium). Second, the model is static, hence it does not trace the time-path of impacts, including various economic cycles associated with employment and investment changes. In addition, the model is constructed through a deterministic approach on the basis of a single base year of data (in contrast to the superior approach of econometric models, which use time series data and have goodness of fit measures), hence it lacks of the ability to incorporate uncertainty. Third, the model has limited power in analyzing regional spillovers given its single-region modeling structure.

Some of these limitations, such as the constraints of regions and temporal interactions, are due to the intrinsic structural characteristics of a single-region CGE model, whereas other limitations, such as the deterministic structure for modeling various interactions among the economy, can be further improved in future research. For instance, one potential area for improvement is to integrate Dynamic Stochastic General Equilibrium (DSGE) modeling procedure into a CGE model to enable policy analysts to conduct impact assessment with a consideration of forward-looking behavior under uncertainty. Rickman (2010) suggested that dynamic fitting of DSGE has great potential to be applied to regional CGE models to improve their empirical basis and lead to a wide utilization. However, numerous challenges remain. For example, issues such as empirical identification, parameterization and verification of DSGE modeling structure need to be further understood before the integration of DSGE into CGE.

5.2.2 Multi-regional CGE

Multi-regional CGE was developed to address the third limitation of the single-region CGE model discussed above. The expansion from a single-region to a multi-region framework is known as regionalization and can be achieved in two ways. The first is called a "top-down" approach in which national results such as gross output, employment and GDP are simulated through a single-region CGE model first, and economic output for different regions are then disaggregated based on certain regional proxy indicators (Klein and Glickman 1977). Alternatively, the other method is known as a "bottom-up" approach in which national results are aggregated based on regional economic outputs that are simulated initially in a multi-regional CGE model. Unlike the single-region CGE or the "top-down" approach of regionalization, a multi-regional CGE model developed through a "bottom-up" approach consists of multiple independent regional accounts and interregional trade involving various commodities and factor flows. Because price and quantities in different regional accounts are determined endogenously by the supply and the demand both interregionally and intraregionally, the model is able to measure distinct regional impacts and associated regional spillover effects caused by a policy simulation. Hence, a multi-regional CGE model is sometimes also called a Spatial CGE model, or SCGE. The multi-regional CGE model has been widely applied for regional economic impacts assessments of infrastructure investment. The model is particularly relevant and critical for the evaluation of regional policies related to regional disparity and regional economic restructuring.

Early applications of the multi-regional CGE model in the transportation sector can be traced back to the 1990s. Buckley (1992) developed a multi-regional CGE model with three regions and five sectors to evaluate the spatial and environmental impacts of transportation services in the U.S in terms of equity and efficiency. His analysis found that transportation costs for both intraregional and interregional trade could be reduced when labor productivity increased. Roson and Dell'Agata (1996) developed a different multi-regional CGE model with 20 regions and 17 sectors for Italy. The model was applied to evaluate environmental and economic impacts of investment in the freight transportation sector. Their study found that traffic congestion could be reduced as a result of an increase in transportation investment. Haddad and Hewings (2001) evaluated the long-run regional impacts of the productivity in the transportation sector in Brazil using a multi-regional CGE model called B-MARIA. The model, which consists of three regions and 40 sectors, was built based on the modeling framework of the MONASH, a multi-regional model for the Australian economy. (Haddad et al. 2010). Bröcker and Mercenier (2011) evaluated the impacts of transportation infrastructure investment for the Trans-European Transport Networks using a SCGE model that consists of 260 European regions. Impacts of new infrastructure links were modeled by reducing transport costs along these links and tracing the effects through the economy. Zhang and Peeta (2011)developed a SCGE model called MINSCGE to evaluate interdependencies of four types of infrastructures: transportation, telecommunication, energy and power.

The Global Trade Analysis Project (GTAP) model developed by Hertel (1997) is one of the well-known multi-regional CGE models. Because regions in the GTAP model are measured as countries or groups of countries, the model has been adopted extensively for policy analysis related to the economy and international trade. The standard GTAP model has a limited application in impact assessment of transportation infrastructures because only transportation margins for international trade are represented in the model. This limitation was reduced in the extended version of GTAP as domestic transportation margins for various transportation modes, such as road, rail, water and air were added to the model (Peterson 2006).

The research team at CoPS is the pioneer of regional CGE modeling. A series of large-scale multi-regional CGE models, such as ORANI (Dixon et al. 1982), FEDERAL (Madden 1990), Monash Multi-Regional Forecasting Model (MMRF) (Adams et al. 2000) were developed and applied for various policy impact analyses. The Enormous Regional Model (TERM) is another MMRF style multi-regional CGE model, but it has an enhanced capacity for regionalization in a "bottom-up" manner (Horridge et al. 2005). Unlike the GTAP model, which is primarily designed for multi-country analysis, TERM is specifically designed for regional impact analysis within a country and the model can handle detailed regional accounts for up to 57 regions and 144 sectors. TERM model is developed based on a "bottom-up" approach and enables researchers to assess regional economic effects of infrastructure investment given that transportation costs are considered explicitly as regional trade margins in the model. Although the model was originally developed for the assessments of the impact of drought on Australian economy, the model has been modified into various versions for over 13 countries.[1] SinoTERM is one example of the modified TERM designed for the Chinese regional economy. The model consists of 31 regions and the number of sectors can be upgraded up to 137 (Horridge and Wittwer 2008).

5.2.3 Dynamic CGE

Dynamization is the other key extension of CGE modeling. It allows for a general equilibrium impact assessment for various years by incorporating time-lagged effects caused by a policy shock into the model. Unlike a static CGE model, a dynamic CGE model provides impact results on GDP, employment, gross output and change of demand for given time periods as well as results for each specific time period. Hence, the modeling framework is more relevant for long-run impacts of policy, such as in the case of infrastructure investment. The detailed modeling mechanism is discussed in Sect. 5.3.

As Dixon and Rimmer (2002) pointed out, the key to upgrading a static CGE model into a dynamic model involves three major modifications: physical capital accumulation, accumulation of financial assets/liabilities and lagged adjustment processes. The first one introduces additional equation systems to the static CGE in order to allow for an accumulation of physical capital. In particular, the flow of annual investment for each sector has to be added to capital stocks. A standard capital accumulation function can be expressed as:

$$K_{i,t+1} = K_{i,t}\left(1 - D_{i,t}\right) + I_{i,t} \tag{5.1}$$

where $K_{i,t}$ denotes the quantity of capital stock available to sector i in year t, $I_{i,t}$ represents the quantity of investment in sector i in year t and $D_{i,t}$ represents the rate of depreciation. The base year quantity of capital stock is normally provided exogenously, which can be retrieved from economic survey or estimated based on private fixed assets.[2] The level of investment is determined by the expected rate of return in sector i in a given time period. The mechanism is also applicable to model capital accumulation at different regions.

The accumulation of financial assets/liabilities is the second key upgrade in converting a static CGE to a dynamic model.[3] This is particularly relevant for

[1]These countries include Brazil, China, Finland, Indonesia, Italy, Japan, Korea, New Zealand, Poland, South Africa, Sri Lanka, Sweden and USA.

[2]As pointed out by one of our reviewers, data scarcity as well as methodological challenges of capital stock estimation should be considered as a caveat.

[3]One should note that applications of the accumulation of financial assets/liabilities in a multi-regional CGE model would require additional specifications or assumptions than that being applied to a single-region CGE in terms of regional balance of payments, For instance, are local assets owned elsewhere and assets outside the region but owned by residents treated modeled in the same

countries with a heavy debt burden. The process requires establishment of linkages between current account flows and net foreign liabilities and then this link feeds into net disposable income and the consumption function relating household spending to disposable income (Dixon and Rimmer 2002). The third aspect of dynamization of CGE modeling involves a lagged adjustment process, which is conducted automatically period-by-period in the model. The process helps to eliminate inconsistencies, for instance between levels of investment and rates of return on the one hand and the theory of investment behavior on the other hand (Dixon and Rimmer 2002).

The dynamic CGE model is often simulated recursively because it traces a time path by sequentially solving a static model, one period at a time. The assumption is that behavior depends only on current and past states of the economy. Alternatively, if the expectations of agents (e.g., producers, consumers, and government) depend on the future state of the economy, the model then requires solutions for all periods simultaneously, leading to a full multi-period dynamic CGE model. Within the latter group, dynamic stochastic general equilibrium models explicitly incorporate uncertainty about the future.

In recent years, there have been burgeoning numbers of empirical studies for impact assessment of transportation infrastructure using a dynamic CGE model. Kim (1998) developed a dynamic CGE model to analyze the economic impact of transportation investment in Korea. Economic impacts were simulated through the shocks of infrastructure investment expenditure and operation services of infrastructure facilities along with the time period. The dynamic mechanism was modeled through a capital accumulation and updates of total labor supply and government policy variables. Rioja (1999) evaluated infrastructure policy using a dynamic CGE model with two sectors for seven Latin American countries. The dynamic mechanism was modeled through the accumulation of capital stock and the study found that more highways and telecommunication infrastructures promote private investment and increased productivity in the private sectors. Seung and Kraybill (2001) evaluated the impacts of infrastructure investment on Ohio's economy using a two-sector dynamic CGE model, in which they found that the magnitude of the stimulus effect was determined by the output elasticity of public capital.

Chen et al. (2015b) evaluated the economic and environmental impacts of rail investment in China using an edited dynamic CGE model based on their earlier static model. The dynamic mechanism follows the approach of Morley et al. (2011), El-Said et al. (2001) and Thurlow (2003) by introducing additional updating equations for all the stock variables (including capital stock, working capital and labor force) as well as dynamic policy shocks. The model is solved recursively period by period with the updated variables. As indicated by Morley et al. (2011), such a dynamic mechanism is a standard method for turning a long-run comparative static CGE model into a tool that gives a time-series solution showing how an

way. The implications to modeling results related to these specifications and assumptions are likely to be substantial.

economy reacts to external shocks or internal changes in policy. The key functions are represented in the following equations:

$$WFKAV_{ft}^{a} = \sum\nolimits_{a}\left[\left(\frac{QF_{fat}}{\sum_{a'} QF_{fa't}}\right)\cdot WF_{ft}\cdot WFDIST_{fat}\right] \tag{5.2}$$

$$INVSHR_{fat}^{a} = \left(\frac{QF_{fat}}{\sum_{a'} QF_{fa't}}\right)\cdot\left(\beta^{a}\cdot\left(\frac{WF_{ft}\cdot WFDIST_{fat}}{WFKAV_{ft}^{a}} - 1\right) + 1\right) \tag{5.3}$$

$$DKAPS_{fat}^{a} = INVSHR_{fat}^{a}\cdot\left(\frac{\sum_{c} PQ_{ct}\cdot QINV_{ct}}{PK_{ft}}\right) \tag{5.4}$$

$$PK_{ft} = \sum_{c} PQ_{ct}\cdot\frac{QINV_{ct}}{\sum_{c'} QINV_{c't}} \tag{5.5}$$

$$QF_{fa,t+1} = QF_{fa,t+1}\cdot\left(1 + \frac{DKAPS_{fat}^{a}}{QF_{fa,t}} - deprate_{f}\right) \tag{5.6}$$

where $WFKAV_{ft}^{a}$: Average capital rental rate of factor f of activity a at time t;
QF_{fat}: Next period sectoral capital stock of factor f of activity a at time t;
WF_{ft}: Wage rate of factor f in time t;
$WFDIST_{fat}$: Wage distortion factor of factor f in activity a and time t;
$INVSHR_{fat}^{a}$: Capital share of factor f of activity a at time t;
β^{a}: Wage-rental ratio parameter. It equals to 1 as default;
$DKAPS_{fat}^{a}$: Gross fixed capital formation of factor f of activity a at time t;
PK_{ft}: Price of capital f at time t;
PQ_{ct}: Composite commodity price of commodity c;
$QINV_{ct}$: Quantity of investment demand of commodity c and time t;
$deprate_{f}$: Capital stock depreciation rate. It equals to 5% in this study.

5.2.4 Dynamic-Recursive Multi-regional CGE

The combination of a dynamic function in a multi-regional CGE modeling framework forms the Dynamic-Recursive Multi-Regional Model. To activate the dynamic mechanisms, additional data is required, which include investment elasticity, rate of depreciation, expected rate of return on investment, and capital growth rate. The model is powerful for regional economic forecasting and policy analysis as it captures dynamic impacts for different regions. The key is to allow physical capital accumulation and lagged adjustments (e.g., wage, employment, and investment) at various rates for different regions.

For instance, FEDERAL-F is a dynamic-recursive multi-regional CGE model developed by Giesecke (2000). A sequence of single-period equilibria is linked via stock-flow functions, and the change of equilibria is computed in response to the value change of stock variables in the model. Specifically, flows in previous periods (such as investment, inter-regional migration, and government borrowings) influence the values for the endogenous variables computed in each period through their contribution to the value of the model's stock variables (such as capital, population, and government debt) in each period (Giesecke 2003, 3).

In addition to FEDERAL-F, the team at CoPS also developed several dynamic multi-regional CGE models with different functions and for different policy analyses. MMRF-GREEN is one of these models designed primarily for regional environmental modeling (Adams et al. 2000). The model was upgraded by incorporating the two dynamic mechanisms of the MONASH model into the comparative-static multi-regional MMRF model: physical capital accumulation and lagged adjustment processes.

TERM-DYN is another dynamic recursive multi-regional CGE model developed by the research team at CoPS and has been applied to analyze the urban water infrastructure project in South-East Queensland (Wittwer 2012). The output of the water and drains sector in the model is considered to be equivalent to the volume of urban water supply. Hence, within a dynamic CGE baseline, water is treated as an exogenous resource, the scarcity of which worsens with economic growth. Conversely, the construction of water infrastructure, such as Dam, improves water supply which, in turn, promotes economic growth.

Other stylized dynamic multi-regional CGE models were also developed for regional economic assessments using similar dynamic mechanisms. For instance, Kim and Kim (2002) evaluated the impacts of regional development strategies on economic growth and equity in Korea using a dynamic and multi-regional CGE model, which consists of six metropolitan areas and eight provinces. The impacts were simulated consecutively through different counterfactual shocks on regional investment expenditures for ten periods, in which a 1995 SAM was treated as the base year. Kim et al. (2004) applied the same model to evaluate the regional economic impacts of highway investment in Korea. The analysis was conducted to determine which highway development deserves priority for investment based on consideration of economic growth and regional economic equity in the long run. Unlike most CGE analysis with interests on measuring regional impacts through the changes in gross output, GDP and employment, Zhang and Peeta (2014) developed a dynamic version of MINSCGE and evaluated interdependencies of four infrastructure types by measuring the change in household utility.

5.3 Key Issues

The modeling framework of CGE has been substantially improved both in terms of regionalization and dynamization in recent decades, which greatly facilitated its application in impact assessment for various policies. Nevertheless, limitations of CGE modeling such as a complex modeling framework and high cost of operation are often criticized (Fæhn 2015). This is because CGE is a complex applied microeconomic/macroeconomic model that contains thousands of equations and variables representing the entirety of economic activities (Donaghy, 2009). The analytical framework is constructed based on numerous assumptions and extensive specifications of parameters. Application of CGE for regional impact assessment can be even more challenging. Partridge and Rickman (1998) pointed out that regional CGE modeling requires adjustment of assumptions designed for a national-level CGE modeling to reflect regional-level activities, such as government fiscal transactions and market structures. Because the number of assumptions on market structure and strategic behavior by government, firm and consumer may potentially lead to different results, sensitivity analysis is considered critical for validation of the model. The following discussion focuses on three major concerns of CGE modeling: parameterization, theoretical underpinnings for policy simulation, and data reliability.

5.3.1 Parameterization

Parameterization remains a major issue of CGE modeling since the previous appraisals by Partridge and Rickman (1998, 2010). As illustrated in Fig. 5.1, CGE analysis requires two sets of input data (a SAM and parameters) to calibrate the simultaneous equation systems through a computerized simulation run. This is a crucial step prior to moving to the implementation of the policy simulation analysis for impact assessment (Sánchez-Cantillo 2004). Key parameters include elasticities of substitution for factor inputs, imports, exports and household consumption. Ideally, these parameters need to be estimated based on data that are consistent with the benchmark data in SAM in terms of sectoral scheme, period and geographic representations. This is because a system-wide econometric estimation of CGE parameters avoids potential simulation errors caused by the inconsistency of benchmark data and various parameters. However, this is a challenging task given the amount of work for estimating various elasticity parameters. This can be so substantial that it becomes prohibitive. Lack of relevant data is another constraint for parameter estimations. Econometric estimation through partial equilibrium models normally requires a sufficient amount of time-series or panel data observations with consistent regional and sectoral schemes. However, researchers often find that data reflecting price and quantity of different sectors and at different geographic aggregation are not always available. Regional level data become even more scant as the geographic scale of analysis narrows.

Given these restrictions on research efforts and data, many CGE studies adopted parameter values from econometric studies as an alternative approach for parameterization. The limitations of such an approach have been widely criticized. For instance, Shoven and Whalley (1992) pointed out that CGE modeling lacks empirical foundations for estimates of behavioral parameters. In fact, the adoption of key parameters from the literature for CGE calibration relies on three underlying assumptions. The first one is that CGE simulations results are insensitive to the specifications of parameters; the second is that parameters are relatively inelastic to time periods and geographic locations, in other words, they are considered both temporally and spatially invariant; third, the parameters are relatively consistent across various related sectors and in various aggregations.

Due to the lack of foundations for parameter identification, there is no consensus on which value to use. Table 5.1 provides an example on the trade elasticity values of some selected sectors used by different CGE models. Except for the GTAP model, which is essentially a multi-country model, all these models were designed specifically for the U.S. with different focuses on policy analysis. The elasticity values were obtained from different studies and are substantially different for most sectors. The inconsistent trade elasticity value being adopted clearly suggests the existence of a potential estimation bias problem due to the various effects of substitution between imports and domestic goods being introduced. Hillberry and Hummels (2013) suggested that one appropriate approach for parameterization is to rely on econometric exercises that employ identifying assumptions and exploit shocks that are similar in nature to those imposed in the model experiment.

Parameterization can be a more severe issue in multi-regional CGE and dynamic CGE models given the involvements of different regional accounts and dynamic

Table 5.1 Comparison of the trade elasticity among different CGE models and studies

Sectors	USITC[a]	GTAP[b]	TERM-USA	USAGE	USCGE	USREP[c]
Grains	5	2.2	5.05	5	2	5
Livestock	3.2	2.8	2.06	5	2	5
Coal	1	2.8	3.05	2.6	0.97	4
Oil and gas	2.8	2.8	5.21	2.6	0.97	4
Other minerals	2	2.8	0.9	2.6	0.97	5
Meat products	2.7	2.2	3.01	3.73	2.5	5
Vegetable fats and oils	5	2.2	3.3	3.73	2	5
Textiles	2.3	2.2	3.74	2.87	1.1	5
Leather products	1.7	4.4	4.05	2.01	1.1	5
Wood products	2.8	2.8	3.4	2.72	3	5
Paper products	3.9	1.8	2.95	3.58	1.1	5
Petroleum and coal products	2.5	1.9	2.1	2.34	2	4
Chemicals rubber and plastic products	2	1.9	3.3	1.93	1.1	5
Metal products	1.9	2.8	3.22	2.35	1.8	5
Transportation equipment	1.7	5.2	3.51	1.5	3	5
Machinery and equipment	2.2	2.8	4.11	2.35	3	5
Electricity	2.8	2.8	4.4	2.8	0.2	0.5
Transport services	1.9	1.9	1.9	1.54	1.1	5

Source: [a]Donnelly et al. (2004), [b]Dimaranan et al. (2006) and [c]Rausch et al. (2011)

mechanisms. New economic geography theory suggests that regional agglomeration and spillover are influenced by two forces: centripetal forces and centrifugal forces, both of which are determined by various regional characteristics such as endowments and transportation costs. Under a multi-regional CGE framework, it is possible that parameters, such as the elasticity of factor substitution, could vary among different regions. For instance, capital and labor may be substituted more easily among the Northeast states in the U.S. than among the states in the Midwest or West due to the concerns of geographic adjacency and homogeneous economic structure. On the other hand, the substitution between capital and labor may occur both intraregionally and interregionally. For instance, the substitution of capital and labor for production activities may occur not only within the New York Metropolitan Statistical Area (MSA), it could also occur between the New York MSA and the Philadelphia MSA given the well-connected infrastructure systems and economic activities. Without capturing the interregional substitution, CGE results of a policy shock could either be underestimated or overestimated due to the adoption of inappropriate elasticity of substitution.

Chen and Haynes (2015) evaluated regional economic impacts of different modes of transportation infrastructure in the U.S. based on a CGE model with elasticities of substitution for factor inputs estimated through spatial econometric models. Their study found that the integration of spatially estimated elasticity of substitution with CGE is important as spatial dependence has been observed among many economic sectors through spatial autocorrelation tests. The elasticity of substitution for factor inputs were found relatively smaller using spatial econometric models than using either OLS or panel estimation. This indicates that the controls of spatial dependence among variables representing quantity and price of labor and capital lead to a high cost penalty to the economy when a policy shock is implemented, which in turn amplifies the impact results. Without considering the issue of spatial dependence in CGE parameter estimation, the various elasticities of substitution are likely to be overestimated using traditional OLS estimation. This can lead to underestimated impact outcomes.

Parametrization for dynamic CGE model is even more challenging as the dynamic recursive mechanism involves additional assumptions and specifications of parameters for physical capital accumulation and investment allocation. For instance, in a dynamic CGE model, depreciation rates for various sectors are required to enable the physical capital accumulation function as indicated in Eq. 5.1. To identify appropriate depreciation rates is important in dynamic CGE models but, unfortunately, relevant data are scarce (Dixon et al. 2013). A higher value for a depreciation rate is likely to underestimate the impacts of a policy shock, whereas a lower value may lead to an overestimation error. Unfortunately, there is a lack of both theoretical and empirical foundations for the selection of appropriate depreciation rates and many studies adopted a rate without explicit justification.

Investment allocation also requires additional information for parameterization. As suggested by Horridge (2002), the mechanism involves two basic assumptions: (1) investment/capital ratios are positively related to expected rates of return and (2), expected rates of return converge to actual rates of return via a partial adjustment mechanism. The two assumptions are represented in Eqs. 5.7 and 5.8, respectively:

$$G = F(E) \tag{5.7}$$

$$G = Q \cdot G_{trend} \cdot \frac{M^{\alpha}}{Q - 1 + M^{\alpha}}. \tag{5.8}$$

where G denotes gross rate of capital growth in the next period and E denotes expected gross rate of return in the next period; M represents the ratio between the expected gross rates of return E and normal gross rates of return R_{normal}; Q denotes (max/trend) investment/capital ratio, and G_{trend} is represented as a function of R_{normal}

Implementation of the first equation assumes that each sector has a long-run or normal rate of return and requires an exogenously determined expected gross rate of return, whereas calibration of the second equations requires specific parameters, such as investment elasticities α, investment/capital ratio G and normal gross rate of return R_{normal}, all of which need to be provided exogenously. It is clear that the information needed for a Dynamic-Recursive Multi-Regional CGE model further increases exponentially given that parameters related to dynamic mechanisms for different regional accounts have to be specified. Despite these issues, empirical applications of dynamic CGE models have emerged rapidly (e.g., Oktaviani et al. 2007; Bohlman 2010; Arndt et al. 2012). However, these issues related to sensitivity and reliability of these parameters for dynamic mechanisms have rarely been discussed or analyzed.

In sum, parameter estimation of CGE models still deserves further attention in future research. Although adopting parameters that were estimated from the literature for CGE analysis has become a normal approach and is widely adopted in many existing studies, the shortfall of such an approach has been generally recognized (Partridge and Rickman 1998; Chen and Haynes 2015). Surprisingly, it is still uncommon to find CGE analysis that is based on a self-estimated parameterization approach rather than depending on estimates from the literature. The lack of incentive to consider estimations of elasticities of substitution as a part of CGE analysis is because the amount of work for estimating elasticities of substitution increases exponentially as the numbers of sectors and regions are added. Another unavoidable fact is that the available data for econometric estimation of parameters become scare as the sectoral structure of CGE is further disaggregated. Nevertheless, given that CGE simulations are found to be sensitive to parameter specifications, additional efforts for parameter validation remain necessary to improve the robustness of CGE analysis.

5.3.2 *Underpinnings of Policy Shock*

In contrast to econometric analysis, CGE modeling is based on a computerized simulation in which macroeconomic impacts in terms of gross output, GDP and employment are calculated as changes before and after a policy shock, which is implemented through adjustments of exogenous variables and/or parameters to reflect the direct effects of a policy reform. Hence, CGE modeling avoids statistical errors such as endogeneity, multicollinearity and heteroscedasticity and has the potential to capture wider economic impacts. On the other hand, CGE modeling also has limitations in the way a policy shock is implemented. Because the direct effects from a policy reform can be modeled in various ways in a CGE model, which approach is most appropriate remains unclear given the lack of theoretical underpinnings for CGE policy shocks. We further elaborate the issue with a focus on impact studies on transportation infrastructure using CGE. We first introduce the various techniques of implementing shocks to reflect changes in equilibrium conditions, such as an increase in investment, and a reduction of stock due to system disruptions. We then discuss the potential caveats of CGE policy simulations using relevant empirical studies as examples.

Economic impact analysis of infrastructure investment and economic consequence analysis of infrastructure disruption under unexpected events, such as natural disaster, terror attack or technological failure are similar in that both can be implemented through CGE simulations. The major difference between these two is that the former represents positive shocks to the economy, whereas the latter is measured as negative shocks. Nevertheless, empirical studies for these two areas can be implemented through four types of shocks in CGE models: capital shock, productivity shock, margin shock and expenditure shock.

A capital shock refers to the approach of measuring economic impacts by altering the quantity of capital input in a CGE model. A positive shock on the quantity of capital input drives up the quantity of production, which thus increases gross output and GDP and vice versa. Many studies evaluate regional economic benefits of transportation investment through such an approach (e.g., Kim 1998; Chen and Haynes 2013). Productivity shock is implemented through adjustments to corresponding productivity parameter in CGE. The shock is normally adopted to reflect improvements or declines of production activities in responses to a status change of economic equilibrium. For instance, a completion of a highway project is expected to increase the productivity of road transportation related services. Hence, an output expansion in the truck sector is expected to contribute to the growth of total gross and GDP. Examples of evaluation of infrastructure investment through a productivity shock can be found in Rioja (1999) and Siegesmund et al. (2008). CGE simulation can also be implemented through a shock on transportation margin as a response of transportation infrastructure improvements. This approach has a requirement for SAM in that trade and transportation margins have to be added in as separate accounts. See examples in Lofgren et al. (2002) and Bröcker et al. (2010).

In addition to these three supply side shocks, macroeconomic impacts of transportation infrastructure can also be measured from the demand side, such as a shock to household expenditure. This is particularly relevant in economic consequence analysis of infrastructure system disruption where behavioral effects play a dominant role. For instance, Chen et al. (2015a) evaluated the economic consequences of aviation system disruption, in which the primary negative impacts from the 9/11 World Trade Center terrorist attack was measured through reductions of travel and tourism related household expenditure.

Most empirical studies using CGE modeling were *de facto* conducted from an *ex ante* perspective based on hypothetical scenarios. Hence, CGE policy shocks generally lack evidence based underpinnings, which can make results difficult to interpret. For instance, Kim et al. (2004) evaluated the regional economic impacts of highway investment in Korea using a multiregional CGE model. Although one of the findings revealed that the selected highway projects mitigate regional disparities in terms of wages with fading impacts over time, the cause of such outcomes remains improbable because several fundamental questions, such as what magnitudes of shocks had been applied for the CGE simulation and how they were determined, were still unclear. Seung and Kraybill (2001) investigated the effects of infrastructure investment on regional output and welfare in Ohio using a regional dynamic CGE model. The policy shocks were implemented through the adjustments of public capital elasticity parameters at three different levels. The approach seems like a hybrid of capital shock and productivity shock but, unfortunately, the specifications of policy shock levels appear to be incomprehensible. Infrastructure investment was found to reduce household welfare instead of increasing it, which seems counterintuitive and raises a red flag for their modeling mechanism.

The lack of evidence based underpinnings also raises concerns on the validity of simulation outcomes for the following two reasons: First, since the magnitude of CGE shocks is generally based on author(s)' arbitrary judgement, the corresponding simulation outcomes have limited power to reflect the real world situation. Second, CGE simulations are likely to involve omission bias due to the lack of scientific procedures to determine appropriate shocks. As a result, empirical CGE studies generally focus on magnitude, direction and distributive patterns instead of interpreting the numeric outcomes. The lack of evidence based underpinnings for CGE shocks further constrains the implications of CGE modeling, and results from CGE analysis can only be used as road maps for policy implications.

It is clear that the underpinnings of policy shock in CGE analysis need to be considered more cautiously and carefully. One potential improvement strategy is to connect the direct shock to external resources. For instance, the classical economic consequence analysis using CGE for natural hazards such as earthquake and tsunami, is to simulate the macroeconomic impacts based on the direct impacts (such as property damages, number of deaths, and losses of trade volume) obtained from other reliable sources, such as government reports and academic research articles. This would be particularly relevant if the focus is on *ex post* impact assessment. In addition, constructing a CGE policy shock scenario using an econometric estimation or side-calculation based on relevant data would also be more pragmatic than a hypothetical scenario that is based on an arbitrary specification. In fact, given that impact drivers for regional CGE modeling assessment can be even more complicated due to the existence of regional heterogeneity, such a data driven or fact driven approach to establish CGE policy shock scenario would be even more critical for regional CGE modeling.

5.3.3 Data Reliability

Data demand for CGE modeling is enormous, and the procedure of data processing is more complicated than econometric analysis. The basic data structure of CGE modeling is a social accounting matrix (SAM), which illustrates the circular process of economic transactions between demand and supply and among different markets. Unlike an Input–Output (I–O) table which shows only the relationship between production accounts and the other accounts (e.g., factor of production, consumption, government, investment, and trade), SAM extends an I–O table by including additional information to reflect the owners of different factor inputs and interrelations between all accounts, such as transfers between household and government, etc. (Rutherford and Paltsev 1999).

Often this information was derived from different resources and not necessarily with consistent reference years. For instance, although information such as trade data, taxes and government transfers are generally available on an annual basis, the I–O tables are updated less frequently. As a result, a SAM table representing a benchmark economic status may reflect an equilibrium condition with multiple reference years and this could make the interpretation of CGE results difficult. One example is the latest GTAP 9 data base used for global trade analysis, which involves three reference years: 2004, 2007 and 2011 and is enormous, containing data for 140 countries/regions and 57 commodities (Narayanan et al. 2015). Some data, such as the macro-economic status, bilateral services trade and energy performance includes three reference years, whereas other data, such as the bilateral merchandise trade, includes one reference year, 2011. Clearly, although GTAP 9 is a gigantic integrated data base, and plays a central role in analyzing important trade policy issues at a global level, the data base has problems when it includes three reference years. The resulting CGE simulations cannot be simply interpreted as a deviation from the initial equilibrium status in a single base year due to a policy shock. Instead, it could only be considered as a change from the benchmark equilibrium status for a given period, since it covers a period between 2004 and 2011.

The process of SAM balancing also raises concerns on data reliability for CGE modeling. A balanced SAM is the foundation of CGE analysis, which requires that all rows and columns must be equal. This means that supply equals demand for all goods and factors, tax payments equals tax receipts, the value of household expenditure equals the value of factor income plus transfers, and the value of government tax revenue equals the value of transfers (Rutherford and Paltsev 1999).[4] The construction of a SAM for CGE analysis requires the balancing of all the data from various sources, such as the I–O tables, tax payment and receipts, government and household transfers. SAM balancing can be achieved in various

[4]It also requires zero profit in production given the assumption of perfect competition. In the situation of non-perfect competition, Mark-up is normally required to be provided exogenously. See Francois (1998).

ways, such as the RAS method, the cross-entropy method or a manual balancing method.

Each method has advantages and disadvantages. For instance, although the RAS is easy to implement, the method lacks an economic foundation and flexibility for specific adjustments. Conversely, the cross-entropy method allows the user to adjust certain cell values while keeping others constant, which helps to maintain economic logic. However, to what extent the adjustment made appropriately is unclear, and the process of balancing is purely based on a mathematical procedure and the personal judgement of modelers. In some cases, accuracy of SAM is sacrificed in order to fit the CGE model. As a result, it becomes unavoidable that information of some specific accounts and transactions in a balanced SAM can be altered substantially from the original data but it is the original data that reflects the real world in the base year.

The issue of data reliability can become very serious if the data is disaggregated for regional CGE modeling. This is because regional economic data, such as regional I–O tables and trade flows, are often incomplete or unavailable. In some cases, the data might be too coarse or inconsistent due to the fact that regional tables may reflect different dates and are in different formats. As a result, regional CGE modeling often requires creation of regional data using various techniques. The classical approach is to create regional data by a "top-down" approach, in which the national accounts are disaggregated into various regional accounts using regional shares as proxies. The approach is based on several assumptions. First, it assumes homogeneity of industrial technology across regions. Second, it assumes a fixed regional share for commodities that are heavily traded between regions. Third, it assumes the outputs of the remaining commodities are adjustable in accordance to regional demand (Horridge 2012). It is clear that such a disaggregation approach relies heavily on these assumptions. While the approach sounds reasonable, its validity is difficult to justify given the lack of empirical evidences and our general knowledge of regional economic differentiation.

In general, although regional CGE modeling has received increasing attention from scholars and practitioners, regional data has to be created based on a variety of assumptions due to the lack of appropriate regional economic information. The lack of appropriate validation for these assumptions also leads to a concern on data reliability since each step adds additional errors and might lead to an imprecise impact assessment in the end. No wonder, as suggested by West (2002), building a CGE model for a small region, while not invalid, may not be a very efficient use of resources in the context of the trade-off between increased complexity and increased data "fuzziness". The evidence and arguments suggest that the input data for CGE modeling should be used with caution. What's more important for future efforts is to support development of more regional data availability for researchers.

5.4 Future Research Directions

Despite some scholars' belief that CGE modeling is inappropriate to be applied to analyze various fields, such as economic sustainability, because its structure is too rigid and stylized to represent the system-wide economic activities (Barker 2004; Scrieciu 2007), it is undeniable that it will continue to be a heavily used approach for economic impact assessment. This is especially true in fields such as impact assessments of infrastructure investment, natural disasters, and analysis of various governmental policy options. Given its strong theoretical underpinnings, its base in empirical analysis, its strong base of consumer support and its wide community of users, it is important for CGE researchers to address the following issues in the near future.

First, the quality of data should be improved to allow for more reliable CGE analysis. In particular, the collection of regional economic data such as regional level IO tables and interregional trade flows should be given more emphasis in regional economic modeling. This will require increased effort and broader and deeper financial support from both the public and private sectors. For instance, an improvement of regional economic data collection through conducting a regional economic survey on a regular basis would be a worthwhile endeavor.

Second, more attention should be paid to CGE parameter estimation. As discussed earlier, the major limitation of most existing CGE models is due to the adoption of parameters that are inconsistent with the analytical data framework. Hence, parameter estimations for CGE models should be given a high priority for research. Spatial dependence of regional factor/commodity substitution is another critical aspect of regional economic modeling, which deserves further attention.

Third, validation of CGE models deserves more research efforts. Currently, the approach to CGE model validation is primarily conducted through various sensitivity tests, which is valuable in identifying the extent of variations in CGE simulation output as a response to the changes in inputs. However, the accuracy of CGE simulation outcomes is uncertain as most existing CGE simulations are based on hypothetical scenarios. There is a lack of validation against reality. One of the major future research endeavors should aim to valid CGE modeling through simulations based on real world data. This also implies that CGE shocks must be conducted cautiously using evidence based data rather than conduct simulations based on arbitrarily specified shocks.

While future research will surely continue to expand economic functions and computational power of CGE modeling from a theoretical perspective, it is also foreseeable that applications of CGE models for impact assessment will be standardized in terms of modeling frameworks. The advantages for using a standardized CGE model are quite clear. First, it improves the efficiency of impact assessment as efforts could be focused primarily on identifying appropriate inputs and parameterization for CGE modeling based on established CGE models. Hence, the cost of CGE analysis can be greatly reduced. Second, given the fact that standardized CGE models are developed by experts and have been run through extensive tests, simulation outcomes will be much more reliable than a self-developed CGE approach.

Last but not the least, the development of CGE with an integration of DSGE will be a worthwhile direction for future research. On the one hand, the development of DSGE models with the levels of sectoral and regional detail found in CGE models would allow researchers to analyze impacts with uncertainty. This type of analysis is particularly relevant and important to help us understand both forward-looking behavior of economic activities and regional and sectoral heterogeneity. On the other hand, the introduction of stochastic optimization specifications into CGE poses a new challenge for computational algorithm development given that the increased modeling size and restrictions will make it much more difficult to solve. Hence, future research for regional CGE modeling should also focus on developing advanced software system to achieve computational efficiency.

References

Adams PD, Horridge MJ, Parmenter BR (2000) Forecasts for Australian regions using the MMRF-Green model. Australas J Reg Stud 6(3):293–322

Arndt C, Chinowsky P, Strzepek K, Thurlow J (2012) Climate change, growth and infrastructure investment: the case of Mozambique. Rev Dev Econ 16(3):463–475

Aschauer DA (1989) Is public expenditure productive? J Monet Econ 23(2):177–200

Arrow KJ, Debreu G (1954) Existence of an equilibrium for a competitive economy. Econometrica 22:265–290

Barker T (2004) The transition to sustainability: a comparison of general–equilibrium and space–time–economics approaches. Tyndall centre working paper (vol 62). University of Cambridge, Cambridge

Bhatta SD, Drennan MP (2003) The economic benefits of public investment in transportation. J Plan Educ Res 22(3):288–296

Bohlman HR (2010) The macroeconomic impact of skilled emigration from South Africa: a CGE analysis. Centre of Policy Studies, Monash University, Melbourne

Boarnet MG (1997) Infrastructure services and the productivity of public capital: the case of streets and highways. Natl Tax J 50(1):39–57

Boarnet MG, Haughwout AF (2000) Do highways matter? Evidence and policy implications of highways' influence on metropolitan development. A Discussion Paper Prepared for The Brookings Institution Center on Urban and Metropolitan Policy. Washington DC

Bröcker J, Korzhenevych A, Schürmann C (2010) Assessing spatial equity and efficiency impacts of transport infrastructure projects. Transp Res B Methodol 44(7):795–811

Bröcker J, Mercenier J (2011) General equilibrium models for transportation economics. In: Palma A, Lindsey R, Quinet E, Vickerman R (eds) A handbook of transport economics. Edward Elgar, Chelenham, pp 21–45

Buckley PH (1992) A transportation-oriented interregional computable general equilibrium model of the United States. Ann Reg Sci 26(4):331–348

Chen Z, Haynes KE (2013) Transportation capital in the United States: a multimodal general equilibrium analysis. Public Works Manag Policy 19(2):97–117

Chen Z, Haynes KE (2015) Spatial impact of transportation infrastructure: a spatial econometric CGE approach. In: Nijkamp P, Rose A, Kourtit K (eds) Regional science matters—studies dedicated to Walter Isard. Springer, Cham, pp 163–186

Chen Z, Rose ZA (2015) Economic resilience to transportation failure: a computable general equilibrium analysis. CREATE working paper. University of Southern California, Los Angeles
Chen Z, Rose A, Prager F, Chatterjee S (2015a) Economic consequences of aviation system disruptions: a reduced-form computable general equilibrium analysis. Available at SSRN 2692177
Chen Z, Xue J, Rose ZA, Haynes KE (2015b) Impact of high speed rail investment on the economy and environment in China: a dynamic CGE analysis. Available at SSRN 2636385
Conrad K (1997) Traffic, transportation, infrastructure and externalities: a theoretical framework for a CGE analysis. Ann Reg Sci 31(4):369–389
Dervis K, de Melo J, Robinson S (1982) General equilibrium models for development policy. Cambridge University Press, Cambridge
Dimaranan BV, McDougall RA, Hertel T (2006) Behavioral parameters. Global trade, assistance, and production: the GTAP, Chapter 18, 1–17
Dixon PB, Parmenter BR, Sutton J, Vincent DP (1982) ORANI: a multisectoral model of the Australian economy. North-Holland, Amsterdam
Dixon P, Rimmer M (2002) Dynamic general equilibrium modelling for forecasting and policy: a practical guide and documentation of MONASH. Elsevier, Amsterdam
Dixon PB, Koopman RB, Rimmer MT (2013) The MONASH style of computable general equilibrium modeling: a framework for practical policy analysis. In: Dixon PB, Jorgenson DW (eds) Handbook of computable general equilibrium modeling, vol 1, pp 23–103
Donaghy KP (2009) CGE modeling in space: a survey. In: Capello R, Nijkamp P (eds) Handbook of regional growth and development theories. Edward Elgar, Chelenham, pp 389–422
Donnelly WA, Johnson K, Tsigas M, Ingersoll D (2004) Revised armington elasticities of substitution for the USITC model and the concordance for constructing a consistent set for the GTAP model, no 2004-01-A. U.S. International Trade Commission, Washington, DC
Duffy-Deno KT, Eberts R (1991) Public infrastructure and regional economic development: a simultaneous equations approach. J Urban Econ 30(3):329–343
El-Said M, Lofgren H, Robinson S (2001) The impact of alternative development strategies on growth and distribution: simulations with a dynamic model. TMD discussion paper 78. International Food Policy Research Institute, Washington, DC
Fæhn T (2015) A shaft of light into the black box of CGE analyses of tax reforms. Econ Model 49:320–330
Fernald JG (1999) Roads to prosperity? Assessing the link between public capital and productivity. Am Econ Rev 89(3):619–638
Francois J (1998) Scale economies and imperfect competition in the GTAP model. GTAP technical papers no 16. Purdue University, West Lafayette
Giesecke JA (2000) The theoretical structure of the FEDERAL-F model. CREA paper no TS-08. Centre for Regional Economic Analysis, University of Tasmania
Giesecke J (2003) Targeting regional output with state government fiscal instruments: a dynamic multi-regional CGE analysis. Aust Econ Pap 42(2):214–233
Gramlich EM (1994) Infrastructure investment: a review essay. J Econ Lit 32(3):1176–1196
Gramlich E (2001) Infrastructure and economic development. Remarks by governor Edward M. Gramlich at the Texas Trade Corridors New Economy Conference, San Antonio
Haddad EA, Hewings GJ (2001) Transportation costs and regional development: an interregional CGE analysis. In: Friedrich P, Jutila S (eds) Policies of regional competition. Nomos, Baden-Baden, pp 83–101
Haddad EA, Hewings GJ, Perobelli FS, Santos RC (2010) Regional effects of port infrastructure: a spatial CGE application to Brazil. Int Reg Sci Rev 33(3):239–263
Harmatuck DJ (1996) The influence of transportation infrastructure on economic development. Logist Transp Rev 32(1):63–76
Hertel TW (1997) Global trade analysis: modeling and applications. Cambridge University Press, Cambridge

Hillberry R, Hummels D (2013) Trade elasticity parameters for a computable general equilibrium model. In: Dixon PB, Jorgenson DW (eds) Handbook of computable general equilibrium modeling, vol 1, pp 1213–1269

Horridge JM (2002) ORANIG-RD: a recursive dynamic version of ORANIG. Melbourne, Centre of Policy Studies, Monash University

Horridge M (2012) The TERM model and its database. In: Wittwer G (ed) Economic modeling of water: the Australian CGE experience. Springer, New York, pp 13–35

Horridge M, Madden J, Wittwer G (2005) The impact of the 2002–2003 drought on Australia. J Policy Model 27(3):285–308

Horridge M, Wittwer G (2008) The economic impacts of a construction project, using SinoTERM, a multi-regional CGE model of China. China Econ Rev 19(4):628–634

Johansen L (1960) A multi-sectoral study of economic growth, vol 82. North-Holland, Amsterdam

Lewis, J. D., Robinson, S., Thierfelder K (2003). Free trade agreements and the SADC economies. Journal of African Economies, 12(2):156–206.

Kim E (1998) Economic gain and loss of public infrastructure investment: dynamic computable general equilibrium model approach. Growth Chang 29(4):445–468

Kim E, Hewings GJ, Hong C (2004) An application of an integrated transport network–multiregional CGE model: a framework for the economic analysis of highway projects. Econ Syst Res 16(3):235–258

Kim E, Kim K (2002) Impacts of regional development strategies on growth and equity of Korea: a multiregional CGE model. Ann Reg Sci 36(1):165–189

Klein LR, Glickman NJ (1977) Econometric model-building at regional level. Reg Sci Urban Econ 7(1):3–23

Lewis JD, Robinson S, Thierfelder K (2003) Free trade agreements and the SADC economies. J Afr Econ 12(2):156–206

Lofgren H, Harris RL, Robinson S (2002) A standard computable general equilibrium (CGE) model in GAMS, vol 5. International Food Policy Research Institute, Washington, DC

Madden JR (1990) FEDERAL: a two-region multisectoral fiscal model of the Australian economy. Doctoral dissertation, University of Tasmania

Mattoon R (2002) Midwest infrastructure: assessing the contribution of basic infrastructure to economic growth. Chicago Fed Letter, 184b(Special Issue December)

McDonald S (2005) The PROVIDE project standard computable general equilibrium model: version 2. Technical Paper Series 15625, PROVIDE Project

Melo PC, Graham DJ, Brage-Ardao R (2013) The productivity of transport infrastructure investment: a meta-analysis of empirical evidence. Reg Sci Urban Econ 43(5):695–706

Morley S, Piñeiro V, Robinson S (2011) A dynamic computable general equilibrium model with working capital for Honduras (no 1130). International Food Policy Research Institute, Washington, DC

Nadiri MI, Mamuneas TP (1996) Contribution of Highway Capital to Industry and National Productivity Growth (No. BAT-94-008). Report Prepared for Apogee Research, Inc., for the Federal Highway Administration Office of Policy Development

Narayanan G, Badri AA, McDougall R (2015) Global trade, assistance, and production: the GTAP 9 data base. Center for Global Trade Analysis, Purdue University

Oktaviani R, Hakim DB, Sahara S, Siregar H (2007) Impact of a lower oil subsidy on Indonesian macroeconomic performance, agricultural sector and poverty incidences: a recursive dynamic computable general equilibrium analysis. SSRN Working paper. Available at http://ssrn.com/abstract=1086380

Partridge MD, Rickman DS (1998) Regional computable general equilibrium modeling: a survey and critical appraisal. Int Reg Sci Rev 21(3):205–248

Partridge MD, Rickman DS (2010) Computable general equilibrium (CGE) modelling for regional economic development analysis. Reg Stud 44(10):1311–1328

Peterson E (2006) GTAP-M: a GTAP model and data base that incorporates domestic margins. GTAP technical papers no 23. Center for Global Trade Analysis, Purdue University

Rausch S, Metcalf GE, Reilly JM, Paltsev S (2011) Distributional impacts of a U.S. greenhouse gas policy. In: Metcalf GE (ed) U.S. energy tax policy. Cambridge University Press, Cambridge, pp 52–112

Rickman DS (2010) Modern macroeconomics and regional economic modeling. J Reg Sci 50(1):23–41

Rioja FK (1999) Productiveness and welfare implications of public infrastructure: a dynamic two-sector general equilibrium analysis. J Dev Econ 58(2):387–404

Rose A, Oladosu G (2002) Greenhouse gas reduction policy in the United States: identifying winners and losers in an expanded permit trading system. Energy J 23(1):1–18

Rose A, Oladosu G, Lee B, Beeler Asay G (2009) The economic impacts of the September 11 terrorist attacks: a computable general equilibrium analysis. Peace Econ Peace Sci Public Policy 15(2):1–28

Roson R, Dell'Agata G (1996) Environmental externalities, transport costs and interregional trade in a general equilibrium model. In: Fossati A (ed) Economic modelling under the applied general equilibrium approach. Avebury, Aldershot

Rutherford T, Paltsev S (1999) From an input–output table to a general equilibrium model: assessing the excess burden of indirect taxes in Russia. University of Colorado

Sánchez-Cantillo MV (2004) Rising inequality and falling poverty in Costa Rica's agriculture during trade reform. A macro–micro general equilibrium analysis. Shaker, Maastricht

Scrieciu SS (2007) The inherent dangers of using computable general equilibrium models as a single integrated modelling framework for sustainability impact assessment. a critical note on Böhringer and Löschel (2006). Ecol Econ 60(4):678–684

Seung CK, Kraybill DS (2001) The effects of infrastructure investment: a two-sector dynamic computable general equilibrium analysis for Ohio. Int Reg Sci Rev 24(2):261–281

Shoven JB, Whalley J (1992) Applying general equilibrium. Cambridge University Press, Cambridge

Siegesmund P, Luskin D, Fujiwara L, Tsigas M (2008) A computable general equilibrium model of the US economy to evaluate maritime infrastructure investments. Transp Res Rec J Transp Res Board 2062:32–38

Thurlow J (2003) A dynamic computable general equilibrium model for South Africa. International Food Policy Research Institute, Washington, DC

West GR (1995) Comparison of input–output + econometric and computable general equilibrium impact models at the regional level. Econ Syst Res 7:209–227

Wittwer G (2012) Economic modeling of water: the Australian CGE experience. Springer, New York

Zhang P, Peeta S (2011) A generalized modeling framework to analyze interdependencies among infrastructure systems. Transp Res B Methodol 45(3):553–579

Zhang P, Peeta S (2014) Dynamic and disequilibrium analysis of interdependent infrastructure systems. Transp Res B Methodol 67:357–381

Zhenhua Chen is an assistant professor in City and Regional Planning at The Ohio State University. His research interest focuses on regional science, risk and resilience, and transportation planning and policy. His dissertation receives a series of awards, including the Benjamin H. Stevens Graduate Fellowship in Regional Science awarded by the North American Regional Science Council, the Vernon E. Jordan, Jr. Fellowship Award awarded by Economic Club of Washington, D.C., and the Best Dissertation Award from the Regional Science Association International (RSAI). Dr. Chen earned the Ph.D. in public policy from George Mason University in 2014.

Kingsley E. Haynes is a University Professor Emeritus of Public Policy at George Mason University. He is also the Ruth D. and John T. Hazel, MD Endowed Chair and Eminent Scholar and Found Dean of the School of Public Policy (now called Schar School of Policy and Government) at the George Mason University. Dr. Haynes has been involved in regional economic development, infrastructure policy, information and transportation, environmental planning and natural resource management since the early 1970s. He has directed international programs for the Ford Foundation's Office of Resources and Environment and EPA. Dr. Haynes earned the Ph.D. in Geography and Environmental Engineering from Johns Hopkins University in 1972.

Chapter 6
Potentials and Prospects for Micro–Macro Modelling in Regional Science

Eveline van Leeuwen, Graham Clarke, Kristinn Hermannsson, and Kim Swales

6.1 Introduction

There is growing interest in regional science, and related fields, in the potential for linking multi-sectoral macro models of economic development and change with micro models of household structures and economic activities. Multi-sectoral macro models, such as input–output (IO), social-accounting matrix (SAM) and computable general equilibrium (CGE), analyse the impact of a major job loss or gain in a region by first exploring the direct impact in terms of changes in variables such as regional GDP and gross/net income. This analysis is then augmented by identifying further multiplier effects on other sectors of the economy. Input–output accounts show the strength of the regional linkage or interaction between different sectors of the economy. Multi-sectoral macro models use this information to estimate the jobs and economic activity generally gained or lost in other sectors of the regional economy through indirect and induced effects. This procedure generates information relating to the dynamics of the regional labour market.

For many applications changes in these regional variables provide the relevant key economic impacts associated with major job changes and are, therefore, of great interest to policy makers. However, there are further interesting questions surrounding potential intra-regional variations. For example, jobs gained or lost

E. van Leeuwen (✉)
Vrije Universiteit Amsterdam, Amsterdam

G. Clarke
School of Geography, University of Leeds

K. Hermannson
School of Educaction, University of Glasgow

K. Swales
Fraser of Allander Institute, University of Strathclyde

R. Jackson, P. Schaeffer (eds.), *Regional Research Frontiers - Vol. 2*,
Advances in Spatial Science, DOI 10.1007/978-3-319-50590-9_6

in location X_1 within region Y may have little impact on the local economy of location X_2 within the same region if these locations are actually very far apart. The proportionate changes in regional economic indicators are in effect averages, potentially masking widespread variations within an individual region. Such models are rarely disaggregated to finer sub-regional geographical scales, although there has always been interest in the potential to do so (see Batey and Madden 2001; Hewings et al. 2001; Jin and Wilson 1993; Ballas and Clarke 2001; Ballas et al. 2006; van Leeuwen 2010; Bourguignon et al. 2010; van Leeuwen et al. 2016).

Microsimulation models can provide a link between regional multisectoral macro models and the individual households and firms that make up the region. By linking households to jobs within the region, we gain the potential to estimate the small-area or local impacts of major changes in the economy. Thus, a major loss of jobs at firm A can be analysed not only by changes to regional GDP, income etc., but also on the basis of which households will be directly impacted in which areas. This, in turn, will allow planners to understand the loss of household incomes, welfare benefit payment changes, multiplier impacts on local shops etc., all at the small-area level.

The aim of this chapter is to demonstrate the benefits of a potential macro–micro model linkage. Our case study region is the Western Islands of Scotland (WIS), an area of considerable interest at the moment given the Scottish Government (2011) target to meet the equivalent of 100% of Scottish gross electricity consumption from renewables by 2020. The work is ongoing and at the time of writing we are a long way from having a fully integrated model. However, it is hoped that we can show sufficient progress to allow the reader to appreciate the advantages of the combined approach. In Sect. 6.2 we discuss the two main modelling approaches in more detail and show ways in which they can be linked. Section 6.3 examines the various components of the modelling exercise as they apply to the Western Isles. An important part of the linkage between the models is the building of a journey to work model, which we also describe in this section. In Sect. 6.4 we demonstrate how two specific investments in employment opportunities in the Western Isles can be modelled using macro and micro models, showing how the framework helps to produce *both* local and regional indicators of economic change. Section 6.5 provides a short conclusion and future road map.

6.2 Linking MSM with IO Modelling

6.2.1 Regional Input–Output Analysis

Regional IO impact analysis is frequently used to capture the total spending effects of institutions, projects or events. This analysis incorporates the multiplier, or "knock-on", impacts of any expenditure injection, obtained by summing the subsequent internal demand feedbacks within the economy. This section briefly outlines the methods adopted in such studies.[1]

Regional demand-driven, multi-sectoral models, including IO, make a basic distinction between exogenous and endogenous expenditures. Exogenous expenditures are determined independently of the level of economic activity within the host economy. In IO studies, many of the elements of final demand, including exports, government expenditure and investment, are typically taken to be exogenous. On the other hand, endogenous expenditures are driven by the overall level of economic activity within the host economy. Specifically, demand for intermediate inputs and often household consumption demands are taken to be endogenous. IO analysis thus identifies a clear causal pathway from exogenous changes in final demand to subsequent adjustments in endogenous economic activity.

These IO and SAM demand-driven models assume that the supply side of the regional economy is entirely passive. Essentially this means that any change in domestic demand is met by a corresponding change in output and no change in prices or wage rates. There are also no physical supply constraints. In the short and medium runs, such a model applies where there is general excess productive capacity and significant regional unemployment. In the long run, supply-side passivity holds where the supply of the primary inputs of labour and capital eventually becomes infinitely elastic, as migration and capital accumulation ultimately eliminate any short-run capacity constraints (McGregor et al. 1996).[2] The lack of a direct modeling of the supply side means that the outputs of a standard IO impact analysis cannot be disaggregated to individual households or small-areas within the study region. Thus, with such models we cannot typically understand the spatial impacts of the changes within the region being modelled (although we again acknowledge attempts to include demand and supply side factors in Batey and Madden 1983, 1999, 2001; Hewings et al. 2001; Jin and Wilson 1993; Ballas and Clarke 2001; Ballas et al. 2006; Hérault 2010; van Leeuwen 2010; Bourguignon et al. 2010).

6.2.2 *Microsimulation Modelling*

Spatial microsimulation (MSM) is a well-established method for estimating the attributes of individuals or households at the small-area level. Although the U.K. population census gives some information on individuals at the small-area level, the detail on interdependency is insufficient for much policy analysis. However, like

[1]For a more detailed account see Armstrong and Taylor (2000), Loveridge (2004) and Miller and Blair (2009).

[2]CGE models allow price flexibility. For example, Learmonth et al. (2007) models the island economy of Jersey. Here a tight labour market combined with institutional restrictions on migration mean that the supply side cannot be treated as passive over any time interval.

many other countries, the U.K. does have some rich survey datasets, which associate many more attributes to individuals. One of these is the Sample of Anonymised Records (SARs). Although rich in attribute data, the level of geography in the SARs is rather coarse—identification is at the regional level at best. However, by effectively cloning individuals in the SAR to match the characteristics of individuals in the census, it is possible to reweight the SARs to provide a detailed set of individuals and their attributes at the small-area level. There are a number of well-known methods for reweighting surveys in this way, which are discussed in Tanton and Edwards (2013) and Hermes and Poulsen (2012).

As far as regional science applications are concerned, MSM has been used to model household income and expenditures, often for input into other economic or spatial interaction models. Some well-known and very policy-relevant economic models are aspatial. These models contain all households in a region or country but are not linked to individual places (for an overview see Bourguignon and Spadaro 2006; Li and O'Donoghue 2013). However, Birkin and Clarke (1989) use a spatial MSM to estimate income in Leeds, U.K., the first of many subsequent spatial analyses of household income and expenditure.

Ballas and Clarke (2001) and Ballas et al. (2006) use a microsimulation model to investigate the detailed spatial impacts of job losses in Leeds following a major factory closure in the east of the city. They do this by linking households to jobs via a journey to work model (see Sect. 6.3.5 below). Then, when jobs are lost, the model can identify the individual households affected in the commuting catchment area of that factory. Households identified as being impacted change their status from employed to unemployed so that subsequent local income and expenditure reductions can be additionally calculated. They further speculate on the impact on other parts of the city that might be affected by the initial factory closure through the forced to close or downsizing of suppliers to that factory. Whilst these papers help to show the advantages of potential macro/micro linkages, the economic changes were hypothesized rather than being formally derived from an IO model. In other words, the second and subsequent round impacts of the factory closures were 'guestimated' rather than derived from the outputs of a multi-sectoral macro model.

6.2.3 *Linking Macro and Micro Models*

6.2.3.1 Top-Down Linkage

When answering questions about the micro effects of a macroeconomic change, top-down linkage is important (Bourguignon et al. 2010). The top-down approach (see Fig. 6.1) builds on insight derived from a multi-sectoral macro model. Information about the way in which sectors are linked with each other and with households (in terms of final demand and/or labour inputs) are translated into multiplier values. A rare example to date is the work of Hérault (2010) who used a CGE model to simulate the changes at the macroeconomic level after a certain policy change,

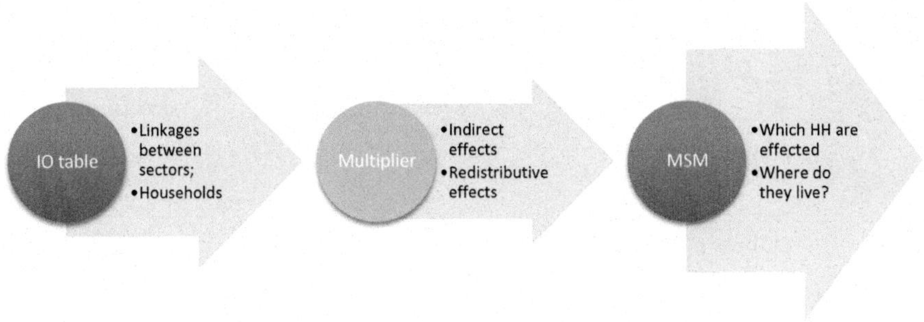

Fig. 6.1 Top-down linkage

which are then passed on to the MSM model. In Ballas and Clarke (2001), the focus is on the type and location of households that are affected by a decline in jobs. In this chapter, we also take a top-down approach.

6.2.3.2 Bottom-Up Linkage

In a bottom-up approach, the modeller starts with behaviour at the individual level, which in a next step is linked to multipliers to show redistributive/indirect effects (see Fig. 6.2). Van Leeuwen (2010), for example, looked at the effect of a new out-of-town shopping centre on the retail sector of a medium-sized Dutch town. Dutch policymakers are often reluctant to allow out-of-town retailing due to possible negative effects on shops in the city center. Lower expenditures in the centre could potentially affect (local) suppliers and result in a loss of jobs. By linking an individual-level spatial shopping model to the simulated population of Nunspeet, van Leeuwen estimated the changing expenditures in the local and wider economic area because of these new larger and, for some, closer shops. Next, the expenditures were combined with the retail multiplier derived from an interregional SAM. Because this multiplier could be decomposed into output, employment and income effects in town and hinterland, the final results showed a range of effects. It confirmed the concerns of local retailers that their sales would fall; however, it also showed to policy-makers that local income will not decrease, and in total more household expenditures will be retained in the local economy (van Leeuwen 2010).

6.2.4 *Enriching IO Tables*

Thirdly, MSM models can enrich empirical regional multi-sectoral macro models. Developing a regional model requires data at the relevant spatial level, which is not always available in exactly the right format (year or scale). Sometimes researchers take a hybrid approach in which carefully collected survey information

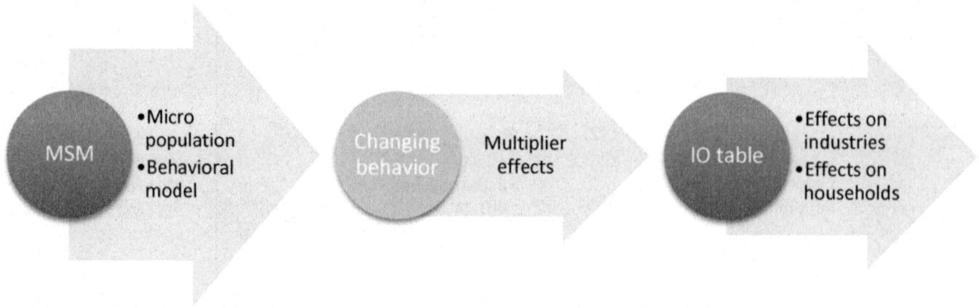

Fig. 6.2 Bottom-up linkage

is reweighted to known totals. However, often a (simple) univariate reweighting is used, which can strongly bias the results. In addition, an important advantage of spatial MSM relates to data linkage (coupling) (van Leeuwen et al. 2016). When there is a link through at least one common attribute, MSM can combine different data sets (for example, questionnaire results and census data at different geographical levels) in the same simulation exercise. Examples are Ballas and Clarke (2001), Lovelace et al. (2014) and van Leeuwen et al. (2016). The results can subsequently be aggregated to whatever level is relevant for the input–output model.

6.3 An Example of a Top-Down Study: Western Isles (WI) in Scotland

6.3.1 Investment in Energy Production in Scotland

To provide a route to sustainability in energy production, the Scottish Government (2011) has set a target that by 2020 the electricity generated in Scotland from renewable energy sources should equal the Scottish gross consumption of electricity. Figure 6.3 shows the electricity generation mix in Scotland between 2000 and 2013. Allan et al. (2011a, b) suggest that the bulk of subsequent increases in renewables will come from on and off-shore wind and that much of this new renewable capacity will be located in peripheral areas. This reinforces current Scottish economic policy which favours development in peripheral areas like the Western Isles.

6.3.2 The Study Area

The Western Isles (Eilean Siar) consist of 36 regions with a working population of around 20,000. Each region has between 400 and 800 workers. Figure 6.4 shows the location of the region in the UK context.

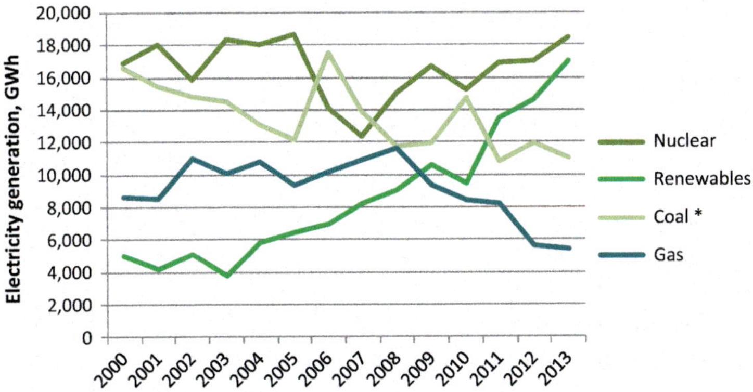

Fig. 6.3 Electricity generation by fuel (GWh) Scotland 2000–2013. Source: Energy in Scotland 2015, Scottish Government (Fig. 3.2)

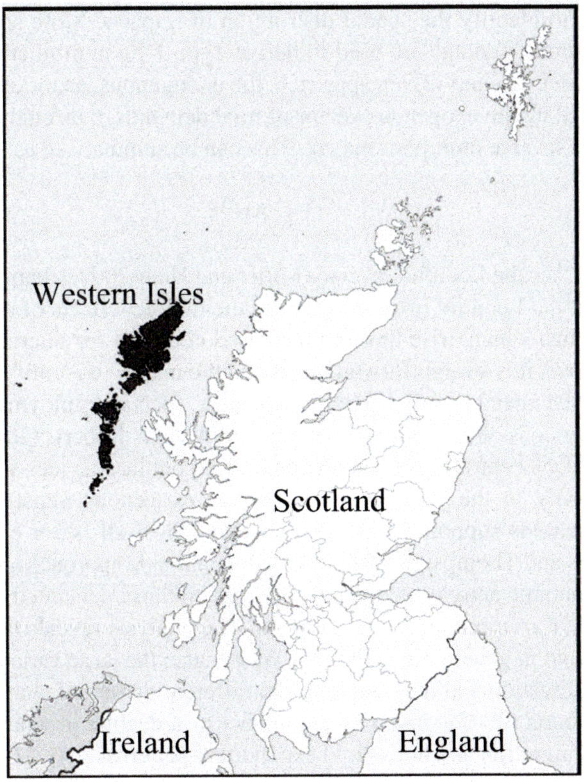

Fig. 6.4 The Western Isles of Scotland

6.3.3 Macro View of the Western Isles Economy

There are already in place good foundations for studying the regional economy of the Western Isles. For a macro view of the economy, we draw on the 2003 Regional Accounts (Roberts 2005), which were commissioned by the local authority and are publicly available.[3] These were constructed by combining secondary data with detailed surveys of business, households and public sector institutions.[4] The regional accounts consist of a SAM, which separately identifies 26 production sectors and three types of households (adults, adults with dependents, retired). Furthermore, the regional accounts comprise an employment occupation matrix, which identifies employment across seven occupation types for each sector.

To align the regional accounts with our microsimulation model, production was aggregated to 12 individual sectors. In this application, we draw on the relative strength of the multisectoral economic accounts in identifying inter-industry linkages. The accounts further inform parameters in the microsimulation model, which is used to identify the spatial distrubtion of impacts. More specifically, the regional economic accounts are used to derive Type-1 IO multiplier values. In the standard Leontief demand-driven approach, the endogenous vector of final outputs, $\boldsymbol{q}$, is determined by the exogenous vector of final demands, f, through the operation of the Leontief inverse multiplier matrix. This can be summarised as:

$$q = (1 - A)^{-1} f \tag{6.1}$$

where $(1 - A)^{-1}$ is the Leontief inverse (Miller and Blair 2009, Chap. 2). In a Type-I specification, the Leontief inverse identifies the indirect effects of any exogenous demand stimulus, which arise through increased demands for intermediate goods. As is well known, it is straightforward to extend the matrix to identify the impact of changes in final demand on other activity indicators, such as employment or income.

In earlier versions of this model based upon 1997 data, Roberts (2003) highlights the importance of central government funding of public services in maintaining economic activity in the Western Isles. Further exogenous transfers of income direct to households support 8% of all jobs and 7% of all factor earnings in the region. Roberts and Thompson (2003) take a demand-side approach and decompose changes in economic activity to distinguish between those generated by changes in technology, local sourcing and final demand. The analysis reveals the importance of export demand in generating activity in WI but also the large variability between sectors in the direction and magnitude of the different drivers of change. Using the 2003 data, Roberts (2005) simulates the impact of a decline in aquaculture and a change in net migration and household expenditure patterns.

[3]The regional accounts can be accessed at: http://www.cne-siar.gov.uk/factfile/economy/regaccounts03/index.asp

[4]For details of the method used in their construction see: http://www.cne-siar.gov.uk/factfile/economy/regaccounts03/methodology.asp

6.3.4 A New Microsimulation Model for WI

For the WI, the SARs provide detailed characteristics on 1500 individuals. We only use individuals living in the WI region, as people living in other parts of Scotland, such as the large cities (Glasgow and Edinburgh for example), might not be suitable for matching in this way. As the principal purpose of the modelling exercise is to examine the economic and social impacts of labour market changes, we simulate the population using age, sex, hours worked and socio economic classification as the main constraint variables. Furthermore, we only take into account individuals aged 16–74. This means, that the 1500 individuals will be reweighted until all known totals of the four variables in the 36 WI regions are met. Table 6.1 gives the definition of the classes.

The constraint variable "age" has been aggregated in such a way that a match could be made between the census and SARS data. The sex and socio-economic class variables were already classified in a similar way in both datasets. For the constraint variable "hours worked", we used the commuting data in which this

Table 6.1 The variables included in the WI microsimulation model

Variables	Classes	Code	Number of respondents
Age	16–24 years	1	136
	25–29 years	2	61
	30–59 years	3	543
	60–64 years	4	81
	65–74 years	5	145
Sex	Male	1	481
	Female	2	485
Hours worked	Full-time (≥30 hours per week)	1	640
	Part-time (<30 hours per week)	2	252
	Not classifiable	3	74
Socio-economic class	Large employers and higher managers	1	11
	Higher professional occupations	2	28
	Lower managerial and professional occupations	3	146
	Intermediate occupations	4	59
	Small employers and own account workers	5	73
	Lower supervisory and technical occupations	6	80
	Semi-routine occupations	7	112
	Routine occupations	8	139
	Never worked/Long-term unemployed	9	42
	FT Students	10	77
	Not classifiable	−9	199

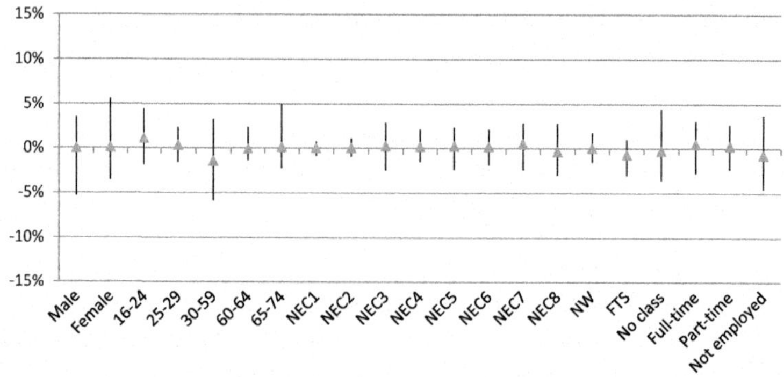

Fig. 6.5 Evaluation of the simulated results: the percentage difference between the error and the total population of the regions

variable is present. We aggregated the available classes of hours worked per week into part-time (<30 hours per week) and full-time (≥30 hours per week).[5] When checking the results with the census data, in particular the number of people in- and out-commuting, we find a precise match, with only small (1%) differences due to rounding.

The deterministic MSM techniques we used were developed by Robin Lovelace and Dimitris Balllas from Sheffield University (Lovelace and Ballas 2013) and are based on an iterative proportional fitting technique (Lovelace et al. 2015).

The results of the simulation for the 36 regions produce a fairly accurate outcome. Figure 6.5 shows the difference between the expected output and the simulated output in average, minimum and maximum terms. It appears that on average we underestimate the number of persons aged between 30–59 by 1%. However, in one area the underestimate is as much as 6%, and in one region there is an overestimate of 3%. Furthermore, although gender is, on average, very well projected, there is one region with 6% overestimate of women. Finally, the social classes (NEC1-8) are very accurately simulated against reality.

The internal evaluation shows reliable results. But what about an external validation? From the SARS, we know whether people work mainly at home, within their local area district (the Western Isles in this case), in the rest of Scotland, or outside Great Britain. When comparing this broad place of work classification with the commuting database, we find very small errors that are all below 2%.

[5]We only assigned the hours worked to those that are economically active according to the SARS.

6.3.5 The Journey to Work Model

The first major task in the linkage process is to allocate individuals in the microsimulation to a place of work. This can be done through the journey to work (JTW) data provided in the census. For every individual, we can use the JTW data to estimate a probability of working in a particular locality, in a particular industry and occupation type. The total number of people aged between 16 and 74 (which we consider as working age) is around 19,500. Of those, almost 11,600 have a full-time or part-time job.

For the journey to work model, we have to link the simulated population to the commuting data at the area level. The external validation showed that the microsimulation predicts commuting behavior extremely well, with an average difference of only 2% (excluding the people who don't have a fixed workplace). To assign the area and sector of work, we use a matching procedure based on zone of residence, gender and hours worked.

First, joint probabilities are calculated using gender and working hours (full-time or part-time). Expression (6.2) defines the probability mass function such that the joint probability is non-negative and Eq. (6.3) states that the sum of the joint probabilities should equal one.

$$f(x, y) \geq 0 \, for \; all \; (x, y) \tag{6.2}$$

$$\sum_x \sum_y (x, y) = 1 \tag{6.3}$$

By multiplying the probabilities by the total number of people with a job in an area, we get the hours worked by men and women.

Secondly, we use the R procedure 'matchby' (Sekhon 2011) to match the commuting data with the simulated micropopulation. We match the two datasets based on gender and hours worked by the individuals grouped by the zone in which they live. By adding a caliper vector to the procedure, we define for different covariates (zone, gender and hours worked) the distances that are acceptable for a match. When all calipers are set to 0 (and only exact matches are allowed), 8791 of the 10,998 are perfectly matched.[6] When allowing for some differences between actual and predicted hours worked (caliper (zone = 0; gender = 0; hours = 3)), 10,474 cases are perfectly matched on zone and gender, but with a different qualification of hours worked in 1800 cases.

The results show an upward bias towards working full-time. This is partly a result of the fact that we used a simple approach to calculate the multivariate variables gender and hours worked. That is to say, we multiply the share of women by the share of full-time jobs in a region to get the share of full-time working women. This does not take into account the fact that women are more likely to be working part

[6] In 650 cases we could assign an exact place of work and sector, without having to use probabilities.

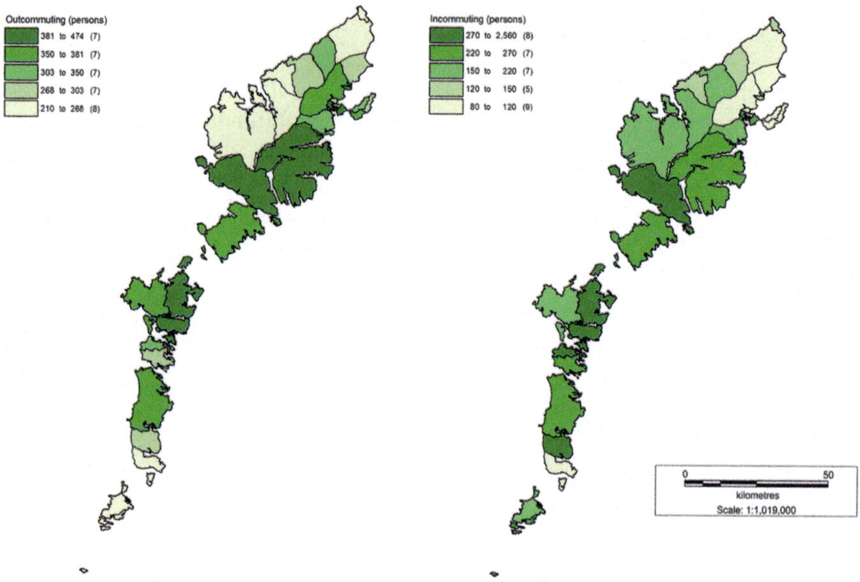

Fig. 6.6 Out and in-commuting predicted by the MSM

time than men. On average, the mismatch is around 3% per zone. In the future, some of these jobs could be allocated to key firms in each locality. Thus, some individuals in the microsimulation model can be allocated not only to a work destination, for example Lerwick, but also to a particular firm, the largest being the most important with which to form a link. The rest can be split between the remaining small firms as appropriate.

Having an establishment database would also allow us in the future to disaggregate the potential employment sector linkages in the input–output model. Figure 6.6 shows the patterns of commuting estimated—the left hand map shows the degree of out-commuting whilst the right hand map shows in-commuting.

6.4 Linking the Macro to the Micro

6.4.1 Input–Output Scenario Analysis

We use the input–output model to explore the impact of two future likely growth scenarios that are illustrative of the nature of the energy investment projects being considered in the Western Isles. First is the production of energy from the anaerobic digestion of seaweed. The second is the operation of a large scale windfarm. We briefly summarise each in turn.

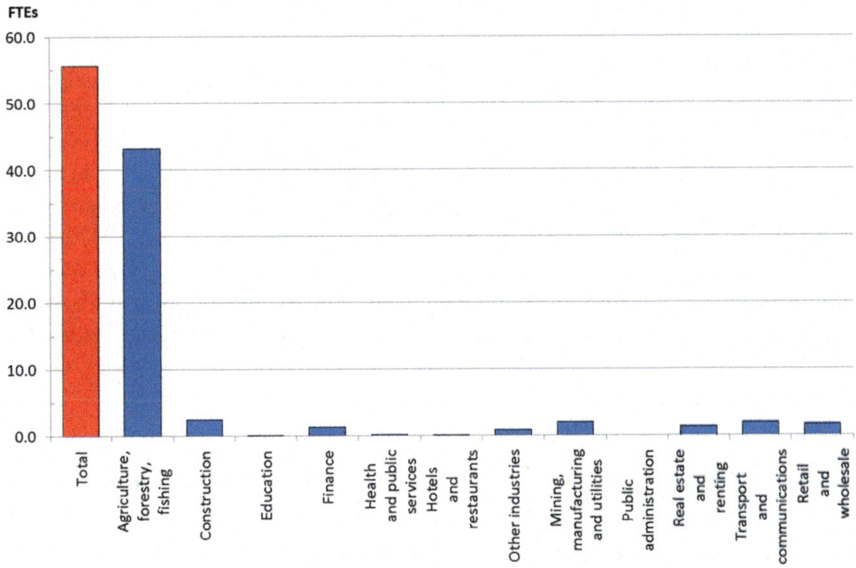

Fig. 6.7 Employment impact of bioenergy scenario in aggregate and by sector (FTEs)

6.4.1.1 Bioenergy from Seaweed

Hermannsson and Swales (2013) appraise the potential economic and environmental impact of harvesting seaweed from the waters around the Western Isles and using it to produce biogas via anaerobic digestion, which in turn is used to produce energy for export to the U.K. national grid. Once in operation this would stimulate final demand for seaweed harvesting in the Western Isles. Based on the energy potential of sustainable seaweed harvests around the WI coastline, electricity prices and subsidies for small scale renewables, Hermannsson and Swales (2013) estimate this could stimulate final demand in the WI to the tune of £2.64 million, which is approximately 0.5% of total final demand in the isles.

In the absence of detail information about the structure of the nascent harvesting sector, we assume for simulation purposes that it can be proxied by the "Agriculture, forestry, fishing" sector in our IO-model, which has a Type-I multiplier of 1.27. Based on this multiplier, the impact on the output of the WI-economy can be estimated at £3.35million. The employment impacts are detailed in Fig. 6.7. The red bar to the left shows the aggregate employment supported across all sectors, whereas the blue bars show employment by individual sectors. Approximately four out of five jobs occur within the "Agriculture, forestry, fishing" sector, whilst other jobs are scattered through service sectors. The spatial distribution of these impacts is likely to depend on how seaweed-harvesting activities will be distributed across the harbours in the Western Isles.

6.4.1.2 Large Scale Windfarm

Plans are under way for a large-scale windfarm just outside the main settlement of Stornoway, the Stornoway Wind Farm (http://www.stornowaywind.com/). Appraisal of this project has suggested that once operational it would support around 75 jobs in the Western Isles.[7] Most of the direct jobs are likely to occur in maintenance and servicing activities in and around Stornoway. According to our IO model, this would be consistent with a final demand stimulus of £1.47million to the "Other industries" sector, which has a Type-I multiplier of 1.12. Based on this multiplier, the output supported directly and indirectly across all sectors can be estimated at £1.64 million. In this case, 96% of the employment occurs within the directly affected sector, as the indirect effects are relatively subdued.

6.4.2 Testing Impacts with the WI Microsimulation Model

The previous discussion identifies the conventional outputs from the WI-IO model. For illustration, we show how the simulation results for the new jobs created in wind farming can be handled by the WI-MSM model. First, the 75 direct new jobs are created in the main wind farm company operating in the WI capital Stornoway. Second, the additional jobs predicted by the IO model in finance and other industries are also assumed to be in locations containing existing concentrations of jobs in those sectors. Existing firms are taken to be more likely to get extra business than new firms entering the labour market, although that assumption could be relaxed in the future. Thus, the jobs predicted through the second order effects are primarily in the main towns. Using the MSM and the journey to work model, unemployed households with the necessary occupation skills from within the commuting catchment areas of the predicted location of the new jobs can be allocated to those new jobs. Again, at the moment this is a very straightforward allocation. In the future it would be useful to include a full labour market model which allocated households to new jobs from a pool of both unemployed and employed households, so as to incorporate job switchers. Figure 6.8 shows the predicted catchment area for the new jobs based on the number of households qualified to be matched to those jobs. If insufficient persons are unemployed, then the model searches in the next zone and so on until all the jobs are allocated. Figure 6.8 shows another important and key result: that the impact of the new jobs is spatially bounded and many areas of the Western Isles will be only slightly impacted by the new job generation.

Finally, we can now change the attributes of the households with the new jobs—they move from a status of unemployed to employed. The most significant change is, therefore, a greater household income. Figure 6.9 shows the increase in

---

[7]http://www.hie.co.uk/about-hie/news-and-media/archive/stornoway-wind-farm-approval-will-support-75-jobs.html#sthash.NFXA7itb.dpbs

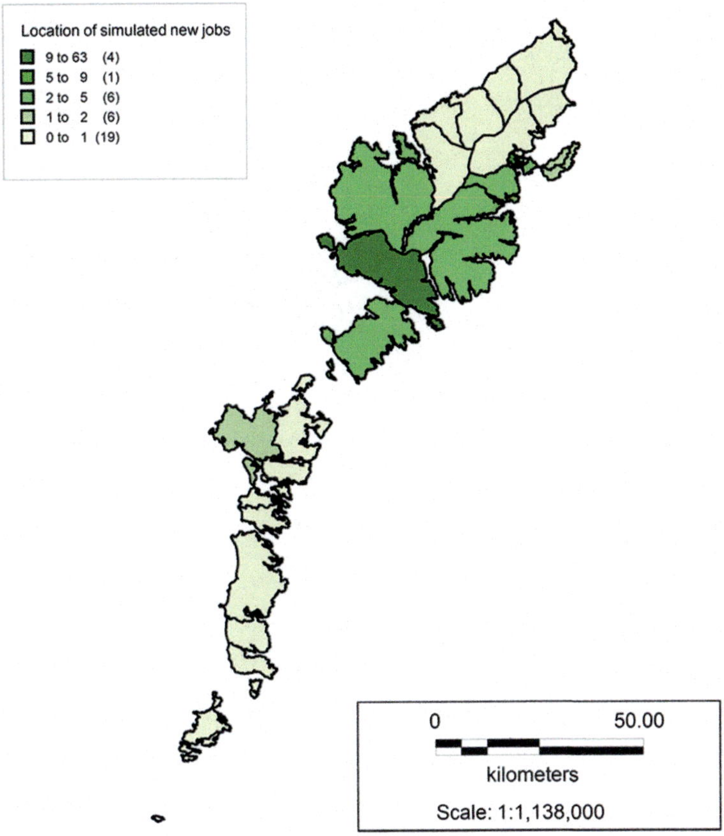

Fig. 6.8 The estimated location of households taking up the new jobs

earned income across the study region—the pattern clearly mirrors the location of households gaining the jobs in Fig. 6.8.

6.5 Conclusions and a Future Road Map

This chapter has presented a case for the greater linkage of macro and micro models in regional science. We believe such a linkage has powerful advantages in terms of the production of both regional and local economic variables associated with job gain or loss. The macro models produce powerful estimates of inter-industry linkage, which the microsimulation models can in turn link to individuals and households at the small-area level. In the preliminary case study analysis presented here illustrates how the combined approach reveals the spatial extent of new job

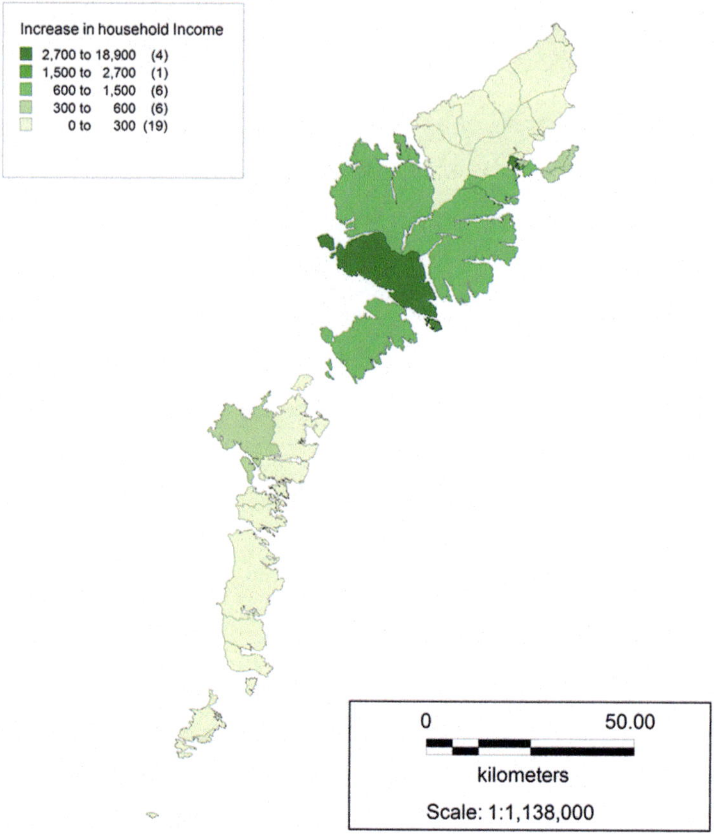

Fig. 6.9 The growth in household income following the new jobs generated

generation estimated by industrial sector. The demand and supply sides of the labour market are now more closely aligned.

The integration of microsimulation and multi-sectoral macro-economic models is driven by the desire to capture the spatial interaction between the population's domestic, work and shopping locations, and the way this is linked to aggregate economic activity. Future enhanced computing power should make such modelling increasingly viable. The improved collection and ease of manipulation of computerized databases (especially in the new era of 'big data') and the increased speed and capacity of model solutions will all aid such analysis.

If the microsimulation model is linked to purely demand-driven IO or SAM economic models, a major advance would be in a more detailed spatial tracking of economic impacts through production (and also consumption) linkages. This could be done in a more sophisticated way than at present through acknowledging spatial gravity effects on inter-industry trade. This would involve running the models in a round-by-round manner specifying both the spatial and sectoral composition of the demand injection in each round. This method would identify the sectoral and spatial ripples of demand economic activity emanating from an original exogenous demand shock.

Such extentions to the macro-modelling would be consistent with simultaneous enhancement of the microsimulation model. In addition to income we could add expenditure by occupation and industry type plus shopping and service-based interactions between home location and supply point. Then, changes between households in and out of work could be supplemented by changes in expenditures at local businesses that might in turn lead to more job losses/gains.

Linking Computable General Equilibrium (CGE), rather than IO or SAM, models to micro-simulation raises a number of challenges. The key-differentiating characteristic of CGE models is that prices (and therefore incomes) are endogenous, determined by market factors. The operation of the labour market is important here. It would be conceptually straightforward to replace the IO model in the Western Isles analysis outlined in Sects. 6.3 and 6.4 with a corresponding CGE model. Moreover regional CGE models typically operate with imperfectly competitive labour markets, where the level of unemployment through a wage-curve specification determines the real wage. However, the operation of the labour market over space would clearly require more thought. A more sophisticated labour market model would incorporate job switching and possibly migration, and progress with this already exists (see, for example, Ballas and Clarke 2000).

One major issue for a fully integrated model is initial model calibration. First data from disparate sources need to be aligned and fully consistent. This is challenging especially if the databases themselves have already been through a data consistency procedure. A related issue is that the integrated model should ideally replicate the base-year values if run forward with no change in exogenous variables. This would mean that the model needs to be parameterized such that both the sectoral and spatial decisions of firms and households are initially in equilibrium. This is a non-trivial task.

We are convinced that future developments in combining the operation of spatial micro-simulation models and multi-sectoral macro-economic models will produce a more spatially nuanced account of the sectoral, demographic and social impact of demand and supply-side economic shocks. Initially this will come from operating the two types of model in tandem, However, increasingly attempts will be made to more fully coordinate and incorporate elements of both into a single model.

Finally, not only what is technically and/or in terms of data-availability feasible, but also what is relevant for other academics or policy-makers should be taken into account. As in most modeling contexts, increased detail comes at a cost. This can be computation time, but also less accurate results. Therefore it is important to decide beforehand what the most appropriate level of analysis and outcome is for specific research questions, but also for the relevant end-users.

References

Allan G, Eromenko I, McGregor P, Swales K (2011a) The regional electricity generation mix in Scotland: a portfolio selection approach incorporating marine technologies. Energy Policy 39:6–22

Allan G, McGregor P, Swales K (2011b) The importance of revenue sharing for the local economic impacts of a renewable energy project: a social accounting matrix approach. Reg Stud 45(9):1171–1186

Armstrong H, Taylor J (2000) Regional economics and policy, 3rd edn. Blackwell, Oxford

Ballas D, Clarke GP (2000) GIS and microsimulation for local labour market analysis. Comput Environ Urban Syst 24:305–330

Ballas D, Clarke GP (2001) Towards local implications of major job transformations in the city: a spatial microsimulation approach. Geogr Anal 33:291–311

Ballas D, Clarke GP, Dewhurst J (2006) Modelling the socio-economic impacts of major job loss or gain at the local level: a spatial microsimulation framework. Spat Econ Anal 1(1):127–146

Batey PW, Madden M (1983) The modelling of demographic-economic change within the context of regional decline. Socio Econ Plan Sci 17:315–328

Batey PW, Madden M (1999) The employment impact of demographic change: a regional analysis. Pap Reg Sci 78(1):69–88

Batey PWJ, Madden M (2001) Socio-economic impact assessment: meeting client requirements. In: Clarke GP, Madden M (eds) Regional science in business. Berlin, Springer, pp 37–60

Birkin M, Clarke M (1989) The generation of individual and household incomes at the small area level using synthesis. Reg Stud 23(6):535–548

Bourguignon F, Bussolo M, Cockburn J (2010) Macro-micro analytics: background, motivation, advantages and remaining challenges. Int J Microsimul 3(1):1–7

Bourguignon F, Spadaro A (2006) Microsimulation as a tool for evaluating redistribution policies. J Econ Inequal 4(1):77–106

Hérault N (2010) Sequential linking of computable general equilibrium and microsimulation models: a comparison of behavioural and reweighting techniques. Int J Microsimul 3(1):35–42

Hermannsson K, Swales K (2013) Economic impact of producing bioenergy from seaweed fraser economic commentary. Special Issue 4:24–30

Hermes K, Poulsen M (2012) A review of current methods to generate synthetic spatial microdata using reweighting and future directions. Comput Environ Urban Syst 36:281–290

Hewings G, Okuyama Y, Sonis M (2001) Creating and expanding trade partnerships within the Chicago Metropolitan area: applications using a Miyazawa Accounting System. In: Clarke GP, Madden M (eds) Regional science in business. Berlin, Springer, pp 11–36

Jin Y-X, Wilson AG (1993) Generation of integrated multispatial input-output models of cities (GIMIMoC): 1: initial stage. Pap Reg Sci 72(4):351–368

Learmonth D, McGregor PG, Swales JK, Turner K, Yin KY (2007) The importance of the regional/local dimension of sustainable development: an illustrative computable general equilibrium analysis of the Jersey economy. Econ Model 24:15–41

Li J, O'Donoghue C (2013) A survey of dynamic microsimulation models: uses, model structure and methodology. Int J Microsimul 6(2):3–55

Lovelace R, Ballas D (2013) Truncate, replicate, sample: a method for creating integer weights for spatial microsimulation. Comput Environ Urban Syst 41:1–11

Lovelace R, Ballas D, Watson M (2014) A spatial microsimulation approach for the analysis of commuter patterns: from individual to regional levels. J Transp Geogr 34:282–296

Lovelace R, Birkin M, Ballas D, van Leeuwen E (2015) Evaluating the performance of iterative proportional fitting for spatial microsimulation: new tests for an established technique. J Artif Soc Soc Simul 18(2):21

Loveridge S (2004) A typology and assessment of multi-sector regional economic impact models. Reg Stud 38(3):305–317

McGregor P, Swales K, Yin YP (1996) A long-run interpretation of regional input–output analysis. J Reg Sci 36:479–501

Miller RE, Blair PD (2009) Input–output analysis: foundations and extensions, 2nd edn. Cambridge University Press, Cambridge

Roberts D (2003) The economic base of rural areas: a SAM-based analysis of the Western Isles 1997. Environ Plan A 35:95–111

Roberts D (2005) The Western Isles Regional Accounts 2003, Final report, University of Aberdeen Business School 2005. Retrieved from the World Wide Web: http://wwwcne-siargovuk/factfile/economy/regaccounts03/indexasp

Roberts D, Thompson K (2003) Sources of structural change in peripheral rural areas: the case of the Western Isles, 1988/89 to 1997. Reg Stud 37:61–70

Scottish Government (2011) 2020 routemap for renewable energy in Scotland, July 2011. Available online at http://wwwscotlandgovuk/Resource/Doc/917/0118802pdf. Accessed 4 Sept 2014

Scottish Government (2015) Energy in Scotland 2015. Available online at http://wwwgovscot/Resource/0046/00469235pdf

Sekhon JS (2011) Multivariate and propensity score matching software with automated balance optimization. J Stat Softw 42(7):1–52

Tanton R, Edwards K (2013) Spatial microsimulation: a reference guide for users. Springer, Dordrecht

Van Leeuwen ES (2010) The effects of future retail developments on the local economy: combining micro and macro approaches. Pap Reg Sci 89(4):691–710

Van Leeuwen ES, Ishikawa Y, Nijkamp P (2016) Microsimulation and interregional input–output modelling as tools for multi-level policy analysis. Environ Plann C Gov Policy 34(1):135–150

Eveline Leeuwen is an associate professor at the department of Spatial Economics at VU Amsterdam. Her research interest lies in modelling complex and interrelated individual behaviour in a spatial context. Her empirical work mainly focuses on agents such as firms, farms and households in small- and medium sized towns. She is a board member of the Dutch-speaking Regional Science Section, and is also active in the European Regional Science Association.

Graham Clarke is Professor of Business Geography at the University of Leeds. He has worked extensively in various areas of GIS and applied spatial modelling, focusing on many applications within urban/social geography. A major research interest has been spatial modelling, especially spatial interaction modelling and spatial microsimulation (especially for estimating small-area patterns of income and wealth, and in applications relating to retail, crime and health). Graham also specialises in retail geography and model development in relation to retail store location planning.

Kristinn Hermannsson is a lecturer in Educational Economics at the Robert Owen Centre for Educational Change in the School of Education, University of Glasgow. A graduate of the University of Strathclyde, his research interest is in the modelling of socio-economic impacts at regional and local levels. He's published on the impacts of education and energy using Input-Output and Computable General Equilibrium models. He is secretary of the British and Irish Section of the Regional Science Association International.

Kim Swales is a Professor (Emeritus) at the Fraser of Allander Institute, Department of Economics, University of Strathclyde. His research interests are primarily in Multi-sectoral economic modelling, regional economic analysis and policy and in energy modelling. He has previously been Director and Research Director of the Fraser of Allander Institute and Head of the Economics Department at the University of Strathclyde.

Part II
Spatial Analysis

Chapter 7
On Deriving Reduced-Form Spatial Econometric Models from Theory and Their Ws from Observed Flows: Example Based on the Regional Knowledge Production Function

Sandy Dall'erba, Dongwoo Kang, and Fang Fang

7.1 Introduction

A number of recent contributions (e.g., Corrado and Fingleton 2012; Pinkse and Slade 2010; McMillen 2012) have called for more attention to two intrinsically related and recurrent issues in spatial econometrics. The first one deals with the common use of diagnostic and goodness-of-fit tests to determine the appropriate form of spatial autocorrelation. However, we demonstrate that spatially explicit reduced form models can be derived from substantive economic theory when the spatial processes at work are motivated theoretically and can be directly embedded in the foundations of the model. A previous application of this approach can be seen in Ertur and Koch (2007), Fischer (2011) and Dall'erba and Llamosas-Rosas (2015) who study the role of inter-regional knowledge externalities in a Cobb-Douglas production function of regional income dynamics.

The second challenge relates to the W matrix of spatial weights being almost consistently based on some degree of geographical proximity as if the strength of inter-regional interactions were to depend on that factor only (Fingleton and Le Gallo 2008). While geographical distance is unambiguously exogenous, it does not

S. Dall'erba (✉)
Department of Agricultural and Consumer Economics and Regional Economics Applications Laboratory, University of Illinois at Urbana-Champaign, Champaign, IL, USA
e-mail: dallerba@illinois.edu

D. Kang
Korea Labor Institute, Sejong, South Korea
e-mail: dwkang1982@kli.re.kr

F. Fang
Graduate Interdisciplinary Program in Statistics and Regional Economics and Spatial Modeling Laboratory, University of Arizona, Tucson, AZ, USA
e-mail: fangfang@email.arizona.edu

R. Jackson, P. Schaeffer (eds.), *Regional Research Frontiers - Vol. 2*,
Advances in Spatial Science, DOI 10.1007/978-3-319-50590-9_7

change with time and does not account for the possible asymmetric nature of the flows between pairs of spatial units. As a result, some contributions have proposed alternatives such as, among many others, the transportation cost (e.g., Conley and Ligon 2002), economic distance (Fingleton 2001, 2008; LeSage and Pace 2008), or technological proximity (Parent and LeSage 2008) across regions. However, such specifications still miss the opportunity to capture the directionality of the flows and their actual magnitude. As such, we advocate that using weight matrices based on observed flows will increasingly become commonplace. Examples of studies that have used this route are Eliste and Fredriksson (2004), Chen and Haynes (2015) who rely on inter-regional trade flows and Kang and Dall'erba (2015) or Sonn and Storper (2008) who use inter-county flows of patent creation-citation.

In order to illustrate the role these increasingly popular approaches will have on the spatial economics literature, we highlight their contribution in the frame of the regional knowledge production function literature (henceforth KPF). While early econometric contributions in this field paid considerable attention to the impact of investments in R&D (Research and Development) on the production of innovation at the firm level (Griliches 1979; Jaffe 1989; Cefis and Orsenigo 2001), the spatial location of the firms as well as the existence of economies of agglomeration and of spatial spillovers were ignored. However, as regional economies try to compete nationally and internationally to attract the factors at the origin of innovation, more recent contributions have adopted a spatial approach, recognized the role of knowledge spillovers and highlighted geographical differences in the dynamics of innovation (Audretsch and Feldman 1996; Crescenzi et al. 2007; Rodríguez-Pose 2001; Acs and Armington 2004; Adams 2002; Ó hUallacháin and Leslie 2007; Sonn and Park 2011; Anselin et al. 1997).

7.2 Traditional Model of Regional KPF and Extensions to the Spatial Case

In his seminal contribution Griliches (1979) formalizes the knowledge production function for each unit i at time t as follows:

$$Y_{it} = A_{it} C_{it}^{\alpha_1} H_{it}^{\alpha_2} L_{it}^{1-\alpha_1-\alpha_2} \tag{7.1}$$

where the production of knowledge Y_{it} is a function of the current state of technical knowledge A_{it} assumed to grow at an exogenous rate similarly experienced in all locations; C_{it} is the level of private reproducible physical capital; H_{it} is the level of human capital and L_{it} reflects the level of labor. As usual in a Cobb-Douglas production function, the coefficients α_1 and α_2 are positive and below 1, thus reflecting the decreasing returns to physical and human capital; and the returns to scale are also assumed decreasing. When rewritten in per capita terms, we get:

$$y_{it} = A_{it} c_{it}^{\alpha_1} h_{it}^{\alpha_2}$$

and applying a log transformation leads to:

$$\ln y_{it} = \ln A_{it} + \alpha_1 \ln c_{it} + \alpha_2 \ln h_{it} + \varepsilon_{it} \text{ with } \varepsilon_{it} \sim N\left(0, \sigma_\varepsilon^2\right) \tag{7.2}$$

As noted earlier, empirical applications of the regional KPF has moved away from the a-spatial model captured in Eq. (7.2) to specifications that allow to explicitly capture the role of knowledge spillovers. They take place when firms, industries, or regions benefit from the knowledge created by other firms, industries or regions without bearing the cost associated to its creation (Fischer et al. 2009). While the role of spillovers in knowledge creation has been well documented in the theoretical literature (Marshall 1920; Jacobs 1969; Jaffe 1986; Glaeser et al. 1992; Fung and Chow 2002; Asheim and Isaksen 2002; Henderson 2003), their appropriate measurement remains a challenge. For instance, a large amount of knowledge spillovers takes place through face-to-face interactions (Jaffe 1986; Jaffe et al. 1993; Audretsch and Feldman 1996; Rodríguez-Pose 2001; Sonn and Storper 2008) and this process is not documented clearly. We do not know how often nor where the agents of one company meet agents from another company to exchange ideas. As a result, regional KPF often deal with this type of undocumented spillovers as if they are limited spatially. Empirical evidence confirms this point. For example, Jaffe et al. (1993) find that patents produced in one state are more likely to be cited within the same state. In addition, when Sonn and Storper (2008) analyze 20 Metropolitan Statistical Areas, they conclude that the proportion of local citations has increased over the 1975–1999 period. At the same time, other contributions indicate that knowledge spillovers may well reach companies located beyond the boundaries of the locality they originate from. For instance, Johnson et al. (2006) show that, in the US, the average distance between patent collaborators has increased from 117 miles in 1975 to 200 miles in 1999. Years earlier, Anselin et al. (1997) uncovered that university research leads to innovation in high technology companies located not only within the same region but also in neighboring ones. The previous study is the first one to have used the formal tools of spatial econometrics to measure these spillovers. Many more have followed since then with applications to many different areas of the world. For instance, Bode (2004) highlights the role of inter-regional knowledge spillovers in West Germany while Parent and LeSage (2008) do so for all the European regions. Recent extensions to spatial panel data models offer the advantage to increase the efficiency of the estimates but are still relatively scarce. To our knowledge, only four have been published so far: Peri (2005) estimates cross-regional citation flows and plugs the estimated fitted values into a spatial weight matrix that captures the diffusion of knowledge flows across a panel of 113 European and North American regions over 22 years. Autant-Bernard and LeSage (2011) examine the spatial spillovers associated with public and private research expenditures by industry from 1992 to 2000 over a sample of 94 French regions. Parent and LeSage (2012) analyze the dynamics of European patenting over 1989–1999 based on a sample of 320 European regions, while Parent (2012) investigates a KPF across the 49 US states over 1994–2005. The latter contribution has the advantage of offering a spatial dynamic panel model so that both spatial and temporal autocorrelations are simultaneously accounted for. We expect that additional spatial panel data models of the KPF will emerge in the coming years given that a growing number of contributions have laid the theoretical (e.g. Baltagi et al. 2003; Kapoor et al. 2007; Elhorst 2014) and methodological foundations (Millo and Piras 2012; Elhorst 2011) for their estimation.

7.3 Selecting the Reduced-Form Spatial Model from Theory

While knowledge spillovers have now been modeled and estimated on numerous occasions, their local (as in a SLX model where the covariates are spatially lagged) or global nature (as in a SAL model where the dependent variable is spatially lagged) is very often the result of the researcher's belief or of a selection through the well-known Lagrange Multiplier tests and their robust version (Anselin et al. 1996). Several contributions have called for more theoretically-grounded foundations in the model selection (Corrado and Fingleton 2012; Pinkse and Slade 2010; McMillen 2012) and one of the most cited example of this approach is Ertur and Koch (2007). Focusing on the issue of regional income inequality, their starting point is also a Cobb-Douglas production function. However, instead of assuming that A_{it} is only exogenously determined, they describe it as the product of three elements as follows:

$$A_{i,t} = \Omega_t c_{i,t}^{\theta_1} h_{i,t}^{\theta_2} \Pi_{j \neq i}^{N} A_{j,t}^{\rho w_{i,j}} \tag{7.3}$$

where Ω_t is the exogenous stock of knowledge that is shared by all entities as proposed by the neoclassical growth model (Solow 1956; Swan 1956); $c_{i,t}^{\theta_1}$ and $h_{i,t}^{\theta_2}$ come from the endogenous growth framework (Romer 1986; Lucas 1988) and indicate that the levels of physical and human capital per worker available in region i increase the stock of knowledge available to all firms in region i by a value θ_1 and θ_2 respectively (with $0 \leq \theta_1, \theta_2$). Finally, the last term captures the knowledge externalities that originate from all the neighboring regions j (with $j \neq i$) and spill over to i (as emphasized in the new economic geography theory: Fujita et al. 1999; Boarnet 1998). The coefficient $\rho (0 < \rho < 1)$ measures the average degree of inter-regional dependence.

After log transformation and some matrix algebra (see Ertur and Koch 2007, for all the successive steps), combining Eqs. (7.2)–(7.3) leads to the following spatial Durbin model:

$$\begin{aligned} \ln y_{i,t} = \ln \Omega_t + \delta_1 \ln c_{it} + \delta_2 \ln h_{it} - \alpha_1 \rho \textstyle\sum_{j \neq i}^{N} W_{i,j} \ln c_{it} - \\ \alpha_2 \rho \textstyle\sum_{j \neq i}^{N} W_{i,j} \ln h_{it} + \rho \textstyle\sum_{j \neq i}^{N} W_{i,j} \ln y_{j,t} \text{ with } \varepsilon_{it} \sim N\left(0, \sigma_\varepsilon^2\right) \end{aligned} \tag{7.4}$$

where $\delta_1 = \theta_1 + \alpha_1$ and $\delta_2 = \theta_2 + \alpha_2$ and the individual coefficients and their significance level can be found through the delta method of Casella and Berger (2002) which builds on the estimated coefficient means and variance-covariance matrix. According to Eq. (7.4), knowledge created in one location would spill over to the rest of the sample with a magnitude that decreases with increasing distance and even feed back to the place of origin (LeSage and Pace 2008). Besides Ertur and Koch (2007), this approach has been used by Fischer (2011) and Dall'erba and Llamosas-Rosas (2015) for the case of regional income growth in Europe and in the US respectively.

As much as deriving the choice of the reduced-form spatial model represents an important contribution compared to the past, one should note that the choice of the initial form taken by $A_{i,t}$ is not neutral. For instance, in the context of the regional KPF, Fang et al. (2016) propose to compare the results obtained using Eq. (7.3) as is versus using the following modified form:

$$A_{i,t} = \Omega_t c_{i,t-1}^{\theta_1} h_{i,t-1}^{\theta_2} \Pi_{j \neq i}^{N} c_{j,t-1}^{\tau_c P_{i,j} + \sigma_c M_{i,j}} h_{j,t-1}^{\tau_h P_{i,j} + \sigma_h M_{i,j}} \tag{7.5}$$

where the first elements are similar to Eq. (7.3), but include the fact that local R&D efforts do not lead instantaneously to the creation of knowledge (Griliches 1979). Furthermore, the latter elements reflect that the spillovers of private and human capital are assumed to originate from two distinct sources: the flows of patent creation-citations $P_{i,j}$ as described in Sonn and Storper (2008) and Kang and Dall'erba (2015) and the flows of educated workers moving from i to j $M_{i,j}$ as in Breschi and Lissoni (2009) and Kerr (2013) although in a different context. Both types of spillovers are assumed to affect the technical knowledge after a one-year period as it is well-known that R&D expenditures take time to produce any innovational output (Griliches 1979). Furthermore, compared to Eq. (7.4) where inter-regional linkages are based on geography only, the obvious advantages are that the elements of $P_{i,j}$ and $M_{i,j}$ identify two different types of spillovers and they are changing from one year to the next. Halleck Vega and Elhorst (2015) provide a list of additional advantages of a SLX model such as Eq. (7.6) compared to models with global spillovers as Eq. (7.4).

Combining (7.5) and (7.2) leads to:

$$\begin{aligned} \ln y_{i,t} &= \ln \Omega_t + \alpha_1 \ln c_{i,t} + \alpha_2 \ln h_{i,t} + \theta_1 \ln c_{i,t-1} + \\ &\theta_2 \ln h_{i,t-1} + \tau_c \textstyle\sum_{j \neq i}^{N} P_{i,j} \ln c_{j,t-1} + \sigma_c \textstyle\sum_{j \neq i}^{N} M_{i,j} \ln c_{j,t-1} + \\ &\tau_h \textstyle\sum_{j \neq i}^{N} P_{i,j} \ln h_{j,t-1} + \sigma_h \textstyle\sum_{j \neq i}^{N} M_{i,j} \ln h_{j,t-1} \text{ with } \varepsilon_{it} \sim N\left(0, \sigma_\varepsilon^2\right) \end{aligned} \tag{7.6}$$

where only the first order places of export or out-migration have an effect on local innovation. Such spillovers are still qualified as "local" even though they are not based on geographical proximity.

When estimating Eqs. (7.4) and (7.6) across the US States, Fang et al. (2016) note that the direct and indirect marginal effects of the inputs (spending in academic and private R&D in their case) correspond for the most part to their expectations in Eq. (7.4). With Eq. (7.6), they find that current and last year's R&D expenditure at universities and colleges support local innovation while private R&D may require more time to show the same effect. Their findings indicate also that past levels of R&D in the states migrants come from benefit the state they move to, hence confirming the transfer of knowledge embedded in labor migration (Almeida and Kogut 1999). The flows of patent creation-citation lead to more novel results by which academic R&D spending that takes place at time *t–1* in the states where patents are originally created is negatively correlated with the production of innovation at time *t* in the states that cite these patents. One possible explanation is that a patent-citing state may, intentionally or not, reduce its marginal spending in academic R&D in its own location if it is known that other states, the patent-creating states, are bearing the costs of academic R&D. Since spending on academic R&D has a positive marginal effect on local innovation, the marginal effect of this "free-rider" behavior leads to a negative effect on local innovation.

7.4 Stepping Away from Proximity-Based Network

The large majority of spatial econometric estimations of the regional KPF define inter-regional interactions based on geographical proximity. The motive is that knowledge spillovers take place through face-to-face interactions (e.g. Jaffe 1986; Jaffe et al. 1993) and thus their spatial extent is geographically limited. For instance, Anselin et al. (1997) and Acs et al. (2002) choose a distance cut-off of 50 miles based on the maximum distance found among US commuting patterns (Rapino and Fields 2013). Their results are robust to 75 miles. In this case, the matrix where regions i and j are separated by distance d_{ij} and a chosen distance cut-off d can be written as:

$$w_{ij} = \begin{cases} 1, 0 < d_{ij} \leq d \\ 0, \quad d_{ij} > d \end{cases}$$

However, the well-established notion of proximity of knowledge spillovers has also been challenged numerous times. The earliest contribution to do so is Jaffe (1986). His focus is on addressing intellectual interactions among regions so that he specifies the knowledge externalities for any considered pair of firms by using a Pearson correlation coefficient. Numerical vectors describing the distribution of firm-level patents over several technological fields are first constructed and the correlation between any pair of vectors is used as a proxy for the firms' interaction. The geographical distance that separates them is thus disregarded. A few years later, Parent and LeSage (2008) have extended his approach by weighting Jaffe's firm-level technology spillovers by GDP_i/GDP_j $\mathrm{GDP_i/GDP_j}$ which captures the output gap between regions i and j. The larger the gap is the larger the asymmetric effect between technological distances w_{ij} and $\mathrm{w_{ji}}$ is assumed:

$$w_{ij} = \left(\frac{GDP_i}{GDP_j}\right)^{1/2} \times \frac{\sum_{k=1}^{m} F_{ki}F_{kj}}{\left(\sum_{k=1}^{m} F_{ki}^2 \sum_{k=1}^{m} F_{kj}^2\right)^{1/2}}$$

where F_{ki} represents the number of patents granted in the technology field k and region i so that regions i and j reduce their technological distance by patenting in the same fields (Jaffe 1986). Other notable exceptions to the general rule of geographically-limited spillover effects are the growing number of contributions based on some measurement of the network of collaboration among innovators. For instance, Autant-Bernard et al. (2007) use a model of cooperation choice to highlight that a firm's position within a network matters more than its geographical location. Their model is based on all the collaborative projects submitted to the European Union 6th Framework Program. Ponds et al. (2010) focus on how networks stemming from university-industry collaborations support the impact of academic research on innovation across Dutch regions and Crescenzi et al. (2016) uncover that among the different types of proximities (geographic, organizational, cognitive, social and cultural-ethnic) they identify and test, U.K. inventors rely more often on social connections while cultural, cognitive and geographic proximity do not matter much.

The above contributions build on the idea of a network of researchers who collaborate in order to create innovative products no matter how far apart they live. To our knowledge, the dataset that has been the most extensively used to measure this network of collaboration is the US Patent and Trade Office (USPTO 2010) as it reports the address of the inventors and the address of the headquarter of the company they work for. The contributions of Jaffe and Lerner (2004) and Crescenzi et al. (2007) are examples of studies that rely on USPTO. In order to allocate spatially the patents that are the fruit of the work of N inventors, they use the fractional counting method suggested by Jaffe et al. (1993) whereby a fraction 1/N of the patent is allocated to each inventor and, as a result, to his/her geographical unit. As such, patent data is not an integer value anymore but a rational number.

7.5 Using Connections that Capture the Directionality of the Flows

However, one element that is missing from the previous approaches is the directionality of the knowledge flows. Indeed, the causality associated to investing inputs in region i to create knowledge output in region j is proxied in various ways but not explicitly captured. For instance, in a spatial network of co-patenting inventors it is impossible to assess who got the idea first. We foresee that the coming decades will offer an increasing number of sources reporting flow data as regional economies become more integrated and the study of spillovers keeps developing. Popular examples of such sources for the U.S. economy are the Commodity Flow Survey that reports data on the movement of goods and the Census Bureau data on migratory flows reported in the Integrated Public Use Microdata Series (IPUMS). When it comes to innovation, the appropriate dataset is the "NBER US Patent Citation Data File" of Hall et al. (2001) as it reports the citation records associated to each patent

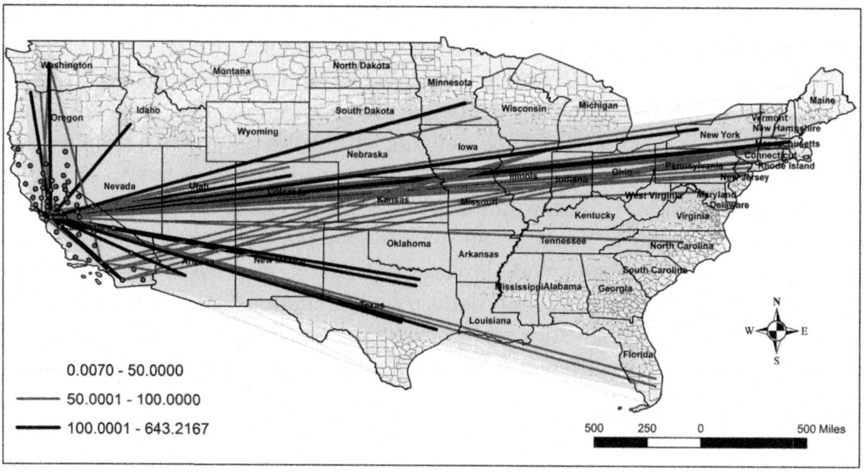

Fig. 7.1 Number and place of creation of the patents cited by California's counties (Kang and Dall'erba 2015)

for the period of 1975–1999, as well as the name of the inventors, assignees and their address. A matrix that clearly stipulates the directionality of the knowledge spillovers from the place(s) of creation of a patent to the place(s) where it is cited for further innovation and patenting can be created from it.

Peri (2005) and Sonn and Storper (2008), among others, have used this approach and an extension of the fractional counting method to origin-destination flows to capture knowledge spillovers. To our knowledge, the most recent application is Kang and Dall'erba (2015) who generate a (3109 $\times$ 3109) patent creation-patent citation flow matrix across US counties. Figure 7.1 below from their manuscript provides a snapshot of their matrix and confirms, as noted earlier, that geographical proximity is not a necessary condition for knowledge spillovers. For example, Santa Clara county (where the Silicon Valley is located) creates new products that are mostly based off of products patented in the East coast, the Midwest and several Southern States. Based on their econometric estimates, the authors conclude that over 1995–1999 the average number of patents created in remote locations (more than 50 miles away) have had a greater role on the US counties' 2003–2005 patenting activities than patents created locally (less than 50 miles away).

Last but not least, we believe that future research interested in capturing the true nature of inter-regional knowledge spillovers requires more efforts in at least three directions: first, interconnections between national and international knowledge spillovers are often disregarded as most studies focus on a single country only. Peri (2005) and Chellaraj et al. (2008) are exceptions to this rule. Second, the list of types of inter-regional spillovers reported above is not exhaustive as Miguélez et al. (2010) indicate that they would also take place through various market transactions, the monitoring of competitors and firm spin-offs. Third, their sectoral heterogeneity has often been ignored but is gaining recognition, as presented next.

7.6 Intra- Versus Inter-Sectoral Knowledge Flows

Most empirical studies in the regional knowledge production function literature use sectorally aggregated data. Sectoral heterogeneity is only partially controlled for by using sectoral dummies (Ponds et al. 2010) or sectoral share in total value added (Bottazzi and Peri 2003). The same lack of evidence applies to the treatment of sectoral heterogeneity among knowledge spillovers. Jaffe (1989) and Anselin et al. (2000) differentiate the localized knowledge spillovers by sector but only capture intra-sectoral spillovers. Autant-Bernard and LeSage (2011) demonstrate the significant impact of inter-sectoral spillovers of private R&D among French metropolitan areas. However, their panel model is averaged across all sectors so that they do not provide an estimate of the marginal effect of such spillover by sector. As a result, if knowledge spillovers are so important for the production of innovation, a deeper understanding of how each sector is likely to benefit from intra- vs. inter-sectoral spillovers and from intra- vs. inter-regional spillovers is necessary. Based on five manufacturing sectors and the sample of US counties, Kang and Dall'erba (2016) show that while both intra-sectoral (MAR) and inter-sectoral (Jacobian) spillovers are significant determinants of knowledge creation, MAR spillovers play a greater role than their corresponding Jacobian spillovers when they take place within the county or across counties (for both short- and long-distance spillovers). Their relative magnitude varies by sector also. For instance, intra-regional private and academic MAR spillovers have a greater elasticity than localized interregional spillovers in the Mechanical, Computer and Electrical sectors.

7.7 Conclusion

Much has already been accomplished in the field of spatial econometrics over the last few decades (Anselin 2010). However, the large majority of applied works does not derive their reduced-form model from a spatially-explicit theoretical framework but from a set of diagnostic tests and goodness-of-fit values. Furthermore, the spatial weight matrix at the core of the spillovers across spatial units is almost always based on some measurement of geographical proximity which does not necessarily capture the true nature, magnitude, asymmetry and directionality of these spillovers (Corrado and Fingleton 2012; Pinkse and Slade 2010; McMillen 2012).

This chapter demonstrates that the seeds to moderate such criticisms have been planted in at least one very active topic of regional science, namely the regional knowledge production function literature. In the period of only three decades, this literature has moved from Griliches' (1979) early work, a seminal contribution in the field but where the spatial organization of the data is completely ignored, to a set of very sophisticated spatial econometric specifications. This chapter shows examples of spatial models that are directly derived from theory, and a long list of weight matrix specifications that go beyond the traditional proximity-based structure is reviewed. Because knowledge spillovers take many forms, we also suggest several venues for the future.

While the regional knowledge production function has been the focus of this chapter, we believe that similar lines of research should be adopted and applied to many of the other exciting topics in regional science.

References

Acs ZJ, Anselin L, Varga A (2002) Patents and innovation counts as measures of regional production of new knowledge. Res Policy 31(7):1069–1085

Acs Z, Armington C (2004) Employment growth and entrepreneurial activity in cities. Reg Stud 38(8):911–927

Adams JD (2002) Comparative localization of academic and industrial spillovers. J Econ Geogr 2(3):253–278

Almeida P, Kogut B (1999) Localization of knowledge and the mobility of engineers in regional networks. Manag Sci 45(7):905–917

Anselin L (2010) Thirty years of spatial econometrics. Pap Reg Sci 89(1):3–25

Anselin L, Bera AK, Florax R, Yoon MJ (1996) Simple diagnostic tests for spatial dependence. Reg Sci Urban Econ 26(1):77–104

Anselin L, Varga A, Acs Z (1997) Local geographic spillovers between university research and high technology innovations. J Urban Econ 42(3):422–448

Anselin L, Varga A, Acs Z (2000) Geographical spillovers and university research: a spatial econometric perspective. Growth Chang 31(4):501–515

Asheim B, Isaksen A (2002) Regional innovation systems: the integration of local 'sticky' and global 'ubiquitous' knowledge. J Technol Transf 27(1):77–86

Audretsch DB, Feldman MP (1996) R&D spillovers and the geography of innovation and production. Am Econ Rev 86(3):630–640

Autant-Bernard C, Billand P, Bravard C, Massard N (2007) Network effects in R&D partnership evidence from the European collaborations in micro and nanotechnologies. Paper presented at the DIME—workshop on "Interdependencies of interactions in local and sectoral innovation systems", IENA, France

Autant-Bernard C, LeSage JP (2011) Quantifying knowledge spillovers using spatial econometric models. J Reg Sci 51(3):471–496

Baltagi BH, Song SH, Koh W (2003) Testing panel data regression models with spatial error correlation. J Econ 117(1):123–150

Boarnet MG (1998) Spillovers and the locational effects of public infrastructure. J Reg Sci 38(3):381–400

Bode E (2004) The spatial pattern of localized R&D spillovers: an empirical investigation for Germany. J Econ Geogr 4(1):43–64

Bottazzi L, Peri G (2003) Innovation and spillovers in regions: evidence from European patent data. Eur Econ Rev 47(4):687–710

Breschi S, Lissoni F (2009) Mobility of skilled workers and co-invention networks: an anatomy of localized knowledge flows. J Econ Geogr 9(4):439–468

Casella G, Berger RL (2002) Statistical inference, 2nd edn. Thomson Learning, Pacific Grove, CA

Cefis E, Orsenigo L (2001) The persistence of innovative activities: a cross-countries and cross-sectors comparative analysis. Res Policy 30(7):1139–1158

Chellaraj G, Maskus KE, Mattoo A (2008) The contribution of international graduate students to US innovation. Rev Int Econ 16(3):444–462

Chen Z, Haynes KE (2015) Spatial impact of transportation infrastructure: a spatial econometric cge approach. In: Nijkamp P, Rose A, Kourtit K (eds) Regional science matters: studies dedicated to walter isard. Springer International Publishing, Cham, pp 163–186

Conley TG, Ligon E (2002) Economic distance and cross-country spillovers. J Econ Growth 7(2):157–187

Corrado L, Fingleton B (2012) Where is the economics in spatial econometrics? J Reg Sci 52(2):210–239

Crescenzi R, Nathan M, Rodríguez-Pose A (2016) Do inventors talk to strangers? On proximity and collaborative knowledge creation. Res Policy 45(1):177–194

Crescenzi R, Rodríguez-Pose A, Storper M (2007) The territorial dynamics of innovation: a Europe–United States comparative analysis. J Econ Geogr 7(6):673–709

Dall'erba S, Llamosas-Rosas I (2015) The impact of private, public and human capital on the US states' economies: theory, extensions and evidence. In: Handbook of research methods and applications in economic geography. Edward Elgar, Cheltenham

Elhorst JP (2011) Matlab software to estimate spatial panels. http://www.regroningen.nl/elhorst/software.shtml. Accessed 11 Apr 2011

Elhorst JP (2014) Spatial econometrics: from cross-sectional data to spatial panels. Springer briefs in regional science. Springer, Heidelberg

Eliste P, Fredriksson PG (2004) Does trade liberalization cause a race-to-the-bottom in environmental policies? A spatial econometric analysis. In: Anselin L, Florax RJGM, Rey SJ (eds) Advances in spatial econometrics: methodology, tools and applications. Springer, Berlin, pp 383–396

Ertur C, Koch W (2007) Growth, technological interdependence and spatial externalities: theory and evidence. J Appl Econ 22(6):1033–1062

Fang F, Dall'erba S, Kang D (2016) On deriving spatial econometric models from theory and W from observations—an application to the US regional knowledge production function. Paper presented at the the 55th annual meeting of the Southern Regional Science Association, Washington, DC, 31 Mar–2 Apr 2016

Fingleton B (2001) Equilibrium and economic growth: spatial econometric models and simulations. J Reg Sci 41(1):117–147

Fingleton B (2008) A generalized method of moments estimator for a spatial model with moving average errors, with application to real estate prices. Empir Econ 34(1):35–57

Fingleton B, Le Gallo J (2008) Estimating spatial models with endogenous variables, a spatial lag and spatially dependent disturbances: finite sample properties. Pap Reg Sci 87(3):319–340

Fischer M (2011) A spatial Mankiw–Romer–Weil model: theory and evidence. Ann Reg Sci 47(2):419–436

Fischer MM, Scherngell T, Reismann M (2009) Knowledge spillovers and total factor productivity: evidence using a spatial panel data model. Geogr Anal 41(2):204–220

Fujita M, Krugman PR, Venables A (1999) The spatial economy: cities, regions and international trade. MIT Press. Available via http://worldcat.org. http://site.ebrary.com/id/10225300

Fung MK, Chow WW (2002) Measuring the intensity of knowledge flow with patent statistics. Econ Lett 74(3):353–358

Glaeser EL, Kallal HD, Scheinkman JA, Shleifer A (1992) Growth in cities. J Polit Econ 100(6):1126–1152

Griliches Z (1979) Issues in assessing the contribution of research and development to productivity growth. Bell J Econ 10(1):92–116

Hall BH, Jaffe AB, Trajtenberg M (2001) The NBER patent citations data file: lessons, insights and methodological tools. NBER woring paper series 8498. National Bureau of Economic Research, Cambridge, MA

Halleck Vega S, Elhorst JP (2015) The SLX model. J Reg Sci 55(3):339–363

Henderson JV (2003) Marshall's scale economies. J Urban Econ 53(1):1–28

Jacobs J (1969) The economy of cities. Random House, New York

Jaffe AB (1986) Technological opportunity and spillovers of R&D: evidence from firms' patents, profits, and market value. Am Econ Rev 76(5):984–1001

Jaffe AB (1989) Real effects of academic research. Am Econ Rev 79(5):957–970
Jaffe A, Lerner J (2004) Innovation and its discontents: how our broken patent system is endangering innovation and progress, and what to do about it. Princeton University Press, Princeton, NJ
Jaffe AB, Trajtenberg M, Henderson R (1993) Geographic localization of knowledge spillovers as evidenced by patent citations. Q J Econ 108(3):577–598
Johnson DKN, Siripong A, Brown AS (2006) The demise of distance? The declining role of physical proximity for knowledge transmission. Growth Chang 37(1):19–33
Kang D, Dall'erba S (2015) An examination of the role of local and distant knowledge spillovers on the US regional knowledge creation. Int Reg Sci Rev. doi:10.1177/0160017615572888
Kang D, Dall'erba S (2016) The role of interregional and inter-sectoral knowledge spillovers on regional knowledge creation across US metropolitan counties. Paper presented at the the 55th annual meeting of Western Regional Science Association, the Big Island, HI, 14–17 Feb 2016
Kapoor M, Kelejian HH, Prucha IR (2007) Panel data models with spatially correlated error components. J Econ 140(1):34
Kerr WR (2013) U.S. high-skilled immigration, innovation, and entrepreneurship: empirical approaches and evidence. National Bureau of Economic Research working paper series no 19377
LeSage JP, Pace RK (2008) Spatial econometric modeling of origin-destination flows. J Reg Sci 48(5):941–967
Lucas RE (1988) On the mechanics of economic development. J Monet Econ 22(1):3–42
Marshall A (1920) Principles of economics. Macmillan, London
McMillen DP (2012) Perspectives on spatial econometrics: linear smoothing with structured models. J Reg Sci 52(2):192–209
Miguélez E, Moreno R, Suriñach J (2010) Inventors on the move: tracing inventors' mobility and its spatial distribution. Pap Reg Sci 89(2):251–274
Millo G, Piras G (2012) Splm: spatial panel data models in r. J Stat Softw 47(1):1–38
Ó hUallacháin B, Leslie TF (2007) Rethinking the regional knowledge production function. J Econ Geogr 7(6):737–752
Parent O (2012) A space-time analysis of knowledge production. J Geogr Syst 14(1):49–73
Parent O, LeSage JP (2008) Using the variance structure of the conditional autoregressive spatial specification to model knowledge spillovers. J Appl Econ 23(2):235–256
Parent O, LeSage JP (2012) Determinants of knowledge production and their effects on regional economic growth. J Reg Sci 52(2):256–284
Peri G (2005) Determinants of knowledge flows and their effect on innovation. Rev Econ Stat 87(2):308–322
Pinkse J, Slade ME (2010) The future of spatial econometrics. J Reg Sci 50(1):103–117
Ponds R, Fv O, Frenken K (2010) Innovation, spillovers and university–industry collaboration: an extended knowledge production function approach. J Econ Geogr 10(2):231–255
Rapino MA, Fields AK (2013) Mega commuters in the U.S.: time and distance in defining the long commute using the american community survey. Working Paper 2013-03. United States Census Bureau
Rodríguez-Pose A (2001) Is R&D investment in lagging areas of Europe worthwhile? Theory and empirical evidence. Pap Reg Sci 80(3):275–295
Romer PM (1986) Increasing returns and long-run growth. J Polit Econ 94(5):1002–1037
Solow RM (1956) A contribution to the theory of economic growth. Q J Econ 70(1):65–94
Sonn JW, Park IK (2011) The increasing importance of agglomeration economies hidden behind convergence. Urban Stud 48(10):2180–2194
Sonn JW, Storper M (2008) The increasing importance of geographical proximity in knowledge production: an analysis of US patent citations, 1975–1997. Environ Plan A 40 (5):1020–1039.
Swan TW (1956) Economic growth and capital accumulation. Econ Rec 32(2):334–361
USPTO (2010) Patents bib: selected bibliographic information from US patents issued 1969 to present. Alexandria, VA

Sandy Dall'erba is Associate Professor, Department of Agricultural and Consumer Economics, and Associate Director of the Regional Economics Applications Laboratory, University of Illinois at Urbana-Champaign. His primary research interests are regional science in general and economic growth, regional development policies, innovation and the economic impact of climate change in particular. Previous faculty position was at the University of Arizona. Dr. Dall'erba earned a PhD in Economics from the University of Pau, France, in 2004.

Dongwoo Kang is associate research fellow, Korea Labor Institute, South Korea. His primary research interests are applied spatial econometrics; regional innovation; regional labor and housing markets. Dr. Kang earned the Ph.D. in geography from the University of Arizona in 2015.

Fang Fang is a PhD candidate in Statistics, University of Arizona, Tucson, Arizona. Her primary research interests are meta analysis in regional econometrics and applied spatial econometrics. Fang will earn the Ph.D in Statistics from the University of Arizona in August, 2016.

Chapter 8
At the Frontier Between Local and Global Interactions in Regional Sciences

Gary Cornwall, Changjoo Kim, and Olivier Parent

8.1 Introduction

Regional scientists have long stressed the importance of spatial spillover effects on local economic outcomes. In his seminal work, Marshall (1890) emphasizes that when economic agents locate in close proximity, they can take advantage of market interactions, knowledge spillovers, and linkages between intermediate and final goods producers. Due to such conveniences, people tend to cluster at specific locations and benefit from the subsequent agglomeration of economies. This clustering not only ends up providing conveniences in markets and economic activity but also fosters, at some level, local growth and development. Measuring the extent to which spillovers are localized remains a key challenge to empirical work in the field. By considering the role of geographic proximity in evaluating spillover effects, LeSage (2014) illustrates the fundamental role of appropriate model specification.

A spatial spillover arises when the decision or outcome of an agent is influenced by a corresponding decision or characteristic of some neighboring agent. Feedback effects are observed when this influence is projected back upon the original agent via a first order reaction to the neighbor's new decision. Spillovers are said to be global when endogenous feedback effects are present.

With the emergence of social network models (Manski 1993; Brock and Durlauf 2001; Bramoullé et al. 2014), researchers have been interested in new forms of

G. Cornwall • O. Parent (✉)
Department of Economics, Carl H. Lindner College of Business, University of Cincinnati, Cincinnati, OH, USA
e-mail: cornwagj@mail.uc.edu; olivier.parent@uc.edu

C. Kim
Department of Geography, University of Cincinnati, Cincinnati, OH, USA
e-mail: changjoo.kim@uc.edu

R. Jackson, P. Schaeffer (eds.), *Regional Research Frontiers - Vol. 2*,
Advances in Spatial Science, DOI 10.1007/978-3-319-50590-9_8

local or group interactions based on spillovers and social distance. Economic agents belonging to the same cluster tend to behave similarly. New spatial econometrics models have been developed to incorporate intragroup interaction (Lee 2007). Similar to the local spatial spillover effect, those models assume that interaction is limited and does not spread across clusters. Interactions between agents do not spill across cluster boundaries, and within a cluster the same weight is often attributed to all individuals leaving aside geographical or social group-wise variations. A clear distinction is made with local spatial spillovers which do not involve endogenous feedback effects. LeSage (2014) discusses in detail the distinction between global and local specifications, advocating respectively for the implementation of the Spatial Durbin (SDM) and the Spatial Durbin Error Models (SDEM).

One of the primary challenges in analyzing interactions amongst economic agents is the inherent complexity in their connectivity structure or network. In standard peer effects models, the local interaction effects represent strategic complementarity in effort across neighboring agents. An agent's incentive to make a particular decision increases as the number of neighboring agents making a similar decision increases. Strategic complementarities correspond to positive partial cross-derivatives. In addition to local complementarities, global interactions across all agents have recently been introduced by Ballester et al. (2006) to reflect strategic substitutability.

Interdependencies can take a variety of forms and little is currently known about their structure. As researchers become more skilled at leveraging geographic information system (GIS) technologies, new types of data will improve the understanding of spatial interactions. Defining a suitable topological structure for network modeling can present a number of GIS challenges and, in general, empirical work has yet to really analyze the transmission of interactions among economic agents. Future research in regional science will greatly benefit from properly specifying the endogenous process that makes economic agents connected. Assuming that connections between agents are mainly explained by exogenous geographical proximity is overly restrictive and could cast serious doubt on causal interpretations of spillover effects. To evaluate the magnitude of local spillover effects, empirical studies in regional science have been exclusively implementing either an SDEM or the so-called SLX model containing exogenous interaction effects. Future research will acknowledge that feedback effects could play an important role in explaining local spillovers effects while being restricted to a limited set of observations or neighborhoods. Moreover, new models will accommodate the possibility that local externalities do not conform to administrative boundaries and will allow for more heterogeneity in the level of spatial dependence.

The remainder of this chapter addresses these challenges as follows. The following section presents modeling issues related to spatial network analysis specifically oriented to GIS. Section 8.3 discusses the limit of a spatial interaction model when regions or groups of society are well delineated. Section 8.4 questions the central issue of endogeneity in the interaction structure. Section 8.5 proposes new spatial mixture models allowing for parameters to be heterogeneous across clusters, and cluster membership is not known to the econometrician. Section 8.6 concludes and points at future work.

8.2 Identifying Networks Using GIS

Regional scientists have long been paying attention to whether agents in close geographical, social, or virtual proximity interact with each other. Their interactions create a conduit by which information is transmitted, and form the fabric of regional development, all of which demands the attention of researchers. The combination of mobile technology and comprehensive datasets have changed how agents interact across space, and new approaches to both local and global interactions will be developed in future regional science research. Today, data is available in exceptional volume and easily accessed over current communication networks more than ever before and has created a new dimension in the study of regional science. In addition to the extended network, GIS has now advanced into new spheres, such as the modeling and analysis of spatio-temporal networks facilitating the understanding of decision making. Despite the great potential, Brugere et al. (2014) consider the intersecting research between spatial networks in GIS and temporal networks in related fields still in its infancy.

Mobile communication tools allow interactive data publishing, which tracks how agents interact with each other and records under what dimensions they are connected. No longer is this data restricted to geographic boundaries and often is contextualized in network structures through social media (i.e. Twitter, Facebook, LinkedIn, etc.). These platforms diminish the importance of traditional measures of distance and, instead, create relationships that may be tangential to those same measures but nevertheless of great importance. Geo-demographics generated in these virtual environments have a great deal of potential when measuring spatial spillover effects. It is now convenient to analyze populations based on who and where under a less restrictive spatial paradigm.

Mobile telecommunications technologies are contributing significantly to the voluminous amount of data being generated by daily online activities. Cameras, phones, and cars have been, and are being, infused with location-aware software designed in some capacity to give producers insights into consumer activities. These devices have, in effect, begun to sense and communicate their absolute and relative positions with locational tags providing a significant medium for organizing, browsing, and retrieving interactions across space. Location-based services have begun to make use of geographic position by identifying the local (global) network of related devices and people across the world.

GIS can also generate social or virtual proximity that could help to detect spatial dependence among individuals beyond physical boundaries as well as geographical proximity. GIS has been playing a significant role in identifying and generating a realistic network of spatial interaction of social processes. With the help of GIS, networks can be developed at the resolution of individual people by their connections. This often requires that large amounts of interaction data are managed and manipulated across scales. Identifying and building a network of massive and hidden connections using GIS is potentially of great value in regional science in providing new tools for advanced model building and in adding spatial dimension and spatial thinking into regional science. Modeling interaction data in both physical and virtual environments will be future challenges in dealing with local and global interactions in regional science.

8.3 Groupwise Spatial Dependence and Spatial Fixed Effects

Researchers have recently recognized the importance of spatial econometric models in identifying and estimating social interaction models. In the empirical literature of regional science, a region, district, or a group of society can be considered a spatial unit whose neighboring units could be defined in terms of a certain socio-economic or physical distance.

One key challenge is to identify the main determinants of the correlation between outcomes of those spatial units who interact with each other. In a seminal work, Manski (1993) points out the difference between endogenous effects capturing the influence of peer behavior and the contextual effects measuring the influence of exogenous peer characteristics. He also mentions the importance of unobserved, correlated effects capturing the likelihood of units to behave similarly due to the similarity of characteristics and/or environment.

Consider some population of n spatial unit for which y_i is the outcome of individual $i = (1 \ldots, n)$. To model how individual units exert some influence on each other, we assume that this influence could be mediated by a network of peer relationships or any socio-economic or physical distances. To constrain those influences, each spatial unit belongs to a group. The interaction between units may occur within a group but not across. For each group $r = (1, \ldots, R)$, we observe n_r units, where $n = \sum_{r=1}^{R} n_r$. As explained in Lee (2007), a group interaction model based on a block diagonal matrix $W = diag(W_1, \ldots, W_R)$ for which each element $w_{ij,r} = 1$ if i and j are direct neighbors or friends, and $W_{ij} = 0$, otherwise.

Lee (2007) and Bramoullé et al. (2009) have rewritten the generic neighborhood effects model described by Manski (1993) as the following Spatial Durbin Autoregressive specification for each group r as:

$$Y_r = \rho W_r Y_r + X_r \beta + W_r X_r \gamma + \iota_{n_r} \alpha_r + \epsilon_r \tag{8.1}$$

where ϵ_r is a n_r-dimensional vector consisting of i.i.d. disturbances with zero mean and a variance σ_2. X_r is an $n_r \times k$ matrix of explanatory variables and Y_r is the n_r-dimensional vector of observation in the rth group.

The spatial weight matrix reflects in principle the structure of the interaction process, and ignoring this process when one is present will induce a misspecified model. The consequence of such a misspecification is that estimates will be biased and inferences will be misleading. To better understand the issue, the reduced form of the spatial lag model can be rewritten as:

$$Y_r = (I_{n_r} - \rho W_r)^{-1}(X_r\beta + W_rX_r\gamma + \iota_{n_r}\alpha_r) + (I_{n_r} - \rho W_r)^{-1}\epsilon_r \quad (8.2)$$

where $(I_{n_r} - \rho W_r)^{-1}\epsilon_r$, is now a spatially correlated and heteroskedastic error term. By using the Taylor's series for the inverse matrix,

$$(I_{n_r} - \rho W_r)^{-1} = I_{n_r} + \rho W_r + \rho^2 W_r^2 + \ldots + \rho^n W_r^n \quad (8.3)$$

Magnitude and significance of spillover effects are assessed via the partial derivatives of the expectation of y_r. LeSage and Pace (2009) show that direct effects are based on the diagonal elements of (8.3), while the off-diagonal elements contain the indirect or spillover effects. An important characteristic of these models is that spillovers only spread within each group or neighborhood r. Unlike a traditional model, they are not global anymore and do not spread across all neighborhoods.

One way to define the neighborhood structure is to assume that all individuals in the same group are neighbors of each other. Each element $w_{ij,r}$ of the spatial weight matrix W is now equal to $1/(n_r - 1)$, and each $n_r \times n_r$-dimensional block matrix W_r can be rewritten as

$$W_r = [1/(n_r - 1)]J_{n_r} - [1/(n_r - 1)]I_{n_r} \quad (8.4)$$

where $J_{n_r} = \iota_{n_r}\iota'_{n_r}$, ι_{n_r} is an $n_r \times 1$ dimensional vector of ones, and I_{n_r} is an identity matrix of dimension n_r. The reduced form of Eq. (8.1) would involve the following inverted matrix for each block r:

$$(I_{n_r} - \rho W_r)^{-1} = \delta_{1,n_r}J_{n_r} + \delta_{2,n_r}I_{n_r}, \quad (8.5)$$

where $\delta_{1,n_r} = \rho/((n_r - 1 + \rho)(1 - \rho))$ and $\delta_{2,n_r} = (n_r - 1)/(n_r - 1 + \rho)$. This model has received substantial attention in the spatial econometric literature for social interaction (Lee 2007). It is important to note that the spatially lagged dependent variable *Wy* asymptotically becomes proportional to the unit vector. In this case, a spatial fixed effects model is asymptotically equivalent to the SDM with group-wise weights. Spatial correlation should disappear by removing the fixed effects.

A spatial fixed effects specification seems appropriate when individual observations are distributed across well-defined groups for which some characteristics α_r are unobserved. However, there are two main issues that are associated with the use of spatial fixed effects. First, the fixed effects are influencing in an identical fashion all observations within a group. If the data were to exhibit heterogeneity or spatial interaction across neighboring individuals within a group, the result would produce correlation in the error term. In this case, the spatial fixed effects would not correct for the presence of spatial correlation, and the model would be misspecified. Second, and more importantly, the spatial delineation of groups or neighborhood is often ambiguous. There is no reason why administrative districts should be used to delineate spatial areas, except as a matter of convenience. Incorrect delineation might exacerbate spatially correlated and heteroskedastic error terms and create additional model misspecification. In other words, unless the structure of the model results in a set of group-wise constants equivalent to the fixed effects, the inclusion of spatial fixed effects will not be robust to the model misspecification.

8.4 Endogeneity in Dependence Within Groups

A key issue with the causal interpretation of estimates in the peer effect Eq. (8.1) is that the connectivity structure between agents may be endogenous. Spatial econometrics has typically been relying on the ad hoc assumption of exogeneity for the spatial weight matrix. This very strong assumption might not be reasonable when assessing the influence of decisions from neighboring agents. In assessing fiscal policy interdependence and budget spillovers across states, Case and Rosen (1993) underline that economic similarities between regions are more likely to exert influence on each other rather that simply sharing a common border. Several subsequent studies have questioned the narrowly defined connectivity structure that relies exclusively on geographical proximity (see Kelejian and Piras 2014). The main concern has become that estimates of regression that do not account for the endogeneity of the spatial weight matrix should suffer from bias, casting doubt on causal interpretations of the peer effects (Qu and Lee 2015).

By modeling group formation, Jackson (2008) makes the assumption that the decision between two agents to form a link is the outcome of two choices. The net utility stemming from the agreement to form a link can be seen as positive. The utility for agent i to form a link with agent j can be defined as $U_i(j)$ and, therefore, the interaction between both agents can be expressed as

$$D_{ij} = \mathbb{1}_{U_i(j)>0} \times \mathbb{1}_{U_j(i)>0} \tag{8.6}$$

In this framework, each potential pair of neighboring agents evaluates the utility of a link between them at the same point in time. The important implication is that those individual utilities depend on the characteristics of the two individuals, conditional on the network at the beginning of the period. Goldsmith-Pinkham and Imbens (2013) propose a Bayesian estimation procedure that separates the likelihood function of the network formation from the likelihood function of the outcome. They find that indirect effects coming at least from the second order neighbors (friends-of-friends) are hard to assess and largely driven by the functional form assumption that ties these indirect effects. The main issue in developing models that allow for endogeneity in the interaction structures between individuals is to define a rule that keeps them separate from each other. As explained by Qu and Lee (2015), estimating a connectivity structure that relies purely on economic distance might be challenging. He underlines the importance of imposing restrictions on the spatial weights, which depend not only on the ad hoc geographical distance but also the magnitude of neighboring effects through socio-economic distance. Interesting extensions would include an examination of how endogeneity over time might change the interaction structure. For all of those situations, the task of properly estimating direct and indirect effects remains daunting.

8.5 Unobserved Dependence Across Groups

It is a common practice in regional science to adopt administrative boundaries for convenience (e.g., census tract or census block boundaries). There is no reason, however, to believe that social interactions will remain within such boundaries. In fact, it is likely that generic neighborhood effects (such as crime, air quality, employment search, etc.) will not conform to such boundaries and will have heterogeneous areas.

As explained in Autant-Bernard et al. (2007), spatial spillovers may occur through collaborative networks (social, scientific, technological, etc.) giving rise to myriad forms of spatial interaction. The geographical dimension of spillover effects appears to be closely related to other mechanisms that are barely measurable. Clusters of individuals should not only rely on geographical proximity. We often observe that across neighboring observations, two individuals might exhibit different patterns or, more specifically, if we consider those patterns to be probabilistic in nature, different distributions. In fact, an aspect that is often overlooked is the considerable heterogeneity of behavior across individuals whether they belong to the same neighborhood or not. Though unobserved heterogeneity across clusters is more difficult to take into account, there is a rapidly growing literature in econometrics using mixture models (see Keane and Wasi (2013) for a review). These models account for unobserved heterogeneity by assuming the data are drawn not from a single distribution but from a finite number of distributions. In fact, they assume, different agents in the population have varying preferences and estimate the proportion of each type.

Cornwall and Parent (2016) consider estimation of spatial data models when the parameters are heterogeneous across groups, and group membership is not known to the econometrician. Thus, they allow parameters to be homogeneous within a group but heterogeneous across groups. This is a form of model-based clustering which partitions a set of data, y_i into G groups according to how near they are to one another. This is easily distinguishable from the aforementioned analysis in which the objective is to understand how the delineated groups differ. It is also important to note that they are allowing the parameters to vary across groups rather than confining themselves to marginal effects, which differ through splitting the sample based on the values of regressors.

Model-based clustering takes as a starting point that a set of data with a group structure is generated by a mixture of distributions such that an observation drawn from sub-population g has density $f_g(y_i|\beta_g, \sigma_g^2)$. If z_i is the identifying label, i.e., $z_i = g$ if unit i belongs to group g, then one can define the dependent variable y_i as being drawn from g different normal distributions with probability $p(z_i = g) = w_g$ and $\sum_{g=1}^{G} w_g = 1$. The normal mixture distribution has means and variances that are different for each group g:

$$P(y_i|\beta, \sigma^2, p_g) = \sum_{g=1}^{G} w_g N(X_i \beta_g, \sigma_g^2). \tag{8.7}$$

We define by $I_g = \{i : z_i = g\}$ the set of agent belonging to the mixture component g and whether an individual belongs to a mixture component g is not known. Cornwall and Parent (2016) develop a spatial extension for which a new dependent variable is defined as $\tilde{y}_{i,r} = y_{i,r} - \rho \sum_{j=1}^{n_r} w_{ij,r} y_{j,r}$, where $w_{ij,r}$ represents the neighborhood structure as defined in (8.4) that is typically based on geographical proximity. This spatial model could be easily extended to the SDM presented in (8.1). In fact, the spatial mixture would then take the following expression:

$$P(\tilde{y}_{i,r}|\beta, \sigma^2, p_g) = \sum_{g=1}^{G} w_g N(\alpha_{r,g} + X_{i,r}\beta_g + \sum_{j=1}^{n_r} w_{ij,r} X_{j,r} \gamma_g, \sigma_g^2). \tag{8.8}$$

Bayesian estimation procedures can be adopted to estimate this model. The introduction of spatial mixtures of distributions relaxes the assumption of independence between observations whether they belong to the same mixture or not. Geographical proximity generates spatial dependence across neighboring individuals even if they exert different behavior and are not part of the same mixture.

8.6 Conclusion

With the increased interest in social interaction, research in regional science has gradually moved from a pure spatial definition of neighboring effects toward a multidimensional measure relying on a different form of socio-economic distances. The emergence of social networking tend to show that agents belonging to a network might not be in close geographical proximity. Moreover, there is no reason why neighborhood effects should be delineated across well-defined groups. It is possible for neighborhood effects to spill over administrative boundaries, and this possibility must be accommodated when modeling such processes. The difficulty in detecting and measuring spillover effects call for a stronger theoretical basis of the interaction structure. Simple weight matrix based on geographical distance might not be enough. Future work will need to rely on the endogeneity of those interactions along with the heterogeneity of behavior that is influenced by physical and socio-economic distance. Promising future direction in regional science will utilize GIS to incorporate data-rich sources from physical and virtual networks to better assess the magnitude of spillover effects.

References

Autant-Bernard C, Mairesse J, Massard N (2007) Spatial knowledge diffusion through collaborative networks. Pap Reg Sci 86:341–350

Ballester C, Calvo-Armengol A, Zenou Y (2006) Who's who in networks. Wanted: the key player. Econometrica 74:1403–1417

Bramoullé Y, Djebbari H, Fortin B (2009) Identification of peer effects through social networks. J Econometrics 150:41–55

Bramoullé Y, Kranton R, D'amours M (2014) Strategic interaction and networks. Am Econ Rev 104:898–930

Brock W, Durlauf S (2001) Discrete choice with social interactions. Rev Econ Stud 68:235–60

Brugere I, Gunturi V, Shekhar S (2014) Modeling and analysis of spatio-temporal social networks. In: Encyclopedia of social network analysis and mining. Springer, New York, pp 950–960

Case AC, Rosen HS (1993) Budget spillovers and fiscal policy interdependence: evidence from the states. J Public Econ 52:285–307

Cornwall GJ, Parent O (2016) Mixture models with spatial dependence. Working paper, University of Cincinnati

Goldsmith-Pinkham P, Imbens GW (2013) Social networks and the identification of peer effects. J Bus Econ Stat 31:253–264

Jackson M (2008) Social and economic networks. Princeton University Press, Princeton

Keane MP, Wasi N (2013) Comparing alternative models of heterogeneity in consumer choice behavior. J Appl Econometrics 28:1018–1045

Kelejian H, Piras G (2014) Estimation of spatial models with endogenous weighting matrices, and an application to a demand model for cigarettes. Reg Sci Urban Econ 46:140–149

Lee LF (2007) Identification and estimation of econometric models with group interactions, contextual factors and fixed effects. J Econometrics 140:333–374

LeSage JP (2014) What regional scientists need to know about spatial econometrics. Rev Reg Stud 44:13–32

LeSage JP, Pace, RK (2009) An introduction to spatial econometrics. Chapman Hall/CRC Press, Boca Raton, FL

Manski C (1993) Identification of endogenous social effects: the reflection problem. Rev Econ Stud 60:531–542

Marshall A (1890) Principles of economics. Macmillan, London

Qu X, Lee LF (2015) Estimating a spatial autoregressive model with an endogenous spatial weight matrix. J Econometrics 184:209–232

Gary Cornwall is a Ph.D. candidate at the University of Cincinnati. His primary research interests are spatial econometrics, income dynamics/inequality, and Bayesian econometric methods. He received his M.A. in Applied Economics from the University of Cincinnati in 2012 and is expected to complete his Ph.D. in 2017.

Changjoo (CJ) Kim is an associate professor, Department of Geography, University of Cincinnati. His research and teaching interests are in geographic information science, transportation, and spatial modeling. His research addresses theoretical and practical questions in urban and economic geography through the application of GIS methods. He investigates a range of urban and economic concerns including travel activity, urban sprawl, commuting, airline industry, retailing, etc. Dr. Kim earned the Ph.D. in geography from The Ohio State University in 2004.

Olivier Parent is associate professor, Department of Economics, at the University of Cincinnati (UC). His primary research interests are spatial econometrics; regional economics; and Bayesian econometrics. Dr. Parent earned the Ph.D. in economics from the University of Saint-Etienne, France, in 2005.

Chapter 9
Hierarchical Spatial Econometric Models in Regional Science

Donald J. Lacombe and Stuart G. McIntyre

9.1 Introduction

Multilevel or hierarchical models (hereafter 'hierarchical') are becoming increasingly important in regional science because the data that are being used are often *nested* in nature and, thus, provide a natural hierarchy to the data. In the United States, over 3000 counties are nested within 50 states. In a UK context, 11 Government Office Regions nest hundreds of local authorities. Indeed, in the UK there is a further administrative hierarchy level with some parts of the country having a two-tiered structure of local government with unitary and district councils. More generally within the European Union (EU), data are regularly released at different levels of the NUTS (Nomenclature of Territorial Units for Statistics) hierarchy. The NUTS classification is designed to be consistent across the EU, for example, having approximately the same number of residents in each NUTS 1, 2, and 3 area in each country and be subject to infrequent revisions. Thus, there is a lot of inherently nested data available to regional scientists.

The nested nature of the data poses problems for standard econometric techniques, such as Ordinary Least Squares (OLS), which assumes that the data are independent. In addition to the clustering issue, there may be hypotheses that are explicitly designed to be answered at the second level of the hierarchy. For example, minimum wage laws in the United States are set at the federal level, but states have the flexibility to institute a minimum wage above the federally mandated minimum

D.J. Lacombe (✉)
Department of Personal Financial Planning, Texas Tech University, 1301 Akron Avenue, Room 234C, Box 41210, Lubbock, TX, 79409-1210, USA
e-mail: donald.lacombe@ttu.edu

S.G. McIntyre
Department of Economics, University of Strathclyde, 199 Cathedral Street, Glasgow G4 0NR, UK
e-mail: s.mcintyre@strath.ac.uk

R. Jackson, P. Schaeffer (eds.), *Regional Research Frontiers - Vol. 2*,
Advances in Spatial Science, DOI 10.1007/978-3-319-50590-9_9

wage.[1] If we have data on youth employment and other variables at the county level but the minimum wage data is at the state level, a standard approach such as OLS with fixed-effects will not be able to be utilised because including state-fixed effects along with a state-level explanatory variable would introduce perfect multicollinearity and a non-invertible design matrix. However, this complication poses no difficulties for the hierarchical methodology, which allows for state-level explanatory variables to be introduced into the model, allowing hypotheses explicitly at the second level of the hierarchy (e.g., states) to be empirically investigated.

Combined with the increasing availability of regional data and improvements in statistical theory, advances in computation have enabled the estimation of a greater variety of hierarchical models. The result has been an increased interest among regional scientists in using hierarchical models in applied work. Most obviously this interest has focussed on combining the advantages of hierarchical modelling with the opportunities provided by spatial econometric methods. A number of recent papers have sought to develop and implement such models (Fingleton 2001; Smith and LeSage 2004; Parent and LeSage 2008; Corrado and Fingleton 2012; Lacombe and McIntyre 2016). In this context, this chapter seeks to chart out areas for future development in the use of hierarchical spatial econometric models.

In the next section, we begin by introducing the core model notation used throughout the chapter. In order to do this, we first introduce the non-spatial hierarchical models, then we present the standard nested spatial econometric models before combining these to produce hierarchical spatial econometric models. We then provide a brief review of the existing literature in this area. Thereafter, we identify a number of areas where we feel further developments and improvements should focus. These are: (1) model comparison improvements, (2) further investigation of heteroskedasticity within these models, (3) the developments of limited dependent variable models, (4) the development of random coefficient models and (5) the development of hierarchical origin-destination models. Improvements in data availability and computational methods are such that the estimation of large and complex econometric models is increasingly straightforward. With these improvements comes huge potential, but also important challenges; not least by imposing a greater responsibility on the applied researcher to select the most appropriate model for their work and to transparently document their results. This chapter is intended to further encourage development of a suite of hierarchical spatial econometric models.

9.2 Introducing Hierarchical Econometric Models

For ease of discussion and explanation in this chapter, we begin by establishing our notation. For our purposes here, we will focus on two-level hierarchical models, although that is not to exclude the inclusion of additional levels in the model.

[1] States must set the wage at or above the federal level but not below that level.

Throughout this chapter, we will refer to the lower level of the hierarchy as 'level 1' and the upper level as 'level 2'. There are two main types of hierarchical models, referred to as random intercept and random coefficient models. In the former, the hierarchy operates as a result of each individual area at the lower level (county, say) having an intercept which is partly driven by the average of the dependent variable for its group (states, say) and partly by something idiosyncratic to that lower level area (hence, the nomenclature of 'random'). In the latter case, the impact of the covariates ('X's') on the dependent variable ('Y') is composed of a group average marginal effect and an idiosyncratic element specific to each lower level unit. For example, in a model of air quality, the marginal impact of traffic congestion on air quality may have a localised impact but also a statewide impact as a result of common state regulation of vehicle emissions.

We now outline formally a non-spatial hierarchical random intercept model (Raudenbush and Bryk 2002):

$$\text{Level 1}: \quad y_{ij} = \alpha_j + X_{ij}\beta + \varepsilon_{ij} \qquad \varepsilon_{ij} \sim N\left(0, \sigma^2\right) \tag{9.1}$$

$$\text{Level 2}: \quad \alpha_j = Z_j\gamma + u_j \qquad u_j \sim N\left(0, \tau^2\right) \tag{9.2}$$

where y is a vector of observations on the dependent variable, α_j are the intercepts that are allowed to vary in the model, X is the matrix of explanatory variables at level 1, and ε is the error term. At level 2 [Eq. (9.2)] we specify the model for the intercepts of the level one model. Here the intercepts α_j are the dependent variable, Z_j is the matrix of explanatory variables (including intercept), and u_j is the error term for this level of the model.

This model can be rewritten in matrix form as:

$$y = X\beta + \Delta\alpha + \varepsilon \tag{9.3}$$

$$\alpha = Z\gamma + u \tag{9.4}$$

$$\varepsilon \sim N\left(0, \sigma^2 I_n\right) \tag{9.5}$$

$$u \sim N\left(0, \tau^2 I_j\right) \tag{9.6}$$

where y is an $N \times 1$ vector of observation on the dependent variable, X is an $N \times K$ matrix of explanatory variables at the first level of the hierarchy, β is a $k \times 1$ vector of coefficients, and ε is an $N \times 1$ vector of disturbances with mean 0 and variance $\sigma^2 I_n$. The symbol Δ represents an $N \times J$ (where N represents the total number of observations and J represents the number of groups) matrix that assigns each level 1 observation to a level 2 group. One can also think of the Δ matrix as the matrix of dummy variables one would use in a standard fixed-effects model. The symbol α represents the $J \times 1$ vector of intercept terms, which is given its own model. The second level of the hierarchical model is the model for the individual intercepts. The dependent variables at this level is the $J \times 1$ vector of intercepts α, Z represents the $J \times m$ vector of explanatory variables (including a constant term), γ is the $J \times m$

vector of coefficients, and u represents the $J \times 1$ vector of disturbances with variance $\tau^2 I_j$ for the level 2 part of the model.

This kind of hierarchical model contains the classical regression model as a special case (Gelman and Hill 2006). The essential idea is that the intercepts, i.e. the α_js, come from a distribution with mean μ_α and standard deviation σ_α. A fully-pooled model ignores any heterogeneity and assumes a common intercept for all upper-level units. In other words, the fully-pooled model assumes a common intercept among the groups. At the other extreme, we can posit a separate intercept for all upper level units which is the so-called "no pooling" model and is operationalised by including a dummy variable for each upper-level unit. However, as noted by Gelman and Hill (2006, p. 258), the level 2 error variance can be estimated from the data and there is "no reason (except for convenience) to accept estimates that arbitrarily set this parameter to one of these two extreme values." In other words, the use of a single intercept (i.e. the "fully-pooled" model) or the individual-intercept model (i.e., the "no-pooling" model) are models that make an assumption that may not be warranted and, thus, a hierarchical model may be appropriate.

Note also that the estimates of the intercepts are a linear combination of the "fully-pooled" and "no-pooling" models. Mathematically, this can be expressed as follows (Subramanian 2010; Luke 2004)

$$\hat{\alpha}_j^{EB} = \lambda_j \hat{\alpha}_j^{NP} + (1 - \lambda_j)\,\hat{\alpha}_j^{FP}$$

$$\lambda_j = \frac{\tau^2}{\left(\tau^2 + \sigma^2 / n_j\right)}$$

where $\hat{\alpha}_j^{NP}$ is the "no-pooling" estimate of the intercept, i.e., the intercept one would obtain if each level 2 group had its own indicator variable; $\hat{\alpha}_j^{FP}$ is the value of the intercept from a "fully-pooled" model, i.e., a model with a single intercept for all level 2 groups and $\hat{\alpha}_j^{EB}$ is the "empirical Bayes" or "shrinkage" estimate of the intercept in the hierarchical model, which is a linear combination of the "no-pooled" and "fully pooled" models. The weights assigned to the "no-pooled" and "fully pooled" are given by λ_j and are a function of the level 2 and level 1 error variance as well as the number of level 1 observation in each level 2 unit, i.e., n_j.[2] There

[2]The empirical Bayes or shrinkage estimates work as follows. If the number of level 1 observations within an individual level 2 group is small (i.e. a small value of n_j) then the estimate of the intercept for that group will be "shrunk" towards the overall intercept in a "fully-pooled" model. As an example, the state of Delaware in the United States has only three counties nested within it and therefore we would expect there to be more shrinkage towards the overall intercept as opposed to the case of the state of Texas, which has 254 counties. In the case of Texas, we would expect the estimate of the intercept to be much more accurate and more weight would be placed on the "no-pooled" estimate of the intercept for the state of Texas. Additional details regarding these empirical Bayes or shrinkage estimates are available in Gelman and Hill (2006), Luke (2004), and Subramanian (2010).

is, therefore, a clear relationship between the hierarchical and non-hierarchical specifications and good reason to question the selection of non-hierarchical models in many regional science applications.

The second main type of hierarchical model is the random coefficient model. This model can be expressed as follows:

$$\text{Level 1}: \quad y_{ij} = \beta_{0j} + (X_{ij} - X_{1j})\beta_{1j} + \varepsilon_{ij} \qquad \varepsilon_{ij} \sim N\left(0, \sigma^2\right) \tag{9.7}$$

$$\text{Level 2}: \quad \beta_{0j} = \gamma_{00} \tag{9.8}$$

$$\beta_{1j} = \gamma_{10} + u_{1j} \tag{9.9}$$

where γ_{00} is the average intercept at the upper level, γ_{10} is the average of the slopes across the upper level areas, and u_{0j} and u_{1j} are the idiosyncratic contribution to the intercept and slope. In this model, the coefficients on the level 1 variables are allowed to vary based on the level 2 units, i.e., the slopes for each explanatory variable are allowed to vary according to the level 2 context. It should be noted that the random intercept and random coefficient models can be combined to allow the intercepts and slopes to vary across the level 2 units as well.

9.3 Introducing Hierarchical Spatial Econometric Models

In extending these non-spatial hierarchical models in the previous section to incorporate spatial relationships, there are a number of different approaches one could take. In this section we will outline four different combinations, which stand out as being potentially very useful to the applied researcher. The decision about which of these models to begin the analysis with will depend upon whether the process under study is one characterised by a local or a global spillover process, consistent with LeSage (2014). A spillover can be defined as where the rth characteristic of the ith entity (local authority say) X_i^r has some influence upon the outcome Y of some neighbouring local authority j, Y_j. With this in mind, a *local* spillover process is one where the impact of X_i^r is limited to impacting Y_j with the js defined based on the specification of the spatial weight matrix. In a *global* spillover process in contrast, the impact of X_i^r not only impacts on Y_j, defined according to the weight matrix, but in turn impacts the Y_k of js neighbours k. Thus the X_i^r impact upon all areas Y_j where j now includes higher order neighbours to i (i.e., neighbours to is neighbours, etc.) We can think of these as endogenous feedback effects and representing system wide change; hence, the *global* nomenclature. Before moving to the hierarchical case, we must first briefly outline the standard non-hierarchical nested spatial econometric models, embodying the *local* and *global* spillover process.

The nested SDEM model, representing a *local* spillover process, can be represented as:

$$y = \alpha + X\beta + WX\theta + u \tag{9.10}$$

$$u = \lambda Wu + \epsilon \tag{9.11}$$

This model nests both the spatial lag of X model (SLX) and the spatial error model (SEM). A test of whether $\theta = 0$ and $\lambda = 0$ would determine whether the nested, or a more specific model should be used. We can see that the SDEM model captures the impact of the *X*s in *j* on the y in *i*, as well as capturing spatial heterogeneity in the error term *u*. However, there is no endogenous feedback in this model. The spatial impacts are limited to each area impacting their immediate neighbours as defined by the spatial weight matrix *W*. No scope is given for impacts on higher order neighbours. In order to capture such impacts, we would use the SDM model, which can be represented as follows:

$$y = \alpha + \rho Wy + X\beta + WX\theta + \epsilon \tag{9.12}$$

This model nests the spatial autoregressive model (SAR) and the SEM and the SLX models. The SDM model incorporates endogenous feedback effects and, thus, a role for changes in one area *i* to impact upon not only its neighbours, *j*, but the neighbours of those areas *k* as well. To see how this happens in this model, one only needs to consider the matrix of partial derivatives of the rth explanatory:

$$\frac{\partial Y}{\partial x_{1r}} \cdots \frac{\partial Y}{\partial x_{nr}} = (I - pW)^{-1}\beta_r \tag{9.13}$$

which, when we expand $(I - pW)^{-1}$ gives: $(I - pW)^{-1} = I + \rho W + \rho^2 W^2 \ldots \rho^n W^n$, with the *direct* effects embodied in the first term and the *indirect* effects embodied in the later terms, specifically the second term embodying the impact on the immediate neighbours, and the third term embodying the impact on the neighbours of those neighbours, and so on.

Thus, model selection in a non-hierarchical spatial econometric setting starts with a decision about which type of spatial process characterises the object of the study, *global* or *local*. In common with non-hierarchical spatial econometric model selection, the choice of which general nested model to begin with will be dictated by the type of spillovers believed to be present. Thereafter, the statistical significance of the spatial parameters can be tested, and the model refined from one of these nested models to a more specific model, if appropriate. For example, if one believes that the phenomenon under study is one characterised by endogenous feedback effects throughout the system, then a global spillover model, the SDM, would be appropriate as a starting point. Similarly, where a local spillover is believed to characterise the process under study, the SDEM model would be the appropriate starting point.

In the hierarchical context, the additional complication is the selection of the appropriate nested model at each level of the hierarchy. The models defined below represent an exhaustive combination of these model combinations for the so-called random intercept model.[3] The four possibilities for extending the hierarchical random intercept model based on the SDEM and SDM specification are:

$$\begin{aligned} &\text{Level 1 : SDM} && y = \rho_1 W_1 y + X\beta + W_1 X\theta + \Delta\alpha + \varepsilon \\ &\text{Level 2 : SDM} && \alpha = \rho_2 W_2 \alpha + Z\delta + W_2 Z\gamma + u \end{aligned}$$

$$\begin{aligned} &\text{Level 1 : SDM} && y = \rho_1 W_1 y + X\beta + W_1 X\theta + \Delta\alpha + \varepsilon \\ &\text{Level 2 : SDEM} && \alpha = Z\delta + W_2 Z\gamma + u \\ & && u = \lambda W u + \eta \end{aligned}$$

$$\begin{aligned} &\text{Level 1 : SDEM} && y = X\beta + W_1 X\theta + \Delta\alpha + \varepsilon \\ & && \varepsilon = \lambda W \varepsilon + \eta \\ &\text{Level 2 : SDM} && \alpha = \rho W_2 \alpha + Z\delta + W_2 Z\gamma + u \end{aligned}$$

$$\begin{aligned} &\text{Level 1 : SDEM} && y = X\beta + W_1 X\theta + \Delta\alpha + \varepsilon \\ & && \varepsilon = \lambda W \varepsilon + \eta \\ &\text{Level 2 : SDEM} && \alpha = Z\delta + W_2 Z\gamma + u \\ & && u = \lambda W u + \nu \end{aligned}$$

where y is an $N \times 1$ vector of observation on the dependent variable, X is an $N \times k$ matrix of explanatory variables at the first level of the hierarchy, β is a $k \times 1$ vector of coefficients, W_1 is an $N \times N$ spatial weight matrix, ρ_1 is the spatial autocorrelation coefficient at level 1, and ε is an $N \times 1$ vector of disturbances with mean 0 and variance $\sigma^2 I_n$. Δ represents an $N \times J$ (where N represents the total number of observations and J represents the number of groups) matrix that assigns each level 1 observation to a level 2 group. The second level of the hierarchical model is the model for the individual intercepts. The dependent variables at this level is the $J \times 1$ vector of intercepts α, Z represents the $J \times m$ vector of explanatory variables (including a constant), γ is the $J \times m$ vector of coefficients, W_2 is a $J \times J$ spatial weight matrix, ρ_2 is the spatial autocorrelation parameter for level 2, and u represents the $J \times 1$ vector of disturbances with variance $\tau^2 I_j$ for the level 2 part of the model.

[3]The models outlined in this section could also be applied in the random coefficient context, which we describe in Sect. 9.6.1.

One of the most important aspects of these hierarchical spatial econometric models is the proper interpretation of the marginal effects. In non-spatial hierarchical models, the coefficients represent how a change in a covariate affects the dependent variable. This is also true in spatial econometric models where one does not have a lagged dependent variable $\rho W_s y$ or $\rho W_s \alpha$ term. Thus, in the final model combination outlined above, the θ and γ coefficients can be interpreted in the traditional manner. However, in the other models, at one or both levels of the hierarchy one must calculate the proper marginal effects estimates following LeSage and Pace (2009) to produce the correct marginal *direct* and *indirect* effects. Having outlined the extension of the non-spatial hierarchical random intercept model, we will now briefly review the existing empirical work in this area. This will help the reader to understand the subsequent section outlining a number of areas for improvement in these models and in their use.

9.4 Existing Work in This Area

The nested structure of much of the data in regional science has long been known, but in the context of spatial econometrics, the literature is not particularly well developed.[4] Anselin and Florax (1995) were the first, to our knowledge, to consider spatial econometric models in an explicitly hierarchical context. They combined a multi-state Kalman Filter approach and a hierarchical random intercept model to take advantage of cross-sectional dependencies to 'borrow strength'. This was then used to backcast school district income tax revenues. Anselin and Florax (1995) did not however incorporate explicitly spatial terms as we now understand them within their model. While the broader spatial econometrics literature had some time ago settled many of the issues around the error component problem, i.e., unobserved heterogeneity, which had first been tackled in a spatial econometric context by Kelejian and Robinson (1993), it was in the context of the modelling of disease rates that the first attempt at embedding spatial terms into a hierarchical model appeared (Langford et al. 1999). Thereafter, mention began to appear in the spatial econometric literature of the overlap between these spatial econometrics approaches and hierarchical modeling, e.g. Anselin (2001, 2002).

It was in Anselin and Cho (2002) that the concepts of hierarchical spatial econometric modelling were first more fully discussed, although as Anselin et al. (2004) subsequently noted, little had been done to incorporate advances in hierarchical modelling into spatial econometric modelling. This began to change

[4]Although not central to our discussion here, it is worth noting that Wheeler and Paez in Fischer and Getis (2009) (eds.), discuss geographically weighted regression methods in a Bayesian hierarchical context. In addition, we note the work of Vanoutrive and Parenti (2009) in comparing spatial econometric and hierarchical modelling approaches (which, perhaps confusingly, they refer to, e.g., in Vanoutrive et al. (2009), as 'spatial' multilevel models), although they did not consider combining these methods and focus instead on motivating the decision on which approach to use.

slightly with the work of Smith and LeSage (2004) who introduced a hierarchical spatial econometric probit model. Their model included upper level fixed-effects, something which was commonly done in the non-spatial econometric literature, but with the innovation that these fixed-effects were not modelled as independent of each other. This enabled the estimation of region specific effects for each state (say), but recognised that each of these state specific effects may depend upon the state specific effect of their neighbours. This approach is a version of the spatial random intercept model introduced above without the inclusion of covariates at the upper level. A similar approach is taken in Parent and LeSage (2007) and in Jensen et al. (2012). Dong et al. (2015) and Dong and Harris (2015) extend this model for use with a continuous dependent variable as a spatial random intercept model but without any covariates at the upper level of the hierarchy. The inclusion of covariates at the upper level of the hierarchy was incorporated into the spatial random effects model provided by Lacombe and McIntyre (2016). LeSage et al. (2007) and LeSage and Llano (2013) took this literature in a slightly different direction by including spatially structured fixed-effects into an origin-destination or 'flow' model and in LeSage et al. (2007), they use it to examine knowledge spillovers using patent data, although again no covariate information was included at the upper level of the hierarchy.

Corrado and Fingleton (2012) restated the case for combining hierarchical econometric models and spatial econometric models, demonstrating that the spatial weight matrix at the heart of spatial econometric approaches is present as part of the structure of hierarchical models. In Elhorst (2014), a short section on multilevel modelling is provided, focussing on a mixed random and fixed coefficients model, essentially enabling coefficients across regions (level 1 units) to vary but to be fixed across countries (level 2 units). This is a presentation of the spatial random coefficient model described earlier. To our knowledge, this short summary includes most—if not all—of the work in this area. Given our earlier presentation of different model ideas and combinations, there appears to be significant scope for further development. These developments have two streams; firstly, developments that apply to all the hierarchical spatial models that have been summarised in this section and those detailed later in this chapter and secondly, model extensions that enhance the ability of the hierarchical spatial econometric models currently available.

9.5 Improvements to Existing Spatial Hierarchical Models

Before discussing research frontiers in this area in more detail in the next section, this section briefly documents some areas for improvements or different uses of the spatial hierarchical models that already exist. While these are more 'housekeeping' items than research frontiers, they are nevertheless important in advancing the development and use of these models.

9.5.1 Model Comparison

The first area for improvement is in formalising the model selection process. While it is true, as LeSage (2014) has argued, that the applied researcher ought to know whether the process they are studying is one characterised by local or global spillovers, the inherently subjective nature of this process can be improved upon. This is particularly important given the presence of two potentially different spillover effects being present in the same hierarchical model with the potential for two different weight matrices to be used.

There are a number of model comparison techniques that can be used and we highlight two possibilities within this section. Each of these needs to be further investigated in an experimental setting to establish their relative performance in selecting the correct model. The first method of choosing amongst the different models is to utilize the Deviance Information Criterion (DIC) statistic first developed by Spiegelhalter et al. (2002). The DIC statistic is calculated for each model, and the model with the lowest DIC value is determined to have the best model fit. The DIC statistic is calculated as follows

$$p_D = \overline{D(\theta)} - D\left(\bar{\theta}\right)$$
$$D(\theta) = -2\log\{p(y|\theta)\} + 2\log\{f(y)\}$$

For each potential hierarchical spatial econometric model under consideration, the DIC statistic could be calculated and the model with the lowest value of this test statistic would then be chosen. Additional details regarding the development of the DIC statistic for model choice is contained in Spiegelhalter et al. (2002).

Another potential avenue that one could take in terms of model comparison would be to adopt a Bayesian perspective and calculate the marginal likelihood for each model. The marginal likelihood for a model M takes the following form

$$p(y|M) = \int p(y|\theta, M)\, p(\theta|M)\, d\theta$$

where $p(y|\theta, M)$ denotes the likelihood for model M and $p(\theta|M)$ denotes the prior distribution for the parameters for model M. Model comparison would involve calculating the marginal likelihood (usually on the computationally convenient log scale) for each model, exponentiate each of these values and then divide each marginal likelihood value by this sum to obtain posterior model probabilities. Model comparison then proceeds apace by choosing the model with the highest posterior model probability. Although this procedure is straightforward to explain, there still remains the difficulty of obtaining the marginal likelihood for these models due to the fact that the integrals involved rarely have closed form solutions and thus calculating the marginal likelihood is non-trivial in most cases. Further details regarding Bayesian model comparison is contained in Koop et al. (2007) and in the specific case of spatial econometric models, LeSage and Pace (2009).

9.5.2 *Heteroskedasticity*

In addition to improving model selection procedures, little empirical investigation of the test for and treatment of heteroskedasticity has so far been present in the hierarchical spatial econometric literature. One potential avenue is to adopt the technique of dealing with heteroskedasticity as outlined in LeSage and Pace (2009). LeSage and Pace (2009, Sect. 5.6.1) make note of the fact that the Bayesian Markov Chain Monte Carlo (MCMC) methodology can be extended "to include variance scalars that can accommodate heteroscedastic and/or outliers". The idea is an extension of the one proposed in Geweke (1993) to various spatial econometric models and involves estimating a set of variance scalars $(v_1, v_2, \ldots, v_n)$ that represent unknown parameters to be estimated as an additional step in the MCMC algorithm. This formulation of the problem allows for the assumption that $\varepsilon \sim N\left(0, \sigma^2 V\right)$, where V is a diagonal matrix containing the parameters $(v_1, v_2, \ldots, v_n)$. The prior distribution for each of the v_i scalar variance terms takes the form of a set of n *i.i.d* $\chi^2(r)/r$ distributions, where r represents the single parameter of the χ^2 distribution. We note that the various hierarchical spatial econometric models that we have discussed so far are readily amenable to this extension of the basic MCMC sampling scheme and, thus, each model is capable of accommodating heteroskedasticity, however—to our knowledge—these are not captured in existing routines.

9.5.3 *Extending Beyond Two Level Models*

In this chapter we have exclusively considered a two level hierarchy. In the statistics literature there are a number of cases where more than two levels of the hierarchy characterise the data. This is something true of regional science also. While we set aside consideration of more than two levels in our discussions earlier in this chapter, we would be remiss not to note that this is one direction of future research that remains both very obvious and potentially very valuable in regional science. In the EU, for instance, NUTS 1, 2, and 3 regions are all nested within each other. Many studies have used these statistical geographies for research in regional science, yet none to our knowledge have recognised the opportunity presented by the nested geographical structure of the underlying data.

9.6 New Directions for Spatial Hierarchical Models

In this section, we describe some extensions to the spatial hierarchical models outlined earlier in this chapter. The number of possible models is quite extensive, and we highlight only those models that have been utilized in a non-hierarchical setting.

The models considered include a random coefficient spatial econometric model, limited dependent variable models, and hierarchical spatial origin-destination models.

9.6.1 Random Coefficient Models

The first, and most obvious, new direction for hierarchical spatial econometric models is in the development of spatial hierarchical random coefficient models. These models would be extensions to the hierarchical random coefficients models detailed earlier in Sect. 9.2. In this case, what would be incorporated in addition to what has already been outlined would be the explicit modelling of the β_{1j} in Eq. (9.9) as a spatial function, for example:

$$\beta_{1j} = \rho W \beta_{1j} + \gamma_{10} + u_{1j} \tag{9.14}$$

This would capture spatial dependence in these upper level average slope coefficients. We realise that the notation is becoming increasingly complex at this stage. With that in mind, a hypothetical tax example may help to provide some intuition to both motivate and explain why this kind of model may be useful. In modelling consumption of a particular good at the county level in the U.S. as a function of,*inter alia*, the tax rates, one is interested in how consumption is affected by changes in the county tax rate. With the proper transformations of the variables, one way to think about this slope coefficient is as the elasticity of consumption to the local tax rate. In the case of a hierarchical spatial model, this level one covariate (tax rate) may be affected by state level decisions or restrictions, motivating the hierarchical dimension and the introduction of a second level (state) into the analysis. In addition, these state (level 2) restrictions or decisions may be determined as a best response to policies in neighbouring states, meaning that the group (level 2) average elasticity of consumption [γ_{10} in Eq. (9.9) above] to the tax rate may be dependent upon the average elasticity of consumption in neighbouring states, motivating the exploration of potential spatial dependence in these slope coefficients using a random coefficients model. It would be possible to extend these effects to considering the nature of the spillover process as discussed earlier in this chapter in the context of the spatial random intercept models.

9.6.2 Limited Dependent Variable Models

A number of applications in regional science make use of limited dependent variable (i.e. probit, logit, tobit) models. In this section, we discuss some extensions to the classic limited dependent variable models to incorporate hierarchical spatial processes. While our focus here is on probit models, other limited dependent variable models could be developed. An obvious starting point in spatial hierarchical limited dependent variable models would be a simple extension to Smith and LeSage (2004) to include covariates at the upper level of the hierarchy. This would produce a model of the following form:

$$y = \rho Wy + X\beta + \Delta\alpha + \varepsilon$$
$$\alpha = Z\gamma + u$$
$$\varepsilon \sim N(0, 1)$$
$$u \sim N\left(0, \tau^2 I_j\right)$$

where y is a binary (0,1) dependent variable and where σ^2 is set to 1 for identification purposes. This model is one of the simplest hierarchical spatial econometric models possible. Beyond the addition of covariates at the upper level to the Smith and LeSage (2004) framework, the next step would be to include the *local* and *global* spillover models into this framework. This would produce hierarchical spatial econometric models which were notationally very similar to those presented earlier for the continuous dependent variable case. To give some idea of the potential usefulness of such a model, the reader is directed to Holloway et al. (2014), who examine passage of the 2001 Farm Bill in the U.S. Congress using a standard spatial autoregressive probit model. However, because congressional districts are nested within states, it is plausible that there are additional state level factors that should be included in this kind of model. Estimation of the kind of model proposed here would enable such an empirical investigation.

9.6.3 Hierarchical Origin-Destination Models

One final area where the further extension of hierarchical econometric models to incorporate spatial processes would be useful is in the context of origin-destination models (sometimes referred to as spatial interaction models). LeSage and Pace (2008), in a highly influential study, developed the origin-destination flow model and LeSage and Thomas-Agnan (2015) outline the special steps that need to be taken in order to properly interpret the marginal effects in these origin-destination models.

The OD Flow Model can also be adapted to handle hierarchical data structures as follows

$$y = \rho_d W_d y + \rho_o W_o y + \rho_w W_w y + X_d \beta_d + X_o \beta_o + \gamma g + \Delta\alpha + \varepsilon \tag{9.15}$$

$$\alpha = Z\theta + u \tag{9.16}$$

which takes into account origin (via W_o), destination (via W_d), and origin-to-destination (via W_w) spatial dependence. The vectors β_d and β_o reflect the effect of covariates at the destination locations and origin locations, respectively, while γ reflects the effect of distance, and ε is the standard $N \times 1$ vector of disturbances. As before, Δ represents an $N \times J$ (where N represents the total number of observations and J represents the number of groups) matrix that assigns each level 1 observation to a level 2 group, while Z is a level 2 covariate.

One area of study in which origin-destination spatial econometric models have become popular is in the migration literature. For example, LeSage and Pace (2009) illustrate the spatial econometric interaction model by examining population migration flows between the 50 largest metropolitan areas from 1995 to 2000 in the United States. The explanatory variables in their model include the population at the origin and destination in 1990, the per-capita income at the origin and destination in 1990, a variable that measures whether people resided in the same house at the origin and destination in 1990, and a distance variable.

Although this group of explanatory variables is fairly exhaustive given the aims of the modeling exercise, there may be factors operating at another level that could affect migration flows. Each of the 50 largest metropolitan areas in the study are nested within their own states; thus, state level factors may play a role in whether or not people decide to migrate from one area to another. For example, it may be that the state income tax rate (or whether or not a state even has an income tax) might affect the decision to migrate. In the hierarchical setup that we posit in this section, a measure of a state's tax burden could be included as an upper-level covariate in the empirical analysis to determine if this affects the migration flow from metropolitan area to metropolitan area. Further empirical examples that use origin-destination flow models are provided in LeSage and Polasek (2008) and Marrocu and Paci (2013).

A final case to be considered here is incorporating these kinds of OD Flow Models into a hierarchical random coefficients model. This would encompass the origin and destination specific effects having two elements, one common to their group, and one idiosyncratic to each flow. There are a number of cases where the effect of a covariate in a flow model in regional science may be different across different groups but also may be characterised by spatial dependence. One example would be in modelling FDI flows where the impact of origin and destination financial development on FDI flows may have both a common effect across members of each group and where this common group effect is characterised by spatial dependence across groups. Development of these kinds of models, combined with the work LeSage and Pace (2008) and LeSage and Thomas-Agnan (2015) who have developed spatial OD flow models and their interpretation, would be a huge value to the field.

9.7 Conclusion

Spatial econometric modelling has increased rapidly in popularity among applied regional scientists in the past few decades. The suite of spatial econometric models and routines now available is large, and improvements to these models are frequent.

Nevertheless, as we outlined in this chapter, the existing suite of hierarchical spatial econometric models currently available is quite small. That is despite the fact that many datasets of interest to regional scientists are nested in nature. While the recent interest in hierarchical modelling is spawning new routines and modelling approaches, there is still much to do. This chapter has sought to provide some areas for immediate and future development and research in this area to bring the advantages of hierarchical modelling and the insights of spatial econometric modelling together more fully. While our suggestions are not exhaustive, we hope that they provide some interesting ideas and stimulate further model development.

References

Anselin L (2001) Spatial effects in econometric practice in environmental and resource economics. Am J Agric Econ 83(3):705–710

Anselin L (2002) Under the hood issues in the specification and interpretation of spatial regression models. Agric Econ 27(3):247–267

Anselin L, Cho WKT (2002) Spatial effects and ecological inference. Polit Anal 10(3):276–297

Anselin L, Florax RJ (1995) Small sample properties of tests for spatial dependence in regression models: some further results. In: New directions in spatial econometrics. Springer, Berlin, pp 21–74

Anselin L, Florax RJ, Rey SJ (2004) Econometrics for spatial models: recent advances. In: Advances in spatial econometrics. Springer, Berlin, pp 1–25

Corrado L, Fingleton B (2012) Where is the economics in spatial econometrics? J Reg Sci 52(2):210–239

Dong G, Harris R (2015) Spatial autoregressive models for geographically hierarchical data structures. Geogr Anal **47**(2):173–191

Dong G, Harris R, Jones K, Yu J (2015) Multilevel modelling with spatial interaction effects with application to an emerging land market in Beijing, China. PLoS ONE 10(6):e0130761

Elhorst JP (2014) Spatial panel data models. In: Spatial econometrics. Springer, Berlin, Heidelberg, pp 37–93

Fingleton B (2001) Equilibrium and economic growth: spatial econometric models and simulations. J Reg Sci 41(1):117–147

Fischer MM, Getis A (2010) Handbook of applied spatial analysis: software tools, methods and applications. Springer, Berlin

Gelman A, Hill J (2006) Data analysis using regression and multilevel/hierarchical models. Cambridge University Press, Cambridge

Geweke J (1993) Bayesian treatment of the independent student-t linear model. J Appl Econometrics 8(S1):S19–S40

Holloway G, Lacombe DJ, Shaughnessy TM (2014) How large is congressional dependence in agriculture? Bayesian inference about “scale” and “scope” in measuring a spatial externality. J Agric Econ 65(2):463–484

Jensen CD, Lacombe DJ, McIntyre S (2012) A Bayesian spatial individual effects probit model of the 2010 UK general election. University of Strathclyde, Discussion Papers in Economics, 12-01

Kelejian HH, Robinson DP (1993) A suggested method of estimation for spatial interdependent models with autocorrelated errors, and an application to a county expenditure model. Pap Reg Sci 72(3):297–312

Koop G, Poirier DJ, Tobias JL (2007) Bayesian econometric methods. Cambridge University Press, Cambridge

Lacombe DJ, McIntyre SG (2016) Local and global spatial effects in hierarchical models. Appl Econ Lett 23(16):1168–1172
Langford IH, Leyland AH, Rasbash J, Goldstein H (1999) Multilevel modelling of the geographical distributions of diseases. J Roy Stat Soc: Ser C (Appl Stat) 48(2):253–268
LeSage JP (2014) What regional scientists need to know about spatial econometrics. Rev Reg Stud 44(1):13–32
LeSage JP, Llano C (2013) A spatial interaction model with spatially structured origin and destination effects. J Geogr Syst 15(3):265–289
LeSage JP, Pace RK (2008) Spatial econometric modeling of origin-destination flows. J Reg Sci 48(5):941–967
LeSage J, Pace RK (2009) Introduction to spatial econometrics. CRC, Boca Raton, FL
LeSage J, Polasek W (2008) Incorporating transportation network structure in spatial econometric models of commodity flows. Spat. Econ. Anal. 3(2):225–245
LeSage JP, Thomas-Agnan C (2015) Interpreting spatial econometric origin-destination flow models. J Reg Sci 55(2):188–208
LeSage JP, Fischer MM, Scherngell T (2007) Knowledge spillovers across Europe: evidence from a Poisson spatial interaction model with spatial effects. Pap Reg Sci 86(3):393–421
Luke DA (2004) Multilevel modeling, vol 143. Sage, Thousand Oaks, CA
Marrocu E, Paci R (2013) Different tourists to different destinations. evidence from spatial interaction models. Tour Manage 39:71–83
Parent O, LeSage JP (2007) A Bayesian spatial model composition analysis of knowledge production. Available from: https://www.researchgate.net
Parent O, LeSage JP (2008) Using the variance structure of the conditional autoregressive spatial specification to model knowledge spillovers. J Appl Econometrics 23(2):235–256
Raudenbush S, Bryk A (2002) Hierarchical linear models: applications and data analysis methods (Advanced quantitative techniques in the social sciences; 1). Sage Publications, Newbury Park
Smith TE, LeSage JP (2004) A Bayesian probit model with spatial dependencies. Adv Econ 18:127–160
Spiegelhalter DJ, Best NG, Carlin BP, Van Der Linde A (2002). Bayesian measures of model complexity and fit. J Roy Stat Soc: Ser B (Stat Methodol) 64(4):583–639
Subramanian S (2010) Multilevel modeling. In: Handbook of applied spatial analysis. Springer, Berlin, pp 507–525
Vanoutrive T, Parenti A (2009) On proximity and hierarchy: exploring and modelling space using multilevel modelling and spatial econometrics. In: ERSA Congress, Papers. European Regional Science Association (ERSA), p 20
Vanoutrive T, Van Malderen L, Jourquin B, Thomas I, Verhetsel A, Witlox F (2009) "Let the business cycle!" A spatial multilevel analysis of cycling to work. Belgeo Revue belge de géographie 2:217–232

Donald J. Lacombe is an Associate Professor of Personal Financial Planning. Dr. Lacombe received his B.A. in Economics from the University of Florida and his Ph.D. in Economics from Florida State University. In addition to teaching Research Methods I, Dr. Lacombe serves on the editorial board of two journals and is Co-Editor of Papers in Regional Science. He has previously held positions at Trinity University, Ohio University, and West Virginia University. Dr. Lacombe's areas of expertise include Spatial Econometrics, Bayesian Econometrics, and Hierarchical Linear Modeling. His research is focused on include the application of various econometric techniques to research questions in all areas of personal financial planning.

Stuart McIntyre is a lecturer, Department of Economics, University of Strathclyde. He is also affiliated with the Fraser of Allander Institute at the University of Strathclyde and the Regional Research Institute at West Virginia University. His research interests are primarily in regional, spatial and energy economics and applied spatial econometrics. Dr McIntyre earned his MSc in Economics through the Scottish Graduate Programme in Economics at the University of Edinburgh, and his PhD in economics from the University of Strathclyde in 2013.

Chapter 10
GIS in Regional Research

Alan T. Murray

10.1 Introduction

Geographic information systems (GIS) have come to be an important component of regional science. This is not particularly surprising given the very characteristics used by Walter Isard in establishing and defining regional science (see Isard 2003). Namely, the use of rigorous analytical methods stemming from multiple disciplines coming together to study real world problems and issues was generally noted as foundational principles of regional science. Such approaches no doubt must be supported by data of some sort, likely with a spatial/geographic orientation. And it turns out that spatial data are messy and complicated, requiring specialized techniques, methods, processes, etc. devoted to the creation and use of such data, but also that spatial data often contain a wealth of implicit knowledge.

With the origins of regional science in the 1950s, there was a rather simplistic view of geographic space and the phenomena associated with it. Computers were just coming onto the scene and were far from accessible. Those fortunate enough to get access encountered major computing limitations due to hard drive, memory and processor components. A simplified view of a region or city was a necessity, with objects of interest often being a point with one or more attributes. Further, the number of points was traditionally kept to a minimum, again because computing capabilities were limited. A common analytical processing need was to derive distance, and often Euclidean (or straight line) travel was deemed sufficiently representative of spatial interaction between two points.

Fast forward to present day. Computing is an afterthought, if even considered at all. Computing power and supporting software exists on laptop and handheld

A.T. Murray
Department of Geography, University of California at Santa Barbara, Santa Barbara, CA, 93106, USA
e-mail: amurray@ucsb.edu

R. Jackson, P. Schaeffer (eds.), *Regional Research Frontiers - Vol. 2*,
Advances in Spatial Science, DOI 10.1007/978-3-319-50590-9_10

devices (and watches). Data are plentiful, often with more data generated in real time than can possibly be synthesized and understood. Monitoring devices are everywhere: woven into clothing; on bracelets and cellular phones; embedded in vehicles; scanning sensors recording purchases and spending behavior; and satellites circling the earth measuring and recording any and everything. Regions and cities are not assumed to be points but rather more complex areal features that can be highly non-homogeneous in terms of attribute characteristics, response and behavior. Limitations on the number of observations considered are substantially relaxed as it is not uncommon to consider thousands or millions of spatial objects, often managed and processed using GIS. And many sorts of geographic interactions may be present, including agglomeration, unique paths of travel between objects, neighboring impacts, etc.

Interests in regional science have evolved accordingly as well. A unit of analysis is no longer assumed to be homogenous and static, but rather varied and changing over time. We have witnessed a systematic reduction of simplifying assumptions and a recognition of the importance of local detail. It is no surprise then that the power and capabilities of GIS have only gained significance. In many ways, GIS is becoming, or has become, a central approach used in regional science. However, there remains a view that GIS is merely a tool for making maps. This chapter sets out to provide an overview of GIS. In doing this, the intent is to highlight current and future capabilities beyond map making, as GIS is specifically designed to deal with geographic data creation and the analysis of this data.

10.2 GIS

A formal definition of GIS is that it is a combination of hardware, software and procedures that support spatial analysis and decision making. GIS necessarily requires capabilities for data capture, management, manipulation, analysis and display associated with spatially referenced data (see Church and Murray 2009; Longley et al. 2015). Collectively, this means that geographic space can be abstracted as layers of information, as suggested in Fig. 10.1, enabling integration and analysis within and across layers.

The process of data capture in GIS involves abstracting the earth, or a portion of it, as a digital representation. This is often done as either a raster or a vector model. The creation of data is possible using many approaches, possibly involving the use of GPS, aerial sensing, drones and/or other sensing devices or may be based on manual digitizing, automated conversion and/or geocoding. The data management component of GIS is concerned with storage, access and query efficiency. The operational response and processing capabilities of GIS software is dependent on managing data efficiency.

The manipulation of data in GIS is necessary for many reasons. Transformation of different layers of information to a consistent frame, or coordinate system, is a very common manipulation approach. Another classic spatial manipulation

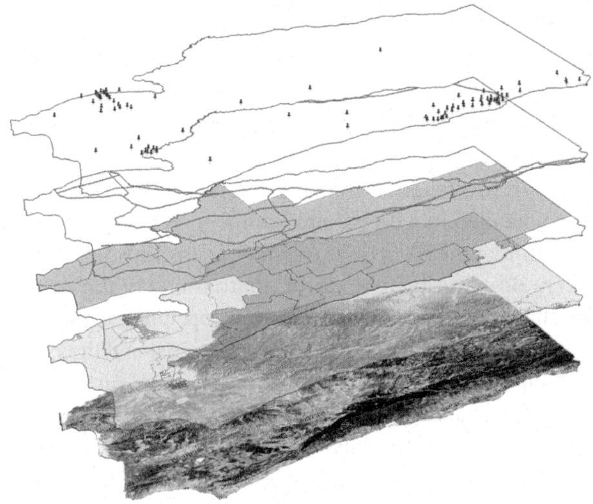

Fig. 10.1 GIS based layers of spatial information

approach is projection of three-dimensional latitude and longitude referenced spatial information into a two-dimensional coordinate system. Processing, calculation and display, as an example, may require two-dimensional representation, depending on the analysis setting. Other contexts, however, may require a three-dimensional depiction. Various spatial manipulation approaches are considered standard in GIS, including simplification and aggregation, among others. Examples of each of these manipulation functions in regional science work is readily found. Simplification may involve the derivation and use of a centroid to represent a county. Aggregation might entail the spatial combination of two (or more) adjacent census tracts in order to form one new polygon that represents the unit of analysis.

The data analysis capabilities of GIS have historically been perceived to be limited. However, this is a bit unfair because there are actually many analytical capabilities, ranging from attribute summary, spatial summary, containment assessment, polygon overlay, map algebra, deriving distance and proximity, buffering, interpolation, cluster detection, etc. In total, there is a wealth of analytical capabilities, but has historically not included advanced statistical, geostatistical, geosimulation and spatial optimization approaches (Anselin and Getis 1992; Fischer and Nijkamp 1992; Goodchild and Haining 2004). The major commercial packages generally include some access to geostatistical and spatial optimization methods, and libraries/software like GeoDA offer advanced spatial statistical methods (Church and Murray 2009; Murray 2010).

Finally, the display of spatial data in GIS has been a mainstay. Often this has entailed the making of a map, either on screen or in a paper form. More contemporary approaches have emerged to support geovisualization in 2D and 3D, but also account for some aspects of temporal variability (see Maciejewski 2014; Rey 2014; Longley et al. 2015).

10.3 Representation and Data

GIS is special and important because the data managed corresponds to activities and observations that exist/occur on the surface of the earth, and also because it has the capability to interact, query, manipulate, etc., associated information in various ways. Accordingly, representation issues are critical as the earth is not regularly shaped nor easily specified in a digital environment. While a convenient, simplified assumption is to consider the earth to be a sphere or an ellipsoid, it actually is neither. Often a geoid may do a reasonable job of approximating the earth. Nevertheless, there are always challenges in accurately and appropriately representing the surface of the earth. Depending on the scale, location of interest and the purpose of a study, a particular representation and associated datum(s) of the earth may be reasonable. Care must therefore be taken with the underlying representation of the earth to ensure that it is satisfactory for intended usages. Associated with an assumed representation of the earth is the need for a referencing system so that attributes and characteristics of places may be encoded, processed and analyzed. Referencing may depend on the abstraction of geographic space, such as whether the interest is in objects or fields. Objects typically consist of points, lines and polygons, whereas a field is generally a regular discretization of continuous space, such as a raster grid surface. The implication of referencing and objects/fields are many, but historically two factors have figured prominently: storage and processing efficiency. System response and computing needs are dependent on how data are stored. Access and query of information are intrinsically linked to the storage of data.

Vast amounts of geographic data exist in a range of formats. Spatial information and sources for obtaining it can be found in Church and Murray (2009) and Longley et al. (2015), among others. Various attempts have been made or exist that bring together publicly available spatial information, and are referred to as Geolibraries or Geoportals (Longley et al. 2015). Some are the byproduct of federal, state and/or local government efforts to ensure public access. An example at the federal level is DATA.GOV. At the state level, California provides public access to geospatial information through the state geoportal, http://gis.ca.gov/, as an example. At a local level, agencies like SANDAG (http://www.sandag.org/) in San Diego provide varying levels of access to certain geospatial data. Other communities, cities and states have policies and data access portals meant for public consumption of geographic information.

Historically the US Census has supplied important data about people and the economy in the United States. To do this, the Census employs an army of people, with primary products being the Decennial Census of Population and Housing (every 10 years), Economic Census (every 5 years), Census of Governments (every 5 years) and the American Community Survey (annually). Of course, a valuable component of census data is that digital records are available for at least a recent history. While a very good source of information, there are issues with the data. These issues can impact data quality, reliability, spatial and temporal accuracy, etc. Particular issues include sampling bias, undercounts, variable ambiguity, conflation, reporting delay/change, as well as others. From a spatial perspective, the fact that census unit boundaries can change presents significant challenges, and most importantly introduces further data uncertainty. Resolving attribute values for reporting units across time periods means that various types of interpolation (intelligent estimation/guessing) are necessary.

A wealth of spatial data now is obtained from sensing based platforms. This includes aerial and ground based equipment ranging from Global Positioning System (GPS), satellites, aircraft and drones to stationary and mobile video, images, road counters and other sensors. While GPS, satellite imagery and aircraft LiDAR are particularly commonplace and accessible, emerging technological capabilities provided by drones offer potential for real time and continuously updated remotely sensed information. On the ground, sensing equipment and technology abounds, from Google Street View vehicles to red light cameras to security video to activity detection devices, there is arguably more continuous sensor data than can be processed and ingested.

Of course, one source of spatial information is to obtain it from private data vendors, typically a byproduct of an assimilation effort on the part of the vendor where various data are brought together through the scraping of digital and print resources. Vendors such as Nokia (HERE), Walls and Associates (National Establishment Time Series), Nielsen (PRIZM), etc. turn raw data into valuable spatial information, often associated with the location of public and private goods or services. Worth noting in particular is a significant reliance on geocoding in the creation of vendor data. An example is National Establishment Time Series produced by Walls & Associates that effectively converts Dun and Bradstreet establishment data into digital, spatially referenced information. This is done by interpreting the establishment/company street address as a global position. This is known as geocoding, the formal process associated with taking a local street address reference and identifying geographic coordinates for that address on the surface of the earth, namely, a latitude and longitude (Murray et al. 2011). While a very common process to produce digital information, there are a range of issues associated with such data. Geocoding works by identifying a successful address match in a street centerline database. Often match rates are high with most commercial software, but not perfect. You can expect 5–10% of the address data to not be successfully matched. Beyond this, a successful match does not necessarily translate to spatially accurate information. The reason for this is that address matching involves interpolation along street centerline segments to estimate the location of an address number. Further, an offset distance is assumed to put the point on the building, hopefully a “rooftop hit”. Ultimately, little is often known about the actual spatial accuracy of geocoded data as the located point may not be precisely on the house, business or building, nor necessarily in the associated land parcel, neighborhood block or census tract. Errors in positional accuracy of a few meters to a few kilometers are not unusual. Worth mentioning is that business address data can be problematic. Often, records reflect headquarters only as a registered place of business, but information on where employees undertake the work is not known.

Another class of spatial information is generated by individuals, possibly solicited or unsolicited. This includes what is widely known as volunteered geographic information (VGI). Websites and software that facilitate VGI include WikiMapia, OpenStreetMap and Map Maker, where individuals create, collect and disseminate spatial data (Longley et al. 2015). Of course, other sources of VGI could include Twitter feeds (when location is disclosed or inferred), Yelp, Urbanspoon, etc. Noteworthy regarding such data is that it may be biased in many ways, not reflective of all opinions, not representative of all social classes, lacking consistency and objectivity, and may not have extensive spatial coverage. Further, data standards and associated metadata often are lacking in many ways. Other sources of user generated data are rather indirect sources, perhaps unknowingly provided by an individual. Spatial location, time and behavior can be obtained through the use of cellular phones and other electronic equipment as well as through the use of customer loyalty card programs, among others. Cell phones are typically GPS enabled, or location can be inferred from cellular towers and satellites. Customer loyalty card programs represent a growing source of data where companies like dunnhumby, Aimia, emnos, Nielsen, Symphony EYC, 5one and Demandtec employ analytics to better understand our collective behavior and trends. While not necessarily publicly available at this time, the data and information extracted by cellular providers and companies with loyalty cards can be purchased and used in various ways without any need for consent on the part of individuals.

10.4 Significance of GIS

There are many implications for regional science in the growth and evolution of GIS. As suggested above, geographic data availability across a range of domains changes how processes may be considered and the detail at which it can be conceived. Not only with respect to more traditional concerns associated with residential location, as an example, but now detailed information on movement and mobility patterns throughout a day or week. Beyond this, there is ubiquitous monitoring by sensors on the ground and in the sky. Big data associated with objects and group or individual activities is generated daily, if not hourly. In many respects this enables details about place and behavior to be considered, and also accounted for explicitly in analysis and modeling efforts. While GIS does have considerable mapping and analytical capabilities, the use of GIS based data and methods to support advanced mathematical and statistical modeling continues to be noteworthy. Thus, what makes GIS special is the ability to create, manage and use data in order to derive and exploit spatial knowledge. Given that the data managed by GIS is geographic in nature, there are many spatial relationships that result. Of note are proximity, adjacency, connectivity, shape, direction, containment, concentration and scale. From a database perspective, GIS is interesting and unique because these spatial relationships are often implicit in that they are not computed in advance and stored as part of the database. Rather, these relationships may be inferred because of geographic location and derived on the fly as needed.

There are many examples of regional science based work that has made use of, or integrated with, GIS in order to carry out advanced modeling of some sort. A prominent area of work is associated with land use planning. Supporting this is the use and development of cellular automata approaches; and more recently, agent-based methods. These approaches operate using some representation of a region, often a regular raster, combined with current and past data on development in each cell. Working on a rather informal model specification, rules are established that reflect observed or inferred growth patterns that can be used to mimic or estimate future patterns. Work in this area includes that of Clarke et al. (1997) and Ward et al. (2000), among many others. What is noteworthy is that spatial relationships have been found to be key to developing good land use transition/change rules. For example, land use around a given area is particularly influential, but also current and future infrastructure is an indicator of likely land use change. The advancement of cellular automata and other related approaches for land use planning/analysis has gone hand-in-hand with GIS proliferation and access to more detailed data. Approaches have pursued greater specification and linkage to regional models (Ward et al. 2003) but now also account for more features of land cover dynamics. Recent discussion of these approaches can be found in Clarke (2014).

A prominent approach in regional science is assessment, evaluation and/or detection of activity concentration. They may represent clusters, hot spots, cold spots, neighborhoods, homogeneous response zones, etc. Identification of such areas could be associated with response correlation or simply detecting agglomeration of some sort. Factors associated with housing prices or foreclosure could be of interest, as an example. Alternatively, one may want to detect whether there are high rates of activity, such as crimes or industry mixes. To support this, a variety of methods have been developed and applied in regional science, including local and global measures of spatial autocorrelation as well as scan statistics. Spatial information critical to most approaches is neighborhood structure, the so called weights matrix or a scanning window depending on the methodological inquiry. This is often based on adjacency or proximity relationships. GIS is invaluable in deriving spatial relationships and details along these lines. A recent review of work in this area can be found in Murray et al. (2014), but of note is the use of GIS to specify more spatially relevant relationship structures, such as AMOEBA and LOSH (Getis 2015) and irregular scan "windows" (Murray et al. 2014).

The direct measurement of proximity is fundamentally important to most regional science based investigations. The interest is generally associated with the spatial interaction between two locations, and in particular the shortest path. This may represent a travel route taken by an individual or a cost/distance to travel. When travel is not restricted to a road network, this is a continuous space problem that offers an infinite number of travel path options. A popular choice for quantifying proximity is to assume straight line (Euclidean) travel between two locations. This can be problematic for many reasons but often does not reflect an actual travel path. Greater realism means that obstacles must be taken into account, such as water

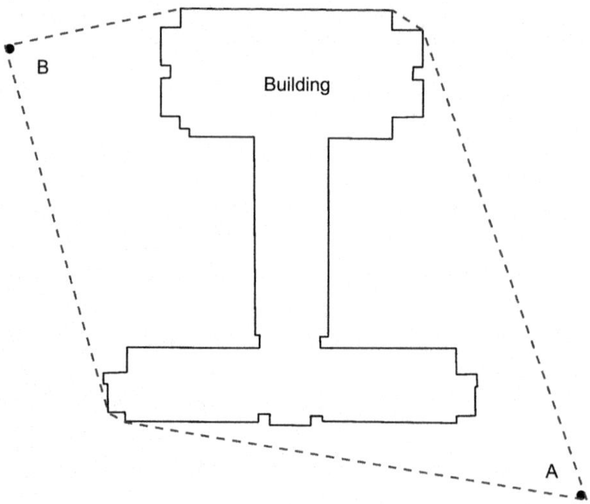

Fig. 10.2 Convex hull for associated objects (A, B and Building)

bodies, bridges, structures, airports, mountains, canyons, etc. When travel is not permitted through obstacles and/or buildings, the most efficient continuous space route is known as the Euclidean shortest path. Of course, travel through geographic space necessarily requires spatial information about movement options between a given origin and destination. Beyond this, however, standard operations in GIS dealing with spatial proximity and computational geometry are critical. In fact, Hong and Murray (2013) report an approach to identify an optimal Euclidean short path based on the use of a convex hull, readily identified using GIS. Figure 10.2 depicts the convex hull associated with three objects, two points and a building. The convex hull represents a minimum length boundary containing all three objects. What Hong and Murray (2013) proved is that the shortest path lies on this boundary, assuming travel from point A to point B, effectively reducing the infinite number of routes possible through continuous space to only a finite number of polygon segments along the convex hull. A technique based on convex hulls was generalized for the case of multiple obstacles. The significance of this is that one can identify an obstacle avoiding shortest path in real-time using GIS, enabling navigation and wayfinding as well as providing an ability to more accurately model travel behavior.

A fairly common spatial analytical method used in regional science is a location model. In particular, coverage models have proven to be invaluable for addressing many types of service situations (Murray 2016). Whether the circumstances involve prescriptive plans or a descriptive understanding of an existing system, location coverage models enable mathematical specification and solution derivation when facilities are to be placed throughout a geographic service area. Facilities could correspond to fire stations, clinics, cell towers, restaurants, etc. The geographic service area may be based on travel time, distance, visibility and/or audibility, and may be regular or irregular in shape. GIS is, therefore, invaluable for helping to

Fig. 10.3 Skeleton of a polygon region

structure proximity, adjacency, contiguity, concentration, etc. in associated models. Beyond this, Murray and Tong (2007), Murray et al. (2008) and Matisziw and Murray (2009) have demonstrated that spatial knowledge and relationships can be exploited through the use of GIS. In particular, Murray and Tong (2007) derived finite dominating sets corresponding to locations where an optimal configuration of facilities would be limited to for continuous space coverage problems. GIS facilitates the identification by systematic evaluation of service areas using overlay functions, reducing an infinite number of siting possibilities to a finite set. Murray et al. (2008) and Matisziw and Murray (2009) proved that an optimal facility site would be located along the skeleton (or medial axis) of a region. As an example, a region is shown in Fig. 10.3 with demand for service distributed throughout. The skeleton for this region is also shown, and we know that service coverage is maximized when the facility is sited somewhere along the skeleton. Again, GIS enables this property of a region with respect to coverage to be exploited in various ways, providing a means of solutions based on the derived property.

10.5 Conclusions

This chapter serves many purposes. One is to provide an overview of GIS within the context of regional science. Beyond this, the hope is that a characterization of GIS based features will result in a greater understanding of what has been and could be done to support regional science using GIS. Finally, a number of recent developments associated with the integrated use of GIS in regional science were discussed. Speculative discussion based on this overview are now possible.

On the analysis side, much of what was discussed in this chapter reflects the continued convergence of GIS and spatial analytics approaches, something noted in Goodchild and Haining (2004), among others. This can be seen in much of contemporary regional science work in that GIS is central to all facets of a study, beginning with data management all the way to the application of particular analytic approaches. The future of regional science will be one where GIS is increasingly more central. This no doubt presents many research challenges for addressing issues of integration as well as efficiently deriving/solving associated models.

What is abundantly clear at this time is that uncertainty abounds. GIS highlights that this is, in fact, the case with digital information. Positional location of objects stored in GIS is rarely without error or uncertainty. Further, even in cases where there is a high level of positional certainty, various manipulation operations could create resident uncertainty. A similar observation holds for model abstractions that attempt to mimic observed regional systems and behavior. The model is a simplification of an actual system, omitting certain features and nuances. Add to this the fact that we may have a limited or biased understanding of systems, processes and behaviors, then collectively there is much potential for all sorts of direct and indirect uncertainty. As a result, this will continue to force researchers to rethink and reevaluate how we approach regional science, and more importantly how we can address issues of uncertainty and bias in the many forms that it may arise. While GIS may highlight how resident uncertainty exists in data, it by no means offers a roadmap on how to take uncertainty into account nor how analysis and planning can be bolstered.

Given technological and computing advances, it is clear that big data will continue to change what is done in regional science and what can be done. The level of detail at which data are collected will necessitate changes in applied analytics, often reflecting a relaxation of simplifying assumptions that have long been relied upon. There is little doubt that this is a good thing as it will enable better modeling and analysis to be carried out. This will translate into better insights, improved plans and superior policy development. The challenges, of course, are how methods will evolve accordingly. This will change perspectives and understanding. As a result, fundamental assumptions likely will prove problematic, thereby needing to be relaxed and/or modified.

While somewhat related to the previous point(s), it is a fact that changes over time are really not well understood in general. Aspects of land use planning may be an exception, as noted above. However, there are significant challenges for relating change over time to actual behavior/response/operation. This can be said for past and current systems, but also for future conditions. What will residential land use patterns look like? To what degree will employment centers continue to decentralize? How will travel and behavior patterns change? What technological advances can we expect and what are implications for regional systems? These questions and others simply highlight that cross sectional work has real limitations, yet with a greater ability to use more detailed spatial information will come an enhanced capability to simulate future conditions.

In closing, this chapter has demonstrated the significance of GIS in regional science to date, but likely this is merely a starting point. What GIS really tells us at this point is that assumptions regarding sufficient data quality are actually very problematic. The impacts and implications for analysis and planning are actually not well understood at all. Couple with this increasing amounts of detailed data from a variety of sources over time, and the suggestion is that there is much left to do in regional science.

References

Anselin L, Getis A (1992) Spatial statistical analysis and geographic information systems. Ann Reg Sci 26:19–33

Church RL, Murray AT (2009) Business site selection, location analysis and GIS. Wiley, New York

Clarke KC (2014) Why simulate cities? GeoJournal 72:129–136

Clarke KC, Hoppen S, Gaydos L (1997) A self-modifying cellular automaton model of historical urbanization in the San Francisco Bay area. Environ Plann B Plann Des 24:247–261

Fischer MM, Nijkamp P (1992) Geographic information systems and spatial analysis. Ann Reg Sci 26:3–17

Getis A (2015) Analytically derived neighborhoods in a rapidly growing West African city: the case of Accra, Ghana. Habitat Int 45:126–134

Goodchild MF, Haining RP (2004) GIS and spatial data analysis: converging perspectives. Pap Reg Sci 83:363–385

Hong I, Murray AT (2013) Efficient measurement of continuous space shortest distance around barriers. Int J Geogr Inf Sci 27:2302–2318

Isard W (2003) History of regional science and the Regional Science Association International. Springer, Berlin

Longley PA, Goodchild MF, Maquire DJ, Rhind DW (2015) Geographic information systems and science, 4th edn. Wiley, New York

Maciejewski R (2014) Geovisualization. In: Fischer MM, Nijkamp P (eds) Handbook of regional science. Springer, Berlin, pp 1137–1155

Matisziw TC, Murray AT (2009) Siting a facility in continuous space to maximize coverage of a region. Socio Econ Plan Sci 43:131–139

Murray AT (2010) Quantitative geography. J Reg Sci 50:143–163

Murray AT (2016) Maximal coverage location problem impacts, significance, and evolution. Int Reg Sci Rev 39:5–27

Murray AT, Grubesic TH, Wei R (2014) Spatially significant cluster detection. Spat Stat 10:103–116

Murray AT, Grubesic TH, Wei R, Mack EA (2011) A hybrid geocoding methodology for spatio-temporal data. Trans GIS 15:795–809

Murray AT, Matisziw TC, Wei H, Tong D (2008) A geocomputational heuristic for coverage maximization in service facility siting. Trans GIS 12:757–773

Murray AT, Tong D (2007) Coverage optimization in continuous space facility siting. Int J Geogr Inf Sci 21:757–776

Rey SJ (2014) Spatial dynamics and space-time data analysis. In: Fischer MM, Nijkamp P (eds) Handbook of regional science. Springer, Berlin, pp 1365–1383

Ward DP, Murray AT, Phinn SR (2000) A stochastically constrained cellular model of urban growth. Comput Environ Urban Syst 24:539–558

Ward DP, Murray AT, Phinn SR (2003) Integrating spatial optimization and cellular automata for evaluating urban change. Ann Reg Sci 37:131–148

Alan T. Murray is professor, Department of Geography, University of California at Santa Barbara (UCSB). His primary research interests are geographic information science; spatial optimization; health informatics; urban growth and development; land use planning; urban, regional, and natural resource planning and development; and, infrastructure and transportation systems. He previously held academic appointments at Drexel University, Arizona State University and Ohio State University. Previous faculty positions were at Drexel University, Arizona State University and Ohio State University. Dr. Murray earned a Ph.D. in geography from the University of California at Santa Barbara in 1995.

Chapter 11
Exploratory Spatial Data Analysis: Tight Coupling Data and Space, Spatial Data Mining, and Hypothesis Generation

Trevor M. Harris

11.1 Introduction

Spatial data analysis and GIS are instrumental components for examining the spatial dimension of regional science. As the role of geographical space has been increasingly recognized in science, social science, and the humanities, partially driven by the explosive growth of GIS, so spatial analysis has become progressively embedded within statistical analysis and modeling. GIS has enabled ever greater access to rapidly expanding quantities of digital spatial data. In a seminal paper Anselin and Getis (1992) made a distinction between confirmatory and exploratory data analysis where, although the edges of both were blurred, the former was largely deductive and theory driven awhile EDA was inductive and data driven. In reality this distinction should be questioned for it suggests that the EDA process begins with little prior theoretical understanding of the problem or of the datasets and is essentially a 'fishing expedition' of available databases before the real deductive theory-driven analysis begins. Few reputable studies proceed in such a fashion. In consigning EDA to a predominantly data driven, atheoretical approach using largely descriptive techniques, the real power and insight that EDA provides is minimized to a lesser role than it deserves. Goodchild (2010) alludes to this point in that the data driven approach and the search for pattern was often viewed as being independent of any theoretical framework and to some degree contributed to the social-theoretic critique of GIS as being essentially not concerned with theory (Pickles 1995).

Importantly, however, Anselin and Getis recognized that to serve the spatial needs of regional science, an integration of spatial analysis and GIS based on computationally intensive approaches and visualizations was required. In recent

T.M. Harris
Department of Geology and Geography, West Virginia University, USA
e-mail: Trevor.Harris@mail.wvu.edu

R. Jackson, P. Schaeffer (eds.), *Regional Research Frontiers - Vol. 2*,
Advances in Spatial Science, DOI 10.1007/978-3-319-50590-9_11

decades, GIS and cyberinfrastructure have brought about a revolution in the availability of spatial data. In addition, the spatial analytical power of GIS and geospatial technologies have contributed to a substantial restructuring of regional science. Many analog map collections have been converted to coordinate-based digital form, and digitally-born spatially referenced data are now available in ever increasing quantities and consumed through spatial data portals and across the Internet. GIS has changed the spatial analysis landscape in many other ways as well, "The geospatial world of today is clearly a much broader domain of data, tools, services, and concepts than the limited GIS world of 1992" (Goodchild 2010, 55). In this respect, the statistical tool box proposed by Anselin and Getis is in many respects a redundant notion, for many software systems are now hybridized and enable sophisticated spatial data analysis to be performed. Significantly, however, Anselin and Getis proposed a dynamic and iterative approach to data analysis in the form of ESDA that, by drawing on EDA and the spatial data management and processing power of GIS, facilitated a tighter interaction between the user and spatial data analysis in a highly interactive and reflexive analytical and graphical environment. Central to shaping ESDA and its spatial extension was the pioneering work of John Tukey (1977).

11.2 Exploratory Data Analysis

Exploratory Spatial Data Analysis advances Tukey's (1977) seminal work on EDA through the tight coupling of geographical space to traditional EDA approaches. While this antecedence to ESDA is often recognized and acknowledged, the unique contribution of EDA to data analysis as espoused by Tukey is sometimes lost in the flurry to examine the spatial dimensions of ESDA. EDA is a critical starting point to research analysis, and there is a tendency to miss this exploratory step in the jump to confirmatory and inferential statistics. Understanding Tukey's work can be valuable to regional scientists, for EDA represents both a philosophical and a methodological approach to data analysis.

EDA stands in some contrast to confirmatory inferential statistics by its emphasis on hypothesis generation rather than on hypothesis testing and confirmation. Tukey's work paved the way for an alternative, yet in many ways a complementary, approach to inferential statistical data analysis. The ideas of Tukey concerning EDA have been pursued and promoted by several authors who provide excellent insight into the essential message of Tukey and his nuanced approach to data analysis through EDA (Chatfield 1985, 1986; Cox and Jones 1981; Hartwig 1979; Hartwig and Dearing 1983; Hoaglin et al. 1983, 1985, 1991; Mosteller 1985; Sibley 1988). At its core, EDA focuses on exploring the properties of data and to use these findings to raise questions, pursue ideas, and generate hypotheses that can be subsequently tested through confirmatory data analysis. Tukey (1977, 1) claimed that, "Exploratory data analysis is detective work . . . numerical detective work . . . or graphical detective work . . . that requires both tools *and* understanding" (italics added). Tukey questioned the ability of inferential statistics to uncover ideas and hypotheses worthy of further investigation and that hypothesis testing alone often ended in a dead-end and provided little guidance as to the directions that a study should proceed. Ideas, Tukey suggested, came from data exploration more often than from "lightning strokes": "Finding the question is often more important than finding the answer" (Tukey 1980, 23–24). Indeed to extend the premise still further, Tukey argued that, "An approximate answer to the *right* problem is worth a good deal more than an exact answer to the wrong question, which can always be made precise" (Tukey 1962, 13).

Much of Tukey's work in EDA represents a critique of traditional confirmatory inferential statistics and a resistance to the Neyman-Pearson approach to confirmatory analysis and a seeming unwillingness to examine the data prior to pursuing inferential statistical analysis (Fernholz and Morgenthaler 2000, 84). Tukey expressed concern from the outset about the 'straight-line paradigm' of confirmatory statistics that seemed to proceed linearly from question, to design, to data collection, to data analysis, and then to answer. One of his primary concerns was that this sequential process neglected how the questions are generated in the first instance. Furthermore, he questioned, how could the research design be guided, or the data collection monitored, or analysis overseen to avoid inappropriate use of statistical models if not by exploring the data before, during and after analysis (Tukey 1980, 23). To pursue confirmatory analysis, he argued, requires substantial exploratory work coupled with quasi-theoretical insight. Tukey suggested reorganizing the early stage of the straight-line paradigm such that a study proceeded from an idea, to an iterative combination of question and analytical (re)design, and thence to data collection, analysis, and outcome (ibid.). In this approach, the formulation of the ideas and questions are critical, yet as Tukey argued, such questions are rarely 'tidy' but rather are inchoate and require extensive exploration of past data (ibid., 24). Tukey saw the essential need for EDA to assist in formulating the questions deserving of subsequent confirmation (ibid., 24). Tukey did not reject confirmatory data analysis in favor of EDA for he argued that each on its own was insufficient: "To try to replace either by the other is madness. We need them both" (ibid., 23). A circular paradigm thus emerges, rather than a linear process, whereby theory defines the problem and EDA provides a feedback loop between the analysis and theoretical formation allowing for subsequent inferential analyses to be pursued or modified in the light of such exploratory work. Thus, analysis and theoretical understanding are enmeshed and not separate stages of an investigation. In this way, EDA emphasizes a constant, but meaningful, return to the data 'honeypot' and, as Tukey remarked, torturing the data until it has revealed all and has no more to confess.

Tukey argued against EDA being seen as comprising wholly descriptive statistics but rather that EDA was an "attitude" and a "flexibility", supported by visual representations and "some helpful techniques" (ibid., 25). Tukey's work in EDA can be seen as providing two primary themes to data analysis. First, he presented a practical philosophy as to how to proceed systematically through a data analysis and especially how to begin that process (Good 1983; Tukey and Wilk 1970). In my experience of teaching ESDA, this practical approach, which can be brought to bear on almost any data analysis, has been of greatest value to many students who balk at where to begin and how to proceed through the data analytical process. The acid test, of course, is to present students with a data set that is known to them and to watch the barrage of inappropriate inferential statistics thrown at the data that invariably fails to generate much understanding or substance from the analysis. EDA more closely replicates the process followed by experienced researchers when analyzing a dataset and provides a practical path through the data analysis process than can be gained from any rigid adherence to standard statistical text books–in geography or otherwise. Despite this valuable practical philosophy, however, Tukey argued against EDA being seen as a kind of theory of data analysis (Fernholz and Morgenthaler 2000, 84).

Second, EDA places considerable emphasis on techniques that are both robust and resistant (Besag 1981; Mosteller 1985; Mosteller and Tukey 1977; Velleman and Hoaglin 1981). Tukey suggested that invariably little is known about the data to which we apply statistical models and, thus, there is a need to explore the data using techniques that minimize prior assumptions about the data and the model (which assumptions he suggested are often violated in practice) and allowed exploration of the data to guide the choice of appropriate questions and analysis. His focus on nonparametric statistics spurred the identification of innovative techniques that were resistant to the effects of extraordinary data values that could unduly influence the results of an analysis, and were robust and lessened the reliance on the assumptions of the data distribution and were essentially distribution free. Thus, the median, interquartile range and percentiles are preferred over the mean and standard deviation because they are more resistant to extreme values and outliers. Creative techniques for univariate, bi-variate, and hypervariate EDA such as boxplots, stem-and-leaf diagrams, q-q plots and Tukey mean-difference plots, parallel coordinate plots, lowess curves and local regression, multi-dot displays, compound filter smoothers of running medians for resistant time series analysis, resistant linear regression, scattergram matrices, and conditional plots provide a diverse mix of robust and resistant EDA techniques that complement more 'fragile' and less resistant measures (Cleveland 1993; Velleman and Hoaglin 1981; Tukey 1977).

A further characteristic of EDA is its focus on the 'five number' summary statistics of minimum, upper and lower quartiles, median, and maximum. This focus on the shape, spread, and central tendency of a distribution and on identifying and examining anomalies, outliers, trends, patterns, and residuals is central to EDA. EDA uses techniques that resist reductionism and summary statistics but tries to keep the original data present at all times. Thus, stem and leaf diagrams are preferred over histograms whose bins 'hide' the original data values. EDA places heavy emphasis on descriptive statistics, and it is here that it battles with the perception that EDA and its statistics are somewhat obvious and trivial and that inferential statistics and progressively more abstract statistical models represent greater legitimacy and intellectual value. To claim a focus on a data distribution curve, for example, may at first sight seem basic, yet the personal story of the eminent paleontologist Jay Stephen Gould represents a powerful example of the importance of just one of these descriptive measures. In the *Median isn't the message,* Gould (1985) recounts being diagnosed with mesothelioma cancer of the abdomen and being told that the median lifespan for people with this disease was 8 months. But Gould's research fascination with variation and his training as a scientist led him to determine that the distribution curve was positively skewed and that for a number of reasons, including good health care, early detection of the cancer, and no other healthcare problems, that he could place himself well into the long tail of the distribution. Indeed Gould survived a further 20 years and died of a different cancer. In a prefatory note Steve Dunn calls Gould's article "the wisest, most humane thing ever written about cancer and statistics" (Dunn 2002).

Thus, EDA resists the allure of the 'magic number syndrome' whereby complex distributions and patterns are reduced to summary numerical form that potentially hides the real pattern or complexity of the data. For this and for other reasons, there is a heavily reliance in EDA on graphical display. As Tukey (1977, vi) contended, "The greatest value of a picture is when it forces us to notice what we never expected to see". The visualization work of Cleveland (1993) and Tufte (1983, 1990) has added considerably to the emphasis on graphical representation in data analysis and to the suite of techniques available in EDA. This focus on exploratory techniques and lessened reliance by EDA on preconceptions and assumptions about data stands in contrast to confirmatory statistics that seeks to make broad conclusions and generalizations about a population based on the inferences drawn from the relationships found in a random sample of that population. The focus of inferential statistics on *a priori* hypothesis testing and probabilistic models and the derivation of estimates and confidence levels points to the need for descriptive statistics of the data as a preliminary step before a statistical model is applied or inferences are generalized about a larger population. EDA is particularly suited to the creative exploration of data and to generating questions and hypotheses, even though it often does not provide definitive answers. As Tukey indicated, EDA is not the whole story but, as he observed, if you took 1000 books on statistics in the 1970s, 999 would be confirmatory. Arguably, the same assessment is probably not that much different today except that to EDA might be added space to create ESDA and its additional focus on understanding the spatial dimensions of data and hypothesis generation.

11.3 Spatial Extensions to Exploratory Data Analysis

If EDA is about using robust, resistant, and graphical techniques to identify, understand, and gain insight into the essential properties of data, then ESDA is an extension to that process that seeks to detect spatial patterns in the data, to formulate hypotheses based on the geography of that data and to assess the appropriateness and assumptions of spatial models (Haining 2009). ESDA utilizes recent and dramatic advances in interactive desktop computer processing and computer graphics to create an exploratory analytical environment capable of linking EDA and spatial data analysis. ESDA provides a powerful idea and hypothesis generation platform with which to undertake complex spatial data analysis and integrates well with recent advances in local spatial statistical techniques, GIS, and geovisualization. The spatial and statistical modeling needs of regional science coupled with ongoing advances in big data and spatial data mining suggests ESDA will be of growing importance in geographical analysis and regional science in the future. The growing availability of hybrid software systems capable of handling spatial data have contributed markedly to the ability to perform ESDA. S-Plus was an early software system equipped with a bridge to ESRI's GIS system though this was subsequently discontinued. Currently there are several analytics systems capable of performing ESDA that include Tableau (www.Tableau.com), Cartovis (cartovis.com), ESRI's Geostatistical Analyst (http://www.esri.com/software/arcgis/extensions/geostatistical), GeoVista (http://www.geovista.psu.edu/), Weave (https://www.oicweave.org/), and within the software environment R (https://www.r-project.org/). Perhaps best well known within geography and regional science is GeoDa (https://geodacenter.asu.edu/).

In addition to the work of Tukey and other researchers in EDA, ESDA has been heavily influenced by the early work of Monmonier (1989) on the geographic brushing of scatterplot matrices, Cleveland's work on data visualization, and Sibley's spatial applications of EDA (1988). In particular, ESDA owes much to the prescient work of Anselin (1993, 1999) who was not only early in identifying the potential for combining advances in GIS and spatial data management with spatial analysis and local spatial analysis but in providing the means to do so through GeoDa. While the linkage between ESDA and GIS has been somewhat tenuous, in reality the tight coupling of space and data analysis as evidenced by the development of the GeoDa software has made the link between GIS and EDA apparent and explicit. ESDA as envisaged by Anselin remains a subset of EDA rather than of GIS and it focuses on exploring the distinguishing characteristics of spatial data through a suite of techniques that specifically focus on spatial autocorrelation and spatial heterogeneity.

ESDA usually contains a similar collection of EDA techniques capable of exploring, describing and visualizing data, but with the additional capability of being able to handle spatial data and mapping. Anselin's particular focus has been to make spatial autocorrelation and spatial heterogeneity central to his ESDA software development and focus (Anselin 2005). As Anselin writes "ESDA is a collection of techniques to describe and visualize spatial distributions, identify atypical locations or spatial outliers, discover patterns of spatial association, clusters or hot spots, and suggest spatial regimes or other forms of spatial heterogeneity. Central to this conceptualization is the notion of spatial autocorrelation or spatial association, i.e., the phenomenon where locational similarity (observations in spatial proximity) is matched by value similarity (attribute correlation)" (Anselin 1999, 79–80). Thus, in addition to many of the robust and resistant techniques to be found in EDA, and as outlined above, in ESDA the analyses are tightly coupled with spatial data and mapping. This tight coupling of spatial and attribute data occurs through brushing and linking of interconnected multiple dynamic window panels. Brushing and linking provides for a powerful exploratory capability not

Fig. 11.1 GeoDa in the Virtual Reality CAVE displaying multiple dynamically linked panels of analyses linked via brushing and linking to each other and to the map display on the floor

just between tables and graphics, but with maps. Currently GeoDa and similar systems dynamically link multiple panes or windows containing various analytical techniques that includes a map display (Fig. 11.1). Anselin provided an important step in ESDA in enabling EDA and spatial analysis to be tightly coupled. In addition to linking panels containing multiple simultaneous analyses, the ability to dynamically 'brush' individual or groups of data items in any panel or map display and to see the corresponding data points or relationships highlighted in the other panels creates a truly powerful exploratory tool. These compelling visual and dynamic displays of multiple analyses are actively and dynamically linked to enable spatial patterns and spatial relationships to be examined, as well as to identify anomalies and outliers. Brushing not only allows for data points selected in one analytical panel to be automatically identified and displayed across all panels, but it is also possible to brush locations on a map to see the respective data displayed in the other panels and vice versa.

In addition to brushing and linking, GeoDa enables both global spatial autocorrelation to be examined using Moran's I, and local spatial autocorrelation using Local Indicators of Spatial Autocorrelation that indicate the specific location and magnitude of spatial autocorrelation to be identified (Anselin 1993). In instances where spatial patterns can be discerned, it is reasonable to assume that the spatial data are related and not independent, and tests for spatial autocorrelation using a spatial weights matrix can be applied based on locational contiguity to test for positive similarity between adjacent spatial units or for negative spatial autocorrelation and dissimilar patterns. A focus on spatial outliers reinforces the exploratory work of Tukey to not ignore anomalies but to embrace their study and the insights that they provide. In tandem with Geographically Weighted Regression (Fotheringham et al. 2002) that identifies the occurrence of spatial non-stationarity and allows relationships to vary over space, the move toward local spatial statistics lends itself well to ESDA. The empirical Bayesian kriging of ESRI's Geostatistical Analyst employs the semi-variogram to identify directional bias in correlations between sample points, and using spatial covariance between data points adjusts the weights of contributing sample points to optimize model interpolators for spatially continuous fields. Thus, within ESDA concepts of distance, adjacency, interaction, and neighborhood spatially enrich the field of statistics that has been relatively insensitive and unsuitable to geographical investigation before the inclusion of the spatial dimension. In overcoming the sampling of data points independent of the characteristics of the data being interpolated, these local spatially adaptive weighting functions are progressively embedding Tobler's (1970) first law of geography into contemporary spatial analysis and within the spatially enabled ESDA in particular.

11.4 Discussion

It is suggested here that EDA, and its spatial counterpart ESDA, provide a powerful, systematic and intuitive approach to spatial data analysis and a necessary precursor to the use of inferential statistics. Despite the embeddedness of these exploratory techniques within ESDA, the extent to which the premises and approaches of Tukey's EDA have been recognized and accepted within regional science as necessary and complementary steps in the spatial data analysis process is not clear. EDA is still seen as 'descriptive' and a 'warm-up exercise' to the real statistical analysis using confirmatory techniques. This perception diminishes the real value of EDA to understand the very nature of a data set. In particular, the potential for ESDA to formulate ideas and hypotheses for pursuit either within the ESDA environment or with confirmatory inferential statistics could represent missed opportunities. ESDA does more than enhance the spatial analytical capabilities of GIS, it represents a powerful approach to gain insight into the heart of the data.

Part of the reason for not fully embracing EDA may be, as others have pointed out (Goodchild 2010; Haining 2009), that in the face of big data and the growing availability of spatial data from GIS, the preference of some is to seek patterns and anomalies automatically. This, of course, flies in the face of Tukey's conception and purpose for EDA. Barnes (2003) in his critique of American regional science arguably suggests that the decline among regional science practitioners could have been avoided. Barnes contends of regional science that, "It is unreflective, and consequently inured to change, because of a commitment to a God's eye view. It is so convinced of its own rightness, of its Archimedean position, that it remained aloof and invariant, rather than being sensitive to its changing local context". The advent of ESDA may be one change that will resonate with regional science and that by drawing on inductive reasoning (and arguably deductive as well through EDAs circular reasoning) provides a reflective and exploratory environment that is creative and open-ended. As Tukey would argue "Exploratory data analysis is an attitude, a flexibility, NOT a bundle of techniques . . . " (Tukey 1980, 23).

In the coming decades, and fueled by a potential avalanche of spatially rich data repositories created from a combination of automatic data sensors and human data generators, regional science will be challenged not only by data storage, curation, search, and query issues, but by how meaningful spatial data analysis of big data will be performed. The profile of ESDA could increase as its philosophy, tools, and techniques are brought to bear on big data to gain an understanding of extremely large and complex spatial datasets. Statistical analyses and visualization technologies struggle with big data in handling the sheer high volume, high velocity, high variety, and increasingly high veracity characteristics of these data assets (Gandomi and Haider 2015). The application of intelligent machine learning approaches replicate some of the early focus of spatial analysis in GIS on automatically detecting patterns from complex data. Amidst assertions that big data will spell the end of theory, a major challenge posed by big data is that little is known about the underlying empirical micro-process that lead to the emergence of the typical network characteristics of big data. And yet, such scenarios continue to beg the question that Tukey laid out nearly four decades ago—how are meaningful questions and hypotheses to be formed without an intensive exploration of the data?

Searching for plausible hypotheses, especially where the spatial pattern is not common knowledge, is problematic. Shekhar and Chawla (2003) proposed the use of interactive exploratory analysis to bring together a number of analytical panels that closely mirror an ESDA approach. Spatial data mining, they suggest, differs from spatial data analysis by its usage of techniques derived from spatial statistics, spatial analysis, machine learning and databases. The output from an iterative spatial data mining process, suggests Shekhar and Chawla, is typically a hypothesis (ibid., 237). One way they suggest to view data mining is as a filter step that occurs before the application of a rigorous statistical tool: "The role of the filter step is to literally plow through reams of data and generate some potentially interesting hypothesis which can then be verified using statistics" (ibid., 240). Thus, a key part of spatial data mining of big data is to comb through big databases in order to identify information that is relevant to building actionable models. As regional science confronts ever larger and more complex spatial databases, these exploratory techniques may take on greater importance in positioning the science and research questions to be pursued.

In the days soon after the publication of Tukey's seminal work, Cox and Jones (1981, 142) made a plea that "it is to be hoped that quantitative geography . . . will be less afflicted than in the past by a craving for the semblance of elegance, exactness, and rigour exuded by inferential ideas, and that geographers will show more willingness to engage in uninhibited exploration of their data, guided but not dominated by the procedures devised by statisticians". Ten years later and following an NCGIA specialist meeting, Fotheringham (1992, 1676) reported that there might be instances "*in certain circumstances*" (italics in the original) where exploratory spatial data techniques within GIS might be appropriate. These circumstances appear to apply to spatial windowing to analyze data on the fly as the window is moved around a set of locations, for detecting spatial outliers, for disaggregating statistics spatially, and to visualize spatial data. In the wake of the GIS revolution, the growing abundance of digital spatial data, the era of big data, the rise of data mining, and the availability of ever more powerful computing and graphical visualization resources and hybrid software solutions, such hopes for ESDA may be closer to reality now than they were three or more decades ago.

References

Anselin L (1993) Exploratory spatial data analysis and geographic information systems. West Virginia University, Regional Research Institute Research Paper #9329

Anselin L (1999) Interactive techniques and exploratory spatial data analysis. In: Longley PA, Goodchild MF, Maguire DJ, Rhind DW (eds) Geographical information systems, vol 1. Wiley, New York, pp 253–266

Anselin L (2005) Exploring spatial data with GeoDa: a workbook. Center for Spatially Integrated Social Science, Urbana-Champaign

Anselin L, Getis A (1992) Spatial statistical analysis and geographic information systems. Ann Reg Sci 26:19–33

Barnes TJ (2003) What's wrong with American regional science: a view from science studies. Can J Reg Sci 1:3–26. Spring

Besag J (1981) On resistant techniques and statistical analysis. Biometrika 68(2):463–469

Chatfield C (1985) The initial examination of data. J R Stat Soc A 148(3):214–253

Chatfield C (1986) Exploratory data analysis. Eur J Oper Res 23:5–13

Cleveland WS (1993) Visualizing data. AT&T Bell Laboratories, Murray Hill, NJ

Cox NJ, Jones K (1981) Exploratory data analysis. In: Wrigley N, Bennett RJ (eds) Quantitative geography: a British view. Routledge and Kegan Paul, London, pp 135–142

Dunn S (2002) Prefatory note to the the median isn't the message by Stephen Jay Gould. http://cancerguide.org/median_not_msg.html. Accessed 30 May 2016

Fernholz LT, Morgenthaler S (2000) A conversation with John W. Tukey and Elizabeth Tukey. Stat Sci 15(1):79–94

Fotheringham AS (1992) Exploratory spatial data analysis and GIS. Environ Plan A 2:1675–1678

Fotheringham AS, Brunsdon C, Charlton M (2002) Geographically weighted regression: the analysis of spatially varying relationships. Wiley, Hoboken, NJ

Gandomi A, Haider M (2015) Beyond the hype: big data concepts, methods, and analytics. Int J Inf Manag 35(2):137–144

Good IJ (1983) The philosophy of exploratory data analysis. Philos Sci 50:283–295

Goodchild MF (2010) Whose hand on the tiller: revisiting spatial statistical analysis and GIS. In: Anselin L, Rey SJ (eds) Perspectives on spatial data analysis. Springer, Berlin, pp 49–59

Gould SJ (1985) The median isn't the message. Discover Magazine, 6 June, pp 40–42

Haining R (2009) Spatial data analysis: theory and practice. Cambridge University Press, New York

Hartwig F (1979) Exploratory data analysis. Sage, Beverly Hills

Hartwig F, Dearing BE (1983) Exploratory data analysis, Sage quantitative applications in the social sciences, 16

Hoaglin DC, Mosteller F, Tukey JW (eds) (1983) Understanding robust and exploratory data analysis. Wiley, New York

Hoaglin DC, Mosteller F, Tukey JW (eds) (1985) Exploring data tables, trends and shapes. Wiley, New York
Hoaglin DC, Mosteller F, Tukey JW (eds) (1991) Fundamentals of exploratory analysis of variance. Wiley, New York
Monmonier M (1989) Geographic brushing: enhancing exploratory analysis of the scatterplot matrix. Geogr Anal 21(1):81–84
Mosteller F (1985) Understanding robust and exploratory data analysis. Wiley, New York
Mosteller F, Tukey JW (1977) Data analysis and regression: a second course in statistics. Addison-Wesley, Reading, MA
Pickles J (ed) (1995) Ground truth: the social implications of geographic information systems. Guilford Press, New York
Shekhar S, Chawla S (2003) Spatial databases: a tour. Englewood-Cliffs, NJ, Prentice Hall
Sibley D (1988) Spatial applications of exploratory data analysis, CATMOG no 49. Geo-Books
Tobler WR (1970) A computer movie simulating urban growth in the Detroit region. Econ Geogr 46:234–240
Tufte ER (1983) The visual display of quantitative information. Graphics Press, Cheshire, CT
Tufte ER (1990) Envisioning information. Graphics Press, Cheshire, CT
Tukey JW (1962) The future of data analysis. Ann Math Stat 33(1):1–67
Tukey JW (1977) Exploratory Data Analysis. Addison Wesley, Reading, MA
Tukey JW (1980) We need both exploratory and confirmatory. Am Stat 34:23–25
Tukey JW, Wilk MB (1970) Data analysis and statistics: techniques and approaches. In: Tufte ER (ed) The quantitative analysis of social problems. Addison-Wesley, Reading, MA, pp 370–390
Velleman P, Hoaglin D (1981) Applications, basics and computing of exploratory data analysis. Duxbury Press, Boston, MA

Trevor M. Harris is professor, Department of Geology and Geography, West Virginia University. His primary research interests are Geographic Information Science; Immersive Geographies; geovisualization and the CAVE; Virtual Reality and Augmented Reality; Geosensory geographies, Spatial Humanities; Qualitative GIS; Spatializing text; GIS and Society, Critical GIS and Participatory GIS; GIS and archaeology; and Exploratory Spatial Data Analysis. He is a faculty research associate in WVU's Regional Research Institute and co-directs the WV State GIS Technical Center and the Virtual Center for Spatial Humanities. Dr Harris earned his Ph.D. in geography from the University of Hull in 1983.

Chapter 12
Location Analysis: Developments on the Horizon

Daoqin Tong and Alan T. Murray

12.1 Introduction

Location analysis has deep roots in regional science and represents a classic method in the discipline. Location analysis, in general, concerns the organization or arrangement of goods, resources, services or activities in space. Such analysis can be used to answer questions of why activities/phenomena occur at certain places and how to best locate goods/services to achieve certain purposes. Early location analysis work can be traced back to Johann Heinrich von Thünen, Walter Christaller, August Lösch, Alfred Weber and Harold Hotelling, among others. von Thünen (1826) proposed a location theory to explain the principles that account for different agricultural land uses by linking locational rent with agricultural production and transportation costs. Focusing on factory location, Weber (1909) was interested in finding the best site on the continuous plane that minimizes transportation costs, equivalent to profit maximization under production, labor supply and demand assumptions. Hotelling (1929) examined the location strategies of two firms and their price setting considering demand distribution, transportation costs and competition. Using a linear city/market, Hotelling showed that with fixed pricing and production costs both firms would ideally locate at the halfway point, with each capturing/serving half the total market. Going beyond a single area or region, Christaller (1933) conceived of human settlements as a system and developed central place theory to explain the spatial organization of villages, towns and cities. Building upon the interrelations of economic activities between places,

D. Tong (✉)
School of Geography and Development, University of Arizona, Tucson, AZ, 85721, USA
e-mail: daoqin@email.arizona.edu

A.T. Murray
Department of Geography, University of California at Santa Barbara, Santa Barbara, CA, 93106, USA

R. Jackson, P. Schaeffer (eds.), *Regional Research Frontiers - Vol. 2*,
Advances in Spatial Science, DOI 10.1007/978-3-319-50590-9_12

this suggests that settlement patterns reflect a hexagon-shaped hierarchy, with centers and their associated hinterlands. Lösch (1941) expanded on central place theory to allow for sophisticated spatial arrangements that considered economies of scale and specialization. These pioneering studies have laid the fundamental foundation for the field of location analysis by connecting locational choices to various economic activities.

Since these pioneering studies, locational analysis has flourished in regional science and beyond. One stream of application and development has sought to verify, extend and refine associated location theory. For example, Alonso (1964) extended von Thünen's agricultural land use theory to the urban setting and developed bid-rent models of land use distribution as a function of the distance from the central business district. Modern agriculture location theory has also evolved to account for more realistic conditions (see Lucas and Chhajed 2004). Similarly, central place theory has been extended to examine city size (Beckmann 1958), hierarchy of villages (von Böventer 1963), and shopping centers (Eaton and Lipsey 1982), and account for customer shopping behavior (Ghosh and McLafferty 1987) and agglomeration effects (Fischer 2011; Mulligan et al. 2012). Models have also been used to interpret, test and/or verify various aspects of central place theory as well as gain insights into underlying processes (Beaumont 1987; Curtin and Church 2007).

Another stream of activity has involved specification and solution of supporting mathematical models. Initial work was devoted to solving and extending the Weber problem (Wesolowsky 1993). Although the Weber problem appears rather simple, solving the problem exactly has been challenging given the continuous nature of the problem, where a firm (or firms) can be sited anywhere in geographic space. Early studies focused on the geometric characteristics of the problem and used a mechanical analogue device known as the Varignon frame. Later, iterative algorithms, including the well-known Weiszfeld algorithm (Weiszfeld 1937), were developed for model solution. Various extensions have also been made to the Weber problem by introducing alternative distance metrics, including multiple facilities, and allowing stochastic demand, among others (Drezner et al. 2002). The Weber problem has also served as the inspiration for a range of contemporary modeling efforts, some of which will be discussed in Sect. 12.2.

Building upon the foundation laid by the above work, location analysis and modeling have evolved in terms of theoretical development and empirical application. While originally focused on descriptive characteristics associated with why and how activities/communities are organized in space, the field has advanced to be more prescriptive in nature through the assistance of making locational decisions for various purposes. A significant number of location models have been developed to support real-world applications at the urban and regional scale for both the public and private sectors. The following section briefly reviews the field with a focus on a selected number of models and applications. This is followed by a discussion of the challenges in location analysis. Looking forward, we highlight future research directions associated with emerging applications, big spatial data and ways to address computational challenges. Finally, concluding remarks are given.

12.2 Analytical Approaches

As suggested above, much of the underlying economic and spatial theory associated with location analysis has historically been descriptive in nature, seeking to develop a better understanding of existing patterns and observed conditions. Examples include bid-rent curves, regression models based on proximity to a city center and distance-decay oriented interaction models. Evolving computing capabilities have enabled description to be carried out using mathematical models, and also support prescriptive decision making about where to best locate goods and services in combination with responsible resource allocation.

12.2.1 Prescriptive Capabilities

While location theory has provided a comprehensive description/explanation of various activities, prescriptive capabilities have come to characterize more contemporary location analysis (Murray 2010). In these studies, determining the best locations for certain services or activities has proven beneficial for achieving overall efficiency. Modern location analysis has therefore been operationalized through development of mathematical models. Over the past few decades, literature on location models and associated applications are prolific. Summaries of work in this area can be found in articles including Chhajed et al. (1993), Brandeau and Chiu (1989), Owen and Daskin (1998); ReVelle and Eiselt (2005), Smith et al. (2009), Murray (2010), as well as books including Love et al. (1988); Drezner (1995), Daskin (1995), Drezner and Hamacher (2002), Church and Murray (2009), Farahani and Hekmatfar (2009), Eiselt and Marianov (2011), Laporte et al. (2015), and Eiselt and Marianov (2015). These reviews have focused on various aspects of the field and major achievements to date. This chapter will be forward-looking with elaboration on important future research areas in the field.

As noted previously, a location model has generally been conceived to be a bid-rent curve, regression model that includes distance and/or an interaction model represented as an equation. The prescriptive approach extends descriptive capabilities to allow for resource allocation and spatial decision making. In this sense, a contemporary location model therefore consists of one or multiple objective function(s) as well as a set of constraints. Objective functions are used to articulate the goal(s) that a particular problem aims to achieve. An objective function may reflect overall investment/operation costs or perhaps service benefits. These would then be optimized accordingly, with decisions made to produce the best objective function outcome. Constraints reflect the problem specific conditions that limit activities in some manner, necessarily establishing a mathematical linkage between decision variables.

Prescriptive oriented location models have been classified into different categories based upon a range of criteria. Categories of particular note include:

continuous space, discrete space, network, stochastic, deterministic, single objective, multiple objective, number of facilities, service capacity, etc. Depending on how space is treated in a location model, it may be considered either continuous, discrete or network. The classic Weber problem is an example of a continuous problem as the factory to be located can be anywhere on the continuous plane. Alternatively, a discrete problem is one where there are only a finite number of candidate sites, identified a priori, and a finite number of objects to be served. Finally, a network problem could be discrete but may also be continuous, depending on whether siting could occur along arcs or if demand is distributed along arcs. Elaboration on these points and others follows in the subsequent sections.

12.2.2 Classic Models

There are a number of noteworthy location models that will serve to illustrate prescriptive capabilities. The location-allocation problem and its variants have arguably been among the most influential and widely relied upon prescriptive models. The location-allocation problem was formally introduced in the seminal work by Cooper (1963), extending the Weber problem to allow for multiple facilities to be sited on the continuous plane. Hakimi (1964) considered a network version of the problem where demand and service provision occur on a network with the objective to minimize the overall travel costs along the network. Demand is assumed to be at nodes, and facilities can be sited anywhere on the network. Although no specific solution method is provided, Hakimi proved that nodes on a network contain at least one optimal solution in the case of a network. Given this, the search for the best configuration of facility can be narrowed to the finite set consisting of only network nodes. This gives rise to the p-median problem: finding p sites among n predetermined points to serve discrete demand such that total travel cost is a minimum. ReVelle and Swan (1970) formulated the p-median problem. Location-allocation problems, especially the p-median problem, have been widely applied and extended to incorporate various problem specific conditions, including facility capacity, hierarchical structure, stochastic demand and competition. A summary of model development and application can be found in Mirchandani (1990), Marianov and Serra (2011), ReVelle et al. (2008), and Daskin and Maass (2015), among others.

Another category of prescriptive location models concerns regional coverage. Critical then is the notion of "coverage", which is often defined based on whether demand can be served within a maximum acceptable travel distance/time. This coverage standard corresponds to the "range" concept introduced in central place theory. In contrast to location-allocation models, covering problems are driven by different performance criteria. Toregas et al. (1971) introduced the location set covering problem (LSCP) seeking to find the minimum number of facilities (and where to locate them) needed to provide complete coverage to a region. Recognizing that in many situations resources are not sufficient to ensure a full coverage of a region, Church and ReVelle (1974) proposed the maximal covering location problem (MCLP) to locate a limited number of facilities in order to achieve the greatest coverage of a region. These two classic covering problems have been extended to incorporate various coverage standards, redundant coverage, cooperative service provision and service capacity. A review of the covering problems and associated applications can be found in Schilling et al. (1993), Murray et al. (2010) and Farahani et al. (2012).

A third category of prescriptive location models is center problems. The concern in this case involves locating one or more facilities/services so that the maximum distance from a demand to its closest sited facility is as short as possible. Differing from other location modeling approaches that focus on cost or system efficiency, center problems seek equality by ensuring that the worst access provided to any individual/place is as good as possible. The p-center problem was introduced by Hakimi (1964) and often assumes that facilities can be located anywhere in a region (continuous space) and that demand is concentrated at discrete points. Various algorithms have been developed to solve center problems, including a Voronoi diagram heuristic (Suzuki and Okabe 1995). The problem becomes a vertex p-center problem if the candidate facility sites are also restricted to predefined sites (Daskin 1995). The p-center problem has also been extended to consider service capacity, continuous demand and backup service provision. Refer to Drezner (2011), Tansel (2011) and Calik et al. (2015) for further discussion of center problems.

A fourth category of prescriptive location models is competitive demand approaches. Following the seminal work of Hotelling (1929), recognition of the need to address competition for service has arisen, with approaches developed to explicitly account for competition among sited facilities. In these problems, the location of additional firms will not only affect new markets but those of the competitors. Early theoretical studies have focused on modifying some of the economic assumptions made in Hotelling (1929) and examining associated equilibrium patterns. Subsequent competitive location models have shifted to account for market share consideration. Various conditions have been explored, including the type of service to be provided (e.g., convenience stores, shopping malls, gas stations, hotels), space (e.g., network, discrete location or continuous region), Nash and Stackelberg equilibria, consumers' choices and market share delineation approaches. Refer to the work of Friesz et al. (1988), Serra and ReVelle (1995), Plastria (2001) and Drezner (2014) for more details.

It is conceivable that listing of categories could continue, likely numbering in the hundreds to account for the significant location model nuances. Rather than continue further, we leave it at the above major categories, but note that issues of dispersion (Goldman and Dearing 1975; Church and Garfinkel 1978; Moon and Chaudry 1984; Kuby 1987; Murray and Church 1995; Verter and Erkut 1995), hubs (O'Kelly 1986; Alumur and Kara 2008), interdiction (Scaparra and Church 2015), etc. are no less important or significant. However, due to space limitations, further review and discussion is not possible.

12.3 Challenges

There are numerous challenges confronting the use and application of location models. One issue noted here has to do with decision making processes unique to particular application contexts. A second issue concerns computing capabilities associated with solving structure models.

12.3.1 Application Contexts

Location analysis and modeling have been applied to solve a wide range of urban and regional problems. These applications include public facility siting (such as libraries, schools, post offices, and police stations), emergency facility placement (fire stations, ambulance), districting (political districting, service districting, police districting), healthcare facility and service planning, network design and routing (telecommunication, transportation), business locations (such as bank branches, retail facilities), military operations, agricultural management (production, storing, processing and distribution of agricultural products) as well as environmental problems (such as nature reserve site selection, wildlife management). A number of classic and modern applications have also been summarized in Lucas and Chhajed (2004) and Eiselt and Marianov (2011).

In a location model, mathematical abstraction is very critical as improper specification of the objective function or constraining relationships/conditions can result in locational decision making that is far from the best. The diverse applications of location analysis present challenges to problem formulation and model construction. Depending on the particular problem of interest, an existing location model might not be applicable, and constructing a new location model is sometimes necessary. Such a new model will involve identifying and formulating one or multiple goals and specifying the associated constraining conditions. Even for problems where an existing location modeling framework applies, oftentimes the existing model may need to be modified to account for application specific goals or constraining conditions, such as different cost functions, specific capacity requirement and special relationships among facilities or between demand and facilities. In other cases, when problems cannot be mathematically articulated or formulated, heuristic approaches will be needed to solve the problems approximately. These heuristic based approaches will be discussed below. Due to problem variety and complexity, constructing location models requires some level of creativity to accurately abstract real-world problems as well as the ability to link components/relationships mathematically.

12.3.2 Problem Solution

Beyond the abstraction process is the need for identifying, comparing and understanding alternative solutions. Decision making often involves a host of constitutes, particular for public section contexts. Different groups or individuals may have differing concerns and objectives. Further, they may have their own ideas about good alternatives to consider. Generating solutions remains a challenge. Understanding strengths and weaknesses and being able to communicate them is essential.

As mentioned earlier, predictive approaches for location analysis and modeling have mainly focused on identifying the best locational decision(s) for serving certain purpose(s). Solving these problems necessitates a search for the best set of locations, either in a continuous region or limited to predefined discrete sites. While a continuous problem usually means it is difficult to solve as there exists an infinite number of candidate sites to select from, searches confined to a finite number of sites may be nontrivial as well. In general, two strategies have been used to solve location models: exact methods and heuristic methods.

Exact methods are those producing a provably optimal solution. That is, solutions identified by these methods can be shown to be superior to all others, found in a process or not found. Enumerating all the possible solutions is sometimes relied upon, enabling identification and evaluation of associated objective function values. The method guarantees the best solution to be identified because all are explicitly considered. However, when the problem size grows in terms of the number of different configurations, solutions to consider, the computational requirements can be prohibitive, making enumeration impractical. Enumeration in the case of continuous space problems is generally infeasible given that an infinite number of siting configurations would need to be considered. For this reason, other exact methods have been developed, including linear programming, integer programming, branch-and-bound, dynamic programming, Lagrangian relaxation based methods as well as specialized algorithms that exploit geometric characteristics of certain problems (Elzinga and Hearn 1972; Matisziw and Murray 2009).

Irrespective of whether we have a continuous or discrete location problem, many are known to be NP-hard (Kariv and Hakimi 1979; Megiddo and Supowit 1984). This means that solving these problems exactly can be difficult or impossible, especially for large sized ones. For these problems as well as problems that are difficult to mathematically formulate, heuristic approaches are widely used for problem solution. Heuristic methods are often rule of thumb, ad-hoc strategies. Compared with exact methods, heuristic approaches can often solve a problem faster but problem solution quality is not known or guaranteed. Various heuristics have been used to solve location models, including the "alternate" method (Cooper 1963; Maranzana 1964), greedy based search (Church and ReVelle 1974), and vertex substitution or interchange (Teitz and Bart 1968). While many early heuristics focus on iterative improvement based on a local search neighborhood, high level modern metaheuristics represent a family of methods that often allow other solution spaces to be considered simultaneously, resulting in solutions less likely to be trapped in local optima (Brimberg et al. 2000). Modern metaheuristics have been widely applied to solve various location problems, including tabu search (Murray and Church 1995; Rolland et al. 1996), simulated annealing (Murray and Church 1996; Chiyoshi and Galvao 2000), and genetic algorithms (Bozkaya et al. 2002).

12.4 Looking Forward

The field of location analysis has evolved tremendously with continued visibility within and outside of regional science. Looking forward, we believe location analysis will continue to be essential for helping address future regional challenges. Future applications may require closer interaction/collaboration of researchers in location analysis with experts in other fields in order to enhance problem understanding and develop efficient problem solution strategies. Although GIS (geographic information system) continues to be recognized as important in location analysis, a wider adoption and integration of GIS into location analysis is expected. The advent of big data has the potential to revolutionize location analysis theoretically and practically. Additional insights gained from big data may help refine existing modeling frameworks and motivate novel solution approaches. With increased complexity and detail in location models due to big data, high performance computing will be an integral component of future analytical frameworks.

12.4.1 New Application Contexts

In years to come, location analysis will be used to help solve emerging challenges and issues at regional and national scales. Closer collaboration with scholars in other disciplines is expected for solving these challenges. One example concerns sustainable development. For example, moving towards a more sustainable environment, US EPA (2015) requires significant annual CO2 reductions: "22%–23% below 2005 levels in 2020; 28–29% below 2005 levels in 2025, and 32% below 2005 levels in 2030". The CO2 reduction goal necessitates an increased use of renewable energy resources to substitute the conventional coal resources for future electricity generation. Solar has been identified as one of the important emerging renewable resources for future energy supply. Location analytical studies of future solar energy power plants and the distribution network presents an important application that will contribute to CO2 reduction goals. However, such an analysis requires collaboration with climate scientists, environmental experts, economists, and geographers to take into account future weather uncertainty, environmental impacts, economic development and population growth. Other likely applications relate to the challenges brought about by climate change. Extreme weather events such as droughts and floods are expected to occur more frequently in some local regions, leading to countless economic losses. Howitt et al. (2015) estimate that the recent drought in California has caused an economic loss of 2.7 billion dollars in 2015. Incorporating location analysis into efficient water allocation and flood mitigation strategies presents a sound way to help mitigate losses due to climate change. Of course, there are many other areas as well.

For some new regional applications, existing modeling frameworks can be used but might need substantial revisions to account for problem complexity. For example, interdiction approaches detailed in Scaparra and Church (2015) provide ways to identify critical components or locations in a region in order to prioritize fortification efforts when preparing for a future disaster. However, such location models have mainly focused on a certain type of service or facility. As for disaster management, many aspects need to be addressed simultaneously (such as lives, properties, transportation infrastructures, communication networks, etc.) and the consequent location analysis can be much more complicated. Scaparra and Church (2015) also noted that even though existing models are already complex, they have not been able to adequately address the interconnection of various components in a system. This also calls for interdisciplinary collaboration for a better understanding and modeling of interdependence and complexity of relevant elements in a region. Driven by the new applications, revisions of existing models or sometimes new modeling frameworks might be needed to address problem specific requirements and complexity. Overall, location analysis as an evolving field will continue to make contributions to regional science and help solve new regional challenges.

12.4.2 GIS

Location analysis often involves various types of data, ranging from demographics (e.g., population distribution), the built environment (e.g., transportation networks, land uses) to the natural environment (e.g., terrain information). Many of these data tend to be spatially explicit, but do give rise to various sorts of implicit information. For example, population is associated with specific cities in a region and roads connect certain places in an area. Given that GIS is a special information system designed to store, manage, process, analyze and display spatial and non-spatial data, there is a natural linkage between GIS and location analysis. In recent years, GIS has been increasingly used to support location analysis and has been widely recognized as important due to its powerful capabilities in data acquisition (as many data are readily available in the GIS form), management and processing. For example, GIS has been directly employed to conduct suitability analysis for various location decisions, including hospitals, roads and utility lines. Murray (2010) also highlighted the critical role of GIS in theoretical development of location analysis that goes beyond simple data support or manipulation. Reviews by Church (1999), Murray (2010) and Bruno and Giannikos (2015) all note the various contributions GIS has made to location analysis.

A wider adoption of GIS by location analysts and modelers will continue to help the field of location analysis advance. The integration of GIS into location analysis can further refine current models and broaden the applications. New location models or variants of existing models better reflecting a problem of interest might emerge due to finer details or alternative representation schemes available in GIS (Murray 2010). Also, GIS can be used to gain insights into the uncertainty associated with spatial data, scale, and modeling practices (Tong and Church 2012). Meanwhile, constructing location models requires certain level of mathematical skills, which is often beyond the knowledge of a general planner or analyst. The incorporation of location models in GIS software helps location analysis and modeling to reach a wider audience. In fact, some GIS commercial software has started to incorporate some of the classic location models. For example, the Esri ArcGIS software provides a location-allocation module that includes the p-median problem, location set covering problem and maximal covering location problem with optional considerations of service capacity and competition.

12.4.3 Big Data

Compared with decades ago when availability of locational data was an issue, big data has revolutionized the amount and detail of information available about human activities and the environment. Such data are collected through a range of technologies, such as cell phones, wearable devices, GPS, social media, cameras and various sensors, and provide an enormous amount of information about people's movement and activities. For example, in 2014 New York City shared with the public the information about 173 million taxi trips. The data provided information about where and when individuals were picked up and dropped off. The unprecedented spatial-temporal coverage, as well as the richness and granularity of big data, allow researchers to gain new knowledge about human activities. It is estimated that big data will have a transformative impact on almost all fields (Shaw 2014). We also expect that the advent of big data will bring about new opportunities to further advance the field of location analysis.

We anticipate that the integration of big data into location analysis will enhance the resolution and accuracy of data input. Conventional data input in many location models relies upon field work or a number of data collection agencies, such as the Census Bureau. Often these data come in an aggregate form, e.g., total population at the census tract level, so how individuals are distributed within the aggregation unit is unknown. Depending on the specific aggregation scheme and scale used in the aggregation, solutions given by a location model may vary substantially (Francis et al. 2009). When continuous regional demand is assumed in location models, uniform or some theoretical distributions are often used, which may differ from where people are in reality. Such a discrepancy will also lead to solutions that may be far from the best. With increased data resolution and accuracy, big data has the potential to help location problems generate better results.

Also, evidence provided by big data will help us revise some assumptions made in existing location models to better reflect the reality. For example, in many location problems demand is often assigned to the closest facility when capacity allows. Insights gained from big data about individuals' preferences can be incorporated into current location models to draw allocations more accurately. Also, in many location models demand is assumed to be fixed (often originate at home). Building on big data, location modelers can also take into account individual level movement dynamics into the locational decisions of the intended service. An incorporation of such travel-activity patterns in location analysis will significantly enhance modeling accuracy. Although in the past few decades a number of empirical studies have been conducted to examine individuals' patron patterns as well as how trips are chained, most were based on conventional data collection such as travel diaries with a very limited number of individuals for a minimal number of days. Big data allows one to do such an examination with a much larger sample size for a longer time period.

While modern location analysis has mainly focused on prescription of the best locations for certain activities/services, the variety of big data offers researchers the opportunity to revisit the location theories. For example, geotagged big data mining can be used to help reveal meaningful distributions of activities or patterns in space. These findings may provide empirical evidence to verify or modify location theories. This also points to a new direction for future research.

Overall, big data will bring new opportunities to advance the field of location analysis. The large scale, fine resolution information provided by big data will help refine exiting models and inspire new approaches/models to better reflect real-world problems. Meanwhile, the large volume of big data will inevitably increase problem complexity tremendously, and solving the associated problems optimally may become extremely difficult. This will necessitate development of new efficient solution approaches and incorporation of high performance computing for problem solution. Additional discussion on these issues will be provided in the following section.

12.4.4 *Efficient Solution Approaches and High Performance Computing*

As mentioned earlier, solving problems exactly presents an important challenge in location analysis. The increased level of model complexity and big data will add more difficulty to problem solution. Solving these problems may involve tremendous amount of computation, especially for large sized problems. One the one hand, novel solution approaches will be needed to solve these problems efficiently. For example, considering that solving small sized problems are often much easier than large sized ones, novel strategies can be developed to decompose certain large problems into smaller sub-problems without sacrificing problem optimality. On the other hand, more efficient and effective heuristics will be continuously sought to solve large sized location problems approximately. Recent development of hybrid metaheuristics combining strengths of metaheuristics and classical solution techniques, such as branch and bound, has shown promise for location problem solution (Blum et al. 2008).

In addition to developing efficient approaches to solve problems, future efforts will be needed to focus on taking the advantage of high performance computing (HPC), which consists of a cluster of computers or processors known as nodes. In HPC, individual nodes can work together to solve complex problems more efficiently than can an individual computer. When solving a problem, workload of solving the entire problem needs to be divided and distributed to a number of nodes simultaneously in a parallel fashion. Using HPC to solve a problem requires an understanding of the computing hardware as the associated parallel architecture may differ, leading to different computational performances. More importantly, in HPC a scalable and efficient procedure is essential for performing the parallel computing. This often involves a customized process of division and synchronization of sub-tasks as well as information interchange (communication) among multiple processors. Studies have started to incorporate HPC to solve large sized, challenging location problems. For example, Redondo (2008) proposed evolutionary algorithm based heuristic approaches in a HPC setting to solve a competitive facility location problem on the continuous plane. However, to achieve the best HPC performance in terms of both solution quality and efficiency, the design of the parallel computing process can be challenging as it often varies with the problem to be solved and the specific solution approaches used. This points to an important area where more research is needed in the future.

12.5 Conclusions

Location analysis represents one of the core fields of regional science. Building upon the classic location theories, location analysis has evolved considerably in the few past decades. While early studies focused on an understanding of the distribution pattern and associated mechanism of human settlements and activities, contemporary location analysis has evolved to assist the locational decision making in various regional problems. Looking forward, the field of location analysis will continue to be relevant and influential in regional science. Location analysis will be used to help solve emerging issues concerning sustainability and environmental challenges. The big data age presents great opportunities for researchers to revisit location theories as well as further advance location modeling frameworks and the applications. We also anticipate a continued integration of GIS into location analysis for data support, model refinement and efficient problem solution. With increased problem complexity, future research will consist of development of computationally efficient solution approaches and an incorporation of high performance computing.

References

Alonso W (1964) Location and land use. Harvard University Press, Cambridge, MA

Alumur S, Kara BY (2008) Network hub location problems: the state of the art. Eur J Oper Res 190:1–21

Beaumont JR (1987) Location-allocation models and central place theory. In: Ghosh A, Rushton G (eds) Spatial analysis and location-allocation models. Van Nostrand Reinhold, New York

Beckmann MJ (1958) City hierarchies and the distribution of city size. Econ Dev Cult Chang 6:243–248

Blum C, Roli A, Sampels M (2008) Hybrid metaheuristics: an emerging approach to optimization. Springer, Berlin

Bozkaya B, Zhang J, Erkut E (2002) An efficient genetic algorithm for the p-median problem. In: Drezner Z, Hamacher H (eds) Facility location: applications and theory. Springer, Berlin

Brandeau ML, Chiu SS (1989) An overview of representative problems in location research. Manag Sci 35:645–674

Brimberg J, Hansen P, Mladenovic N, Taillard ED (2000) Improvements and comparison of heuristics for solving the uncapacitated multisource Weber problem. Oper Res 48:444–460

Bruno G, Giannikos I (2015) Location and GIS. In: Laporte G, Nickel G, Saldanha da Gama F (eds) Location science. Springer International Publishing, Berlin

Calik H, Labbé M, Yaman H (2015) p-Center Problems. In: Laporte G, Nickel G, Saldanha da Gama F (eds) Location science. Springer International Publishing, Berlin

Chhajed D, Francis RL, Lowe TJ (1993) Contributions of operations research to location analysis. Locat sci 1(4):263–287

Chiyoshi F, Galvao RD (2000) A statistical analysis of simulated annealing applied to the p-median problem. Ann Oper Res 96:61–74

Christaller W (1933) Die zentralen Orte in Süddeutschland. Fischer, Jena. Translation: Baskin CW (1966) Central places in southern Germany. Prentice-Hall, Englewood Cliffs

Church RL (1999) Location modelling and GIS. In: Longley PA, Goodchild MF, Maguire DJ, Rhind DW (eds) Geographical information systems. Wiley, New York

Church RL, Garfinkel RS (1978) Locating an obnoxious facility on a network. Transp Sci 2:107–118

Church RL, Murray AT (2009) Business site selection, location analysis and GIS. Wiley, New York

Church RL, ReVelle C (1974) The maximal covering location problem. Pap Reg Sci 32:101–118

Cooper L (1963) Location–allocation problems. Oper Res 11:331–343

Curtin KM, Church RL (2007) Optimal dispersion and central places. J Geogr Syst 9:167–187

Daskin MS (1995) Network and discrete location: models algorithms and applications. Wiley, New York

Daskin MS, Maass KL (2015) The p-median problem. In: Laporte G, Nickel G, Saldanha da Gama F (eds) Location science. Springer International Publishing, Berlin

Drezner Z (ed) (1995) Facility location: a survey of applications and methods. Springer, New York

Drezner Z (2011) Continuous center problems. In: Eiselt HA, Marianov V (eds) Foundations of location analysis. Springer, New York

Drezner T (2014) A review of competitive facility location in the plane. Logist Res 7:114

Drezner Z, Hamacher HW (eds) (2002) Facility location: applications and theory. Springer, Berlin

Drezner Z, Klamroth K, Schobel A, Wesolowsky GO (2002) The Weber problem. In: Drezner Z, Hamacher HW (eds) Facility location: applications and theory. Springer, Berlin

Eaton BC, Lipsey RG (1982) An economic theory of central places. Econ J 92(365):56–72

Eiselt HA, Marianov V (eds) (2011) Foundations of location analysis. Springer, New York

Eiselt HA, Marianov V (eds) (2015) Applications of location analysis. Springer International Publishing, Gewerbestrasse

Elzinga J, Hearn DW (1972) Geometrical solutions for some minimax location problems. Transp Sci 6:379–394

Farahani RZ, Hekmatfar M (eds) (2009) Facility location: concepts, models, algorithms and case studies. Physica Verlag, Heidelberg

Farahani RZ, Asgari N, Heidari N, Hosseininia M, Goh M (2012) Covering problems in facility location: a review. Comput Ind Eng 62(1):368–407

Fischer K (2011) Central places: the theories of von Thünen, Christaller, and Lösch. In: Eiselt HA, Marianov V (eds) Foundations of location analysis. Springer, New York

Francis RL, Lowe TJ, Rayco MB, Tamir A (2009) Aggregation error for location models: survey and analysis. Ann Oper Res 167:171–208

Friesz TL, Miller T, Tobin R (1988) Competitive network facility location models: a survey. Pap Reg Sci 65:47–57

Ghosh A, McLafferty SL (1987) Location strategies for retail and service firms. Lexington Books, Lexington, MA

Goldman AJ, Dearing PM (1975) Concepts of optimal location for partially noxious facilities. Bull Oper Res Soc Am 23:1–31

Hakimi SL (1964) Optimal locations of switching centers and the absolute centers and medians of a graph. Oper Res 12:450–459

Hotelling H (1929) Stability in competition. Econ J 39:41–57

Howitt R, MacEwan D, Medellín-Azuara J, Lund J, Sumner D (2015) Economic analysis of the 2015 drought for California agriculture. UC Davis Center for Watershed Sciences, Davis

Kariv O, Hakimi SL (1979) An algorithmic approach to network location problems: the p medians. SIAM J Appl Math 37:539–560

Kuby M (1987) Programming models for facility dispersion: the p-dispersion and maxisum dispersion problems. Geogr Anal 19(4):315–329

Laporte G, Nickel G, Saldanha da Gama F (eds) (2015) Location science. Springer International Publishing, Berlin

Lösch A (1941) Die raumliche Ordnung der Wirtschaft. Fischer, Jena. Translation: Woglom WH, Stolper WP (1954) The economics of location. Yale University, New Haven

Love R, Morris JG, Wesolovsky GO (1988) Facilities location: models and methods. North Holland, New York

Lucas MT, Chhajed D (2004) Applications of location analysis in agriculture: a survey. J Oper Res Soc 55:561–578

Maranzana F (1964) On the location of supply points to minimize transport costs. Oper Res Q 15:261–270

Marianov V, Serra D (2011) Median problems in networks. In: Eiselt HA, Marianov V (eds) Foundations of location analysis. Springer, New York

Matisziw TC, Murray AT (2009) Siting a facility in continuous space to maximize coverage of continuously distributed demand. Socioecon Plann Sci 43:131–139

Megiddo M, Supowit KJ (1984) On the complexity of some common geometric location problems. SIAM J Comput 13:182–196

Mirchandani PB (1990) The p-median problem and generalizations. In: Mirchandani PB, Francis RL (eds) Discrete location theory. Wiley, New York

Moon ID, Chaudry SS (1984) An analysis of network location problems with distance constraints. Manag Sci 30:290–307

Mulligan GF, Partridge MD, Carruthers GI (2012) Central place theory and its reemergence inregional science. Ann Reg Sci 48:405–431

Murray AT (2010) Advances in location modeling: GIS linkages and contributions. J Geogr Syst 12:335–354

Murray AT, Church RL (1995) Heuristic solution approaches to operational forest planning problems. OR-Spektrum 17(2):193–203

Murray AT, Church RL (1996) Applying simulated annealing to location-planning models. J Heuristics 2:31–53

Murray AT, Tong D, Kim K (2010) Enhancing classic coverage location models. Int Reg Sci Rev 33(2):115–133

O'Kelly ME (1986) The location of interacting hub facilities. Transp Sci 20:92–106

Owen SH, Daskin MS (1998) Strategic facility location: a review. Eur J Oper Res 111:423–447
Plastria F (2001) Static competitive facility location: an overview of optimisation approaches. Eur J Oper Res 129:461–470
Redondo JL (2008) Solving competitive location problems via memetic algorithms. High performance computing approaches. Ph.D. dissertation, University of Almería
ReVelle CS, Eiselt HA (2005) Location analysis: a synthesis and survey. Eur J Oper Res 165:1–19
ReVelle CS, Swain RW (1970) Central facilities location. Geogr Anal 2:30–42
ReVelle CS, Eiselt HA, Daskin MS (2008) A bibliography for some fundamental problem categories in discrete location science. Eur J Oper Res 184:817–848
Rolland E, Schilling DA, Current JA (1996) An efficient tabu search procedure for the p-median problem. Eur J Oper Res 96:329–342
Scaparra MP, Church RL (2015) Location problems under disaster events. In: Laporte G, Nickel G, Saldanha da Gama F (eds) Location science. Springer International Publishing, Berlin
Schilling DA, Jayaraman V, Barkhi R (1993) A review of covering problems in facility location. Locat Sci 1(1):25–55
Serra D, ReVelle C (1995) Competitive location in discrete space. In: Drezner Z (ed) Facility location: a survey of applications and methods. Springer, New York
Shaw J (2014) Why "big data" is a big deal. Harvard Magazine. http://harvardmagazine.com/2014/03/why-big-data-is-a-big-deal. Last Accessed 3 Feb 2016
Smith HK, Laporte G, Harper PR (2009) Locational analysis: highlights of growth to maturity. J Oper Res Soc 60:140–148
Suzuki A, Okabe A (1995) Using Voronoi diagrams. In: Drezner Z (ed) Facility location—a survey of applications and methods. Springer, New York
Tansel BÇ (2011) Discrete center problems. In: Eiselt HA, Marianov V (eds) Foundations of location analysis. Springer, New York
Teitz M, Bart P (1968) Heuristic methods for estimating the generalized vertex median of a weighted graph. Oper Res 16:955–961
Tong D, Church RL (2012) Aggregation in continuous space coverage modeling. Int J Geogr Inf Sci 26(5):795–816
Toregas C, Swain R, Revelle C, Bergman L (1971) The location of emergency services facilities. Oper Res 19:1363–1373
US Environmental Protection Agency (US EPA) (2015) Clean Power Plan for Existing Power Plants, U.S. Environmental Protection Agency. Available at http://www2.epa.gov/cleanpowerplan/clean-power-plan-existing-power-plants#CPP-final. Last Accessed 3 Feb 2016
Verter V, Erkut E (1995) Hazardous materials logistics: an annotated bibliography. In: Haurie A, Carraro C (eds) Operations research and environmental management. Kluwer Academic, Boston, MA
von Böventer E (1963) Toward a unified theory of spatial economic structure. Pap Reg Sci 10:163–187
von Thünen JH (1826) Der isolierte Staat in Beziehung auf Landwirtschaft und Nationalökonomie. Fischer, Jena
Weber A (1909) Über den Standort der Industrien, Tübingen, J.C.B. Mohr—English translation: The Theory of the Location of Industries. Chicago University Press, Chicago, 1929
Weiszfeld E (1937) Sur le point pour lequel la somme des distances de n points donn_ees est minimum. TMJ 43:355–386
Wesolowsky GO (1993) The Weber problem: history and procedures. Locat Sci 1:5–23

Daoqin Tong is associate professor, School of Geography and Development, University of Arizona (UA). Her primary research interests are spatial optimization; location analysis and modeling; GIS; spatial data uncertainty; food security and food assistance services; transportation. She is a faculty member of the UA's Graduate Interdisciplinary Program in Statistics and has a courtesy appointment in the Department of Systems and Industrial Engineering. Dr. Tong earned the Ph.D. in geography from the Ohio State University in 2007.

Alan T. Murray is professor, Department of Geography, University of California at Santa Barbara (UCSB). His primary research interests are geographic information science; spatial optimization; health informatics; urban growth and development; land use planning; urban, regional, and natural resource planning and development; and, infrastructure and transportation systems. He previously held academic appointments at Drexel University, Arizona State University and Ohio State University. Previous faculty positions were at Drexel University, Arizona State University and Ohio State University. Dr. Murray earned a Ph.D. in geography from the University of California at Santa Barbara in 1995.

Chapter 13
Structural Decomposition and Shift-Share Analyses: Let the Parallels Converge

Michael L. Lahr and Erik Dietzenbacher

13.1 Introduction

Both structural decomposition analysis (SDA) and shift-share analysis (SSA) have been widely applied in multi- and inter-regional input-output (I-O) studies. This paper shows how elements from SSA can be integrated in SDA. This adds a novel spatial perspective to decomposing the change over time in an endogenous variable into the changes in its constituent exogenous factors.

Rose and Casler (1996) forwarded the idea that the structural decomposition of input-output (I-O) tables was not unlike shift-share analysis (SSA). (Incidentally, they likened it to growth accounting and index number analysis as well.) Intuitively, structural decomposition analysis (SDA) demonstrates strong similarities to SSA. Both examine the effects of industry shifts due to growth (or decline) and some sort of difference in industry shares. But SSA works its shares across space while SDA works its shares again across industries via technology change (fabrication effects). Interestingly, using a set of multiregional I-O tables from Spain over 6 years and without drawing parallels to either SSA or SDA, Oosterhaven and Escobedo-Cardeñoso (2011) demonstrated that regional I-O tables can be forecasted fairly well. One innovation they applied was lagging the "remainder" from the biproportional adjustment technique. This remainder looks remarkably like the "regional component" (also termed the "competitive effect") in SSA. More recently, Arto and Dietzenbacher (2014) performed what might be termed a "dynamic" SDA to examine the effect of trade changes on the growth of global CO_2 emissions. This harkens parallels to dynamic SSA (Thirlwall 1967; Barff and Knight 1988).

M.L. Lahr (✉)
EJB School of Planning and Public Policy, Rutgers University, New Brunswick, NJ, USA
e-mail: lahr@rutgers.edu

E. Dietzenbacher
Professor of Economics, University of Groningen, The Netherlands

R. Jackson, P. Schaeffer (eds.), *Regional Research Frontiers - Vol. 2*,
Advances in Spatial Science, DOI 10.1007/978-3-319-50590-9_13

Suffice it to say, SDA and SSA are related and this chapter formally combines the two disparate strands of literature. In particular, it shows how changes in regional growth differentials can be included into a structural decomposition analysis. Moreover, the present availability of a large number of I-O table panels appears to enable the detection of even more parallels between the two approaches. Between the formalization of the SSA-SDA relationship and the available I-O data, a wide range of new, policy-relevant empirical applications is possible. The method proposed in this chapter may be useful for several avenues of research.

13.2 Background

The notion of shift-share analysis (SSA) has been around since at least Creamer (1943).[1] SSA disaggregates regional change by industry (on a particular economic measure, generally employment) in order to identify the relative influence of components of that change. It is roughly predicated on the concept of regional comparative advantage. Consequently, it is used to decompose growth into (a) general national trends, (b) nationwide industry deviations from that general trend, and (c) some remainder that is identified as the "regional component" of the industry's change. Occasionally, when the region of focus is a very small geographic unit, some interim political-geography growth trend differentials—both regional and industrial—are also applied. Key points of the continued popularity of the approach are its minimal data requirements and technical simplicity. Of course, it helps that despite these potential oversimplifications, SSA tends to do a fairly good job in identifying the relative importance of factors that influence industrywise change in a region's economy (Nazara and Hewings 2004).

In addition to the parallels drawn by Rose and Casler (1996) between SDA and SSA, SDA has been used to disaggregate economic change, more generally, into its proximate change components. The larger count of economic indicators available in I-O tables, as opposed to the SSA convention of using just employment or wage data, enables more variation in the analyses. But the lower frequency and time delay of I-O table production for a fixed geographic space has made available fewer data points of analysis. At its outset, SDA controlled for the three components of change—activity level and industry mix (as in SSA), plus technology change. But as many as 14 different components of change have been analyzed simultaneously using the approach (Rose and Chen 1991). And while regional and multiregional SDAs have been performed, both have only used pairs or multiple pairs of regional or multiregional tables to perform the analysis.

[1] Victor Fuchs (1959), Edgar Dunn (1960), Lowell Ashby (1964), and Anthony Thirlwall (1967) were major players in the technique's early development, and the prominence of these authors in the field of regional science and planning certainly induced SSA's popular appeal.

In summary, while SSA and SDA have similar roots, it is clear that, as yet, no SDA analysis has examined how a regional economy differs from its nation parent over time, which is the point of SSA. The purpose of the present paper is to lay out an SDA approach for performing such an analysis.

Of course, this then begs the question of why it might be desirable to perform SSA in an SDA context. SSA reveals how well a region's industries are performing relative to the nation, or other economy that contains the salient region, along a dimension of change. Thus, the focal quantities of SSA are the "regional components," which show the distance of actual regional performance from expectations. The expected values are derived by assuming regional industries grow at the national average rates. In this vein, the actual and relative distances from expectations for industries can help reveal a region's competitive strengths and weaknesses relative to national performance. This feature can be important in developing strategic regional development initiatives. As presently formulated, SDA does not offer this sort of result.

While the above explains why SSA-type findings are of value, it does not explain why performing them in an SDA context could be worthwhile. Analysts have used the myriad of different economic indicators available in input-output accounts to good effect. But less sophisticated sets of indicators have generally been applied within SSA. Still, theoretical underpinnings of SSA, as articulated by Casler (1989), have been extended by Graham and Spence (1998) to unfold employment-based SSA's "regional component" further into partials related to input-price- and technology-related trends by using regression analysis to develop a productivity-growth decomposition within SSA. But their approach requires a panel of regional data on wage rates and output as well as employment, albeit a shallow one. And most countries do not release such panels of data by region. Meanwhile, an SDA equivalent would demand similar data for any region that is analyzed, but for only two points in time. Moreover, only data for the focal region and the nation of the analysis are needed. That is, given that I-O tables pre-exist, the data needs of SDA-based SSA should be far less demanding than that of the standard, regression-based SSA with equivalent complexity insofar as the array of applied indicators is concerned. Recall that, along with its intuitive implications, SSA's low data requirements have been key to its popularity. It would seem that SDA could minimize data requirements in certain shift-share settings and yet enable sophistication in the approach's theoretical underpinnings.

Yet another feature of SDA over conventional SSA is that it is able to measure the contribution of indirect (spillover and feedback) effects across regions. This is not to say that such effects cannot be measured by SSA. Indeed, Nazara and Hewings (2004) account for three components of change rather than the conventional two: the first is the usual national average growth component, and the second accounts for national sectoral growth differentials, and the third accounts for differential between the national sectoral growth and the weighted average sectoral growth rate for neighboring regions. Some spatial statistical approaches also have been applied to examine interregional spillover effects of shift-share components (Le Gallo and Kamarianakis 2011; Li and Haynes 2011).

A limitation of conventional shift–share that parallels the interregional aspect mentioned above is its omission of intersectoral relationships. Using an approach parallel to that of Nazara and Hewings (2004), Ramajo and Márquez (2008) and Màrquez et al. (2009) have suggested an extension that accounts for such interindustry shift-share contributions to change. That is, they define and use as the interindustry structure component for a particular referent industry the difference between the weighted-average growth rate of all other industries within the referent region and the weighted-average national growth rate of those same industries.

But like most extensions of SSA, the data demands for each additional component can be quite extensive. Moreover, as more variables are added, more degrees of freedom are consumed by the analysis, which in turn require more data observations (years and regions). This is not so much the case on SDA.

In summary then, in this paper we undertake a sort of technical reconnaissance into the potential of SDA for performing SSA. SDA can simultaneously account for interregional and interindustry effects while also accounting for nationwide and industrywide trends. In the original vein of SSA, SDA also has the potential to provide solid insight using few data points. But SDA has not examined regional trends in light of national trends in the manner that SSA does. We hope we sufficiently demonstrate how such an approach might be formulated. We conclude by pointing out the myriad types of analyses that might follow based on the SDA-based SSA that we formulate.

13.3 The Input-Output Framework

SDA works on I-O accounts. So let us start with an interregional I-O table. For our purposes, we use the accounts shown in Table 13.1, which are for a country with three regions (R, S, and T).[2]

Here, $\mathbf{Z}^{RS}$ is an $n \times n$ matrix and its element z_{ij}^{RS} gives the intermediate deliveries from industry i in region R to industry j in region S; $\mathbf{f}^{RS}$ is an n-element (column) vector with typical element f_i^{RS} indicating the final demand (including household consumption, private investments, and government expenditures) by region S for the produce of industry i in region R; $\mathbf{e}^R$ is an n-element (column) vector with typical element e_i^R indicating the exports by industry i in region R; $\mathbf{x}^R$ is an n-element (column) vector with typical element x_i^R indicating the output of (or total amount of production by) industry i in region R; $(\mathbf{v}^R)'$ is an n-element (row) vector with typical element v_j^R indicating the value added generated in industry j in region R; and $(\mathbf{m}^R)'$ is an n-element (row) vector with typical element m_j^R indicating the imports of industry j in region R. In addition, information from satellite accounts is often available. For example, the use of labor

[2] There is no reason this could not be four or even more regions. But three regions typically takes any analysis beyond a trivial case.

Table 13.1 An interregional input-output table

	Intermediate deliveries			Final demands				Total
	R	*S*	*T*	*R*	*S*	*T*	*Exp*	
R	$\mathbf{Z}^{RR}$	$\mathbf{Z}^{RS}$	$\mathbf{Z}^{RT}$	$\mathbf{f}^{RR}$	$\mathbf{f}^{RS}$	$\mathbf{f}^{RT}$	$\mathbf{e}^{R}$	$\mathbf{x}^{R}$
S	$\mathbf{Z}^{SR}$	$\mathbf{Z}^{SS}$	$\mathbf{Z}^{ST}$	$\mathbf{f}^{SR}$	$\mathbf{f}^{SS}$	$\mathbf{f}^{ST}$	$\mathbf{e}^{S}$	$\mathbf{x}^{S}$
T	$\mathbf{Z}^{TR}$	$\mathbf{Z}^{TS}$	$\mathbf{Z}^{TT}$	$\mathbf{f}^{TR}$	$\mathbf{f}^{TS}$	$\mathbf{f}^{TT}$	$\mathbf{e}^{T}$	$\mathbf{x}^{T}$
VA	$(\mathbf{v}^{R})'$	$(\mathbf{v}^{S})'$	$(\mathbf{v}^{T})'$					
Imp	$(\mathbf{m}^{R})'$	$(\mathbf{m}^{S})'$	$(\mathbf{m}^{T})'$					
Total	$(\mathbf{x}^{R})'$	$(\mathbf{x}^{S})'$	$(\mathbf{x}^{T})'$					
Labor	$(\mathbf{c}^{R})'$	$(\mathbf{c}^{S})'$	$(\mathbf{c}^{T})'$					

(say in hours worked). In that case, $(\mathbf{c}^R)'$ is an n-element (row) vector with typical element c_j^R indicating the use of labor in industry j in region R.

Following the recent discussion on global value chains and trade in value added (or trade in emissions), one of the questions at the regional level is: "Who works (or emits) for whom?" (Serrano and Dietzenbacher 2010; Koopman et al. 2014). That is, how much labor is (directly and indirectly) necessary in region R for the final demand bundle of region T? Using an interregional I-O model, the answer is given by the element π_{RT} of the 3×3 matrix $\mathbf{\Pi}$, which is defined as

$$\mathbf{\Pi} = \begin{bmatrix} \pi_{RR} & \pi_{RS} & \pi_{RT} \\ \pi_{SR} & \pi_{SS} & \pi_{ST} \\ \pi_{TR} & \pi_{TS} & \pi_{TT} \end{bmatrix} = \mathbf{HF} = \begin{bmatrix} (\mathbf{h}^{RR})' & (\mathbf{h}^{RS})' & (\mathbf{h}^{RT})' \\ (\mathbf{h}^{SR})' & (\mathbf{h}^{SS})' & (\mathbf{h}^{ST})' \\ (\mathbf{h}^{TR})' & (\mathbf{h}^{TS})' & (\mathbf{h}^{TT})' \end{bmatrix} \begin{bmatrix} \mathbf{f}^{RR} & \mathbf{f}^{RS} & \mathbf{f}^{RT} \\ \mathbf{f}^{SR} & \mathbf{f}^{SS} & \mathbf{f}^{ST} \\ \mathbf{f}^{TR} & \mathbf{f}^{TS} & \mathbf{f}^{TT} \end{bmatrix} \quad (13.1)$$

Note that $\mathbf{H}$ is a 3×3n matrix with labor multipliers and $\mathbf{F}$ is a $3n \times 3$ matrix with regional final demands. The elements of the matrix $\mathbf{H}$ are obtained as follows

$$\mathbf{H} = \begin{bmatrix} (\mathbf{d}^R)' & 0 & 0 \\ 0 & (\mathbf{d}^S)' & 0 \\ 0 & 0 & (\mathbf{d}^T)' \end{bmatrix} \begin{bmatrix} \mathbf{L}^{RR} & \mathbf{L}^{RS} & \mathbf{L}^{RT} \\ \mathbf{L}^{SR} & \mathbf{L}^{SS} & \mathbf{L}^{ST} \\ \mathbf{L}^{TR} & \mathbf{L}^{TS} & \mathbf{L}^{TT} \end{bmatrix} \quad (13.2)$$

The vector $(\boldsymbol{d}^R)'$ contains the direct labor input coefficients and is defined as $(\boldsymbol{d}^R)' = (\boldsymbol{c}^R)'\left(\widehat{\boldsymbol{x}}^R\right)^{-1}$ or $d_j^R = c_j^R/x_j^R$. The second matrix on the right-hand side of (13.2) gives the partitioned Leontief inverse, i.e., $\mathbf{L} = (\mathbf{I} - \mathbf{A})^{-1}$. $\mathbf{A}$ is the $3n \times 3n$ matrix with input coefficients, which in partitioned form is given by

$$\mathbf{A} = \begin{bmatrix} \mathbf{A}^{RR} & \mathbf{A}^{RS} & \mathbf{A}^{RT} \\ \mathbf{A}^{SR} & \mathbf{A}^{SS} & \mathbf{A}^{ST} \\ \mathbf{A}^{TR} & \mathbf{A}^{TS} & \mathbf{A}^{TT} \end{bmatrix}$$

where the input coefficients are defined as $\mathbf{A}^{RS} = \mathbf{Z}^{RS}\left(\widehat{x}^S\right)^{-1}$ or $a_{ij}^{RS} = z_{ij}^{RS}/x_j^S$.

Note that the jth element of the vector $(\mathbf{h}^{RS})'$, i.e. h_j^{RS}, gives the total amount of labor used in region R that is necessary for one dollar of final demand for product j from region S. The scalar $(\mathbf{h}^{RS})'\mathbf{f}^{ST}$ then gives the total amount of labor used in region R that is embodied in the final demand of region T for products from region S. The element $\pi_{RT} = (\mathbf{h}^{RR})'\mathbf{f}^{RT} + (\mathbf{h}^{RS})'\mathbf{f}^{ST} + (\mathbf{h}^{RT})'\mathbf{f}^{TT}$ then gives the total amount of labor used in region R that is necessary for all final demands by region T.

Observe that our calculations take indirect linkages and interregional feedback effects into account, as far as they are national. For example, final demands in T require inputs from S that require inputs from R. Indirectly, final demands in T require production and, therefore, labor use in R. What is not included in our analysis are feedback effects that run through foreign countries. Exactly the same example can be used with region S replaced by a foreign country.

13.4 Adding Shift-Share Elements

The next step is to introduce shift-share elements into the equation (13.1). To this end write the first n rows of the $3n \times 3$ matrix $\mathbf{F}$ as follows.

$$\begin{aligned}\begin{bmatrix}\mathbf{f}^{RR} & \mathbf{f}^{RS} & \mathbf{f}^{RT}\end{bmatrix} &= \left\{\mathbf{T}^{R} \otimes \mathbf{\Sigma} \otimes \mathbf{S}\right\} \mathbf{R} f^{NAT} \\ &= \left\{\begin{bmatrix}\mathbf{t}^{RR} & \mathbf{t}^{RS} & \mathbf{t}^{RT}\end{bmatrix} \otimes \begin{bmatrix}\boldsymbol{\sigma}^{R} & \boldsymbol{\sigma}^{S} & \boldsymbol{\sigma}^{T}\end{bmatrix} \otimes \begin{bmatrix}\mathbf{s}^{NAT} & \mathbf{s}^{NAT} & \mathbf{s}^{NAT}\end{bmatrix}\right\} \begin{bmatrix} r^{R} & 0 & 0 \\ 0 & r^{S} & 0 \\ 0 & 0 & r^{T}\end{bmatrix} f^{NAT}\end{aligned} \tag{13.3}$$

Going through the equation from right to left, the scalar f^{NAT} indicates the total amount of national final demand. That is, $f^{NAT} = \sum_{I=R,S,T} \sum_{J=R,S,T} \sum_{i=1}^{n} f_i^{IJ}$, the sum of all elements in the matrix $\mathbf{F}$. The diagonal elements of the 3×3 matrix $\mathbf{R}$ give the share of the regional total final demand in the national final demand. For example, $r^R = \sum_{I=R,S,T} \sum_{i=1}^{n} f_i^{IR} / f^{NAT}$ and observe that $r^R + r^S + r^T = 1$. The $n \times 3$ matrix $\mathbf{S}$ consists of three times the vector s^{NAT} with the national final demand mix. Note that the final demand mix does not distinguish between the region of origin; it matters, for example, what households consume of (domestically produced) good i, not where the consumer goods come from. That is, $s_i^{NAT} = \sum_{I=R,S,T} \sum_{J=R,S,T} f_i^{IJ} / f^{NAT}$ and note that the shares add to one (i.e. $\sum_{i=1}^{n} s_i^{NAT} = 1$).

The regional final demand shares (for example for region R) are obtained as $\sum_{I=R,S,T} f_i^{IR} / \sum_{I=R,S,T} \sum_{i=1}^{n} f_i^{IR}$. The discrepancies between the regional and the national shares of final demands are given by the elements of the $n \times 3$ matrix $\mathbf{\Sigma}$. That is, $\sigma_i^R = \sum_{I=R,S,T} f_i^{IR} / \left(s_i^{NAT} \sum_{I=R,S,T} \sum_{i=1}^{n} f_i^{IR}\right)$. The operator $\otimes$ stands for the Hadamard product of elementwise multiplication. The element in row i and column R of the matrix $\mathbf{\Sigma} \otimes \mathbf{S}$ thus equals $\sigma_i^R s_i^{NAT} = \sum_{I=R,S,T} f_i^{IR} / \sum_{I=R,S,T} \sum_{i=1}^{n} f_i^{IR}$, the share of good i in the total final demands of region R. Finally, the elements of the

$n \times 3$ matrix $\mathbf{T}^R$ give the trade coefficients, indicating the share of a region's final demand for product i that originates from region R. For example, the ith element of the n-element vector $\mathbf{t}^{RS}$ yields $t_i^{RS} = f_i^{RS} / \sum_{I=R,S,T} f_i^{IS}$.

The expression for the full $3n \times n$ matrix $\mathbf{F}$ then becomes

$$\mathbf{F} = \begin{bmatrix} \mathbf{f}^{RR} & \mathbf{f}^{RS} & \mathbf{f}^{RT} \\ \mathbf{f}^{SR} & \mathbf{f}^{SS} & \mathbf{f}^{ST} \\ \mathbf{f}^{TR} & \mathbf{f}^{TS} & \mathbf{f}^{TT} \end{bmatrix} = \left\{ \begin{bmatrix} \mathbf{T}^R \\ \mathbf{T}^S \\ \mathbf{T}^T \end{bmatrix} \otimes \begin{bmatrix} \boldsymbol{\Sigma} \\ \boldsymbol{\Sigma} \\ \boldsymbol{\Sigma} \end{bmatrix} \otimes \begin{bmatrix} \mathbf{S} \\ \mathbf{S} \\ \mathbf{S} \end{bmatrix} \right\} \mathbf{R} f^{NAT} = \left\{ \mathbf{T} \otimes \bar{\boldsymbol{\Sigma}} \otimes \bar{\mathbf{S}} \right\} \mathbf{R} f^{NAT} \tag{13.4}$$

A similar distinction can be made for the $3 \times 3n$ matrix $\mathbf{H}$ with labor multipliers. That is,

$$\mathbf{H} = \begin{bmatrix} (\mathbf{h}^{RR})' & (\mathbf{h}^{RS})' & (\mathbf{h}^{RT})' \\ (\mathbf{h}^{SR})' & (\mathbf{h}^{SS})' & (\mathbf{h}^{ST})' \\ (\mathbf{h}^{TR})' & (\mathbf{h}^{TS})' & (\mathbf{h}^{TT})' \end{bmatrix} =$$

$$\begin{bmatrix} (\boldsymbol{\gamma}^{RR})' & (\boldsymbol{\gamma}^{RS})' & (\boldsymbol{\gamma}^{RT})' \\ (\boldsymbol{\gamma}^{SR})' & (\boldsymbol{\gamma}^{SS})' & (\boldsymbol{\gamma}^{ST})' \\ (\boldsymbol{\gamma}^{TR})' & (\boldsymbol{\gamma}^{TS})' & (\boldsymbol{\gamma}^{TT})' \end{bmatrix} \otimes \begin{bmatrix} (\mathbf{h}^{NAT,R})' & (\mathbf{h}^{NAT,S})' & (\mathbf{h}^{NAT,T})' \\ (\mathbf{h}^{NAT,R})' & (\mathbf{h}^{NAT,S})' & (\mathbf{h}^{NAT,T})' \\ (\mathbf{h}^{NAT,R})' & (\mathbf{h}^{NAT,S})' & (\mathbf{h}^{NAT,T})' \end{bmatrix} = \boldsymbol{\Gamma} \otimes \mathbf{H}^{NAT} \tag{13.5}$$

The elements of the matrix $\mathbf{H}^{NAT}$ give the national labor multipliers. For example, $h_j^{NAT,S}$ gives the total amount of labor that is used nationally for the final demand of one dollar of good j produced by region S. This amount equals the sum of the labor use in each region, i.e., $h_j^{NAT,S} = \sum_{I=R,S,T} h_j^{IS}$. The elements of the matrix $\boldsymbol{\Gamma}$ then give the shares of the national labor use that take place in each of the regions. That is, $\gamma_j^{RS} = h_j^{RS} / h_j^{NAT,S}$ and note that the shares add to one $\left(\sum_{I=R,S,T} \gamma_j^{IJ} = 1, \text{ for } J = R, S, T \text{ and } j = 1, \ldots, n \right)$.

Combining Equations (13.1), (13.4), and (13.5), yields

$$\boldsymbol{\Pi} = \left[\boldsymbol{\Gamma} \otimes \mathbf{H}^{NAT} \right] \left[\mathbf{T} \otimes \bar{\boldsymbol{\Sigma}} \otimes \bar{\mathbf{S}} \right] \mathbf{R} f^{NAT} \tag{13.6}$$

13.5 The Structural Decomposition

Structural decomposition analysis splits the growth in some variable (here, the matrix $\boldsymbol{\Pi}$) into the contributions of the growth in its components (here, the matrix $\boldsymbol{\Gamma}$ is one of these components). That is, one decomposes $\Delta \boldsymbol{\Pi} = \boldsymbol{\Pi}_1 - \boldsymbol{\Pi}_0$, the change in $\boldsymbol{\Pi}$ between two points in time, indicated by 0 and 1. Its element $\Delta \pi_{RS}$ gives the change in labor usage in region R that is embodied in the final demands in region S.

One possible decomposition is

$$\begin{aligned}
\Delta\boldsymbol{\Pi} &= \boldsymbol{\Pi}_1 - \boldsymbol{\Pi}_0 \\
&= \left[\boldsymbol{\Gamma}_1 \otimes \mathbf{H}_1^{NAT}\right]\left[\mathbf{T}_1 \otimes \bar{\boldsymbol{\Sigma}}_1 \otimes \bar{\mathbf{S}}_1\right]\mathbf{R}_1 f_1^{NAT} - \left[\boldsymbol{\Gamma}_0 \otimes \mathbf{H}_0^{NAT}\right]\left[\mathbf{T}_0 \otimes \bar{\boldsymbol{\Sigma}}_0 \otimes \bar{\mathbf{S}}_0\right]\mathbf{R}_0 f_0^{NAT} \\
&= \left[\boldsymbol{\Gamma}_1 \otimes \mathbf{H}_1^{NAT}\right]\left[\mathbf{T}_1 \otimes \bar{\boldsymbol{\Sigma}}_1 \otimes \bar{\mathbf{S}}_1\right]\mathbf{R}_1 f_1^{NAT} - \left[\boldsymbol{\Gamma}_0 \otimes \mathbf{H}_1^{NAT}\right]\left[\mathbf{T}_1 \otimes \bar{\boldsymbol{\Sigma}}_1 \otimes \bar{\mathbf{S}}_1\right]\mathbf{R}_1 f_1^{NAT} \\
&+ \left[\boldsymbol{\Gamma}_0 \otimes \mathbf{H}_1^{NAT}\right]\left[\mathbf{T}_1 \otimes \bar{\boldsymbol{\Sigma}}_1 \otimes \bar{\mathbf{S}}_1\right]\mathbf{R}_1 f_1^{NAT} - \left[\boldsymbol{\Gamma}_0 \otimes \mathbf{H}_0^{NAT}\right]\left[\mathbf{T}_1 \otimes \bar{\boldsymbol{\Sigma}}_1 \otimes \bar{\mathbf{S}}_1\right]\mathbf{R}_1 f_1^{NAT} \\
&+ \left[\boldsymbol{\Gamma}_0 \otimes \mathbf{H}_0^{NAT}\right]\left[\mathbf{T}_1 \otimes \bar{\boldsymbol{\Sigma}}_1 \otimes \bar{\mathbf{S}}_1\right]\mathbf{R}_1 f_1^{NAT} - \left[\boldsymbol{\Gamma}_0 \otimes \mathbf{H}_0^{NAT}\right]\left[\mathbf{T}_0 \otimes \bar{\boldsymbol{\Sigma}}_1 \otimes \bar{\mathbf{S}}_1\right]\mathbf{R}_1 f_1^{NAT} \\
&+ \left[\boldsymbol{\Gamma}_0 \otimes \mathbf{H}_0^{NAT}\right]\left[\mathbf{T}_0 \otimes \bar{\boldsymbol{\Sigma}}_1 \otimes \bar{\mathbf{S}}_1\right]\mathbf{R}_1 f_1^{NAT} - \left[\boldsymbol{\Gamma}_0 \otimes \mathbf{H}_0^{NAT}\right]\left[\mathbf{T}_0 \otimes \bar{\boldsymbol{\Sigma}}_0 \otimes \bar{\mathbf{S}}_1\right]\mathbf{R}_1 f_1^{NAT} \\
&+ \left[\boldsymbol{\Gamma}_0 \otimes \mathbf{H}_0^{NAT}\right]\left[\mathbf{T}_0 \otimes \bar{\boldsymbol{\Sigma}}_0 \otimes \bar{\mathbf{S}}_1\right]\mathbf{R}_1 f_1^{NAT} - \left[\boldsymbol{\Gamma}_0 \otimes \mathbf{H}_0^{NAT}\right]\left[\mathbf{T}_0 \otimes \bar{\boldsymbol{\Sigma}}_0 \otimes \bar{\mathbf{S}}_0\right]\mathbf{R}_1 f_1^{NAT} \\
&+ \left[\boldsymbol{\Gamma}_0 \otimes \mathbf{H}_0^{NAT}\right]\left[\mathbf{T}_0 \otimes \bar{\boldsymbol{\Sigma}}_0 \otimes \bar{\mathbf{S}}_0\right]\mathbf{R}_1 f_1^{NAT} - \left[\boldsymbol{\Gamma}_0 \otimes \mathbf{H}_0^{NAT}\right]\left[\mathbf{T}_0 \otimes \bar{\boldsymbol{\Sigma}}_0 \otimes \bar{\mathbf{S}}_0\right]\mathbf{R}_0 f_1^{NAT} \\
&+ \left[\boldsymbol{\Gamma}_0 \otimes \mathbf{H}_0^{NAT}\right]\left[\mathbf{T}_0 \otimes \bar{\boldsymbol{\Sigma}}_0 \otimes \bar{\mathbf{S}}_0\right]\mathbf{R}_0 f_1^{NAT} - \left[\boldsymbol{\Gamma}_0 \otimes \mathbf{H}_0^{NAT}\right]\left[\mathbf{T}_0 \otimes \bar{\boldsymbol{\Sigma}}_0 \otimes \bar{\mathbf{S}}_0\right]\mathbf{R}_0 f_0^{NAT}
\end{aligned}$$

Or, in more compact form:

$$\begin{aligned}
\Delta\boldsymbol{\Pi} &= \boldsymbol{\Pi}_1 - \boldsymbol{\Pi}_0 \\
&= \left[(\Delta\boldsymbol{\Gamma}) \otimes \mathbf{H}_1^{NAT}\right]\left[\mathbf{T}_1 \otimes \bar{\boldsymbol{\Sigma}}_1 \otimes \bar{\mathbf{S}}_1\right]\mathbf{R}_1 f_1^{NAT} \\
&+ \left[\boldsymbol{\Gamma}_0 \otimes (\Delta\mathbf{H}^{NAT})\right]\left[\mathbf{T}_1 \otimes \bar{\boldsymbol{\Sigma}}_1 \otimes \bar{\mathbf{S}}_1\right]\mathbf{R}_1 f_1^{NAT} \\
&+ \left[\boldsymbol{\Gamma}_0 \otimes \mathbf{H}_0^{NAT}\right]\left[(\Delta\mathbf{T}) \otimes \bar{\boldsymbol{\Sigma}}_1 \otimes \bar{\mathbf{S}}_1\right]\mathbf{R}_1 f_1^{NAT} \\
&+ \left[\boldsymbol{\Gamma}_0 \otimes \mathbf{H}_0^{NAT}\right]\left[\mathbf{T}_0 \otimes (\Delta\bar{\boldsymbol{\Sigma}}) \otimes \bar{\mathbf{S}}_1\right]\mathbf{R}_1 f_1^{NAT} \\
&+ \left[\boldsymbol{\Gamma}_0 \otimes \mathbf{H}_0^{NAT}\right]\left[\mathbf{T}_0 \otimes \bar{\boldsymbol{\Sigma}}_0 \otimes \left(\Delta\bar{\mathbf{S}}\right)\right]\mathbf{R}_1 f_1^{NAT} \\
&+ \left[\boldsymbol{\Gamma}_0 \otimes \mathbf{H}_0^{NAT}\right]\left[\mathbf{T}_0 \otimes \bar{\boldsymbol{\Sigma}}_0 \otimes \bar{\mathbf{S}}_0\right](\Delta\mathbf{R}) f_1^{NAT} \\
&+ \left[\boldsymbol{\Gamma}_0 \otimes \mathbf{H}_0^{NAT}\right]\left[\mathbf{T}_0 \otimes \bar{\boldsymbol{\Sigma}}_0 \otimes \bar{\mathbf{S}}_0\right]\mathbf{R}_0 \left(\Delta f^{NAT}\right)
\end{aligned}$$

The change variable in parentheses (recognized by the Δ symbol) identifies how to interpret each of the seven terms or components.

(1) The first shows the contribution from the change in regional shares in national labor use ($\boldsymbol{\Gamma}$),
(2) the second reveals the effects of change in the national labor multipliers (**H**),
(3) the third identifies the effects due to changes in the supplying region's share of the regional final demand (**T**),
(4) the fourth reveals the effects due to changes in the differences between regional and national final demand mixes $\left(\bar{\boldsymbol{\Sigma}}\right)$,
(5) the fifth reports the effects due to changes in the national final demand mix $\left(\bar{\mathbf{S}}\right)$,
(6) the sixth shows the effects due to changes in the shares of regional total final demand in the national final demand (**R**), and
(7) the seventh reports the effects due to changes in total national final demand (f^{NAT}).

Clearly, this is just one decomposition of many that we could have expressed to analyze the change. Fortunately, Dietzenbacher and Los (1997) have noted that a simple average of the above decomposition and its polar opposite very reasonably represents the average of all possible decompositions. Moreover, while the above is an additive decomposition, multiplicative decompositions are also conceivable (e.g., Dietzenbacher et al. 2001; Dietzenbacher et al. 2004).

A key point to be made here is that the sort of analyses we suggest here would be hampered by roughly estimated final demand accounts. If performed using a multiplicative approach, the analyses could be even more compromised if value added components were also only roughly estimated. In this vein, analyses of national I-O tables, for which extraordinary care has been taken to formulate the input-output accounts, within a broader framework of nations—perhaps those sharing a trade agreement (e.g., EU, BRICs, NAFTA)—could be ideal targets for the sort of SSA-SDA analyses that we are suggesting. In such instances, regions R and S in the framework described above would be representative of countries in the trade group to be analyzed (effectively the "nation" in our framework), and T reflecting relationships with countries outside of it. Still, the basic form of the equations would remain the same, but the number of regions and, hence, partitions composing the matrices would generally be greater and require specific adaptations.

13.6 Conclusions

Structural decomposition analysis (SDA) and shift-share analysis (SSA) bear some similarities. In essence, both disaggregate economic change into its proximate change components. Both examine the effects of industry shifts due to economic change as well as differences in industry shares. But SSA works its shares across space while, to date at least, SDA has worked its shares across industries via technology change. Literature on the two approaches suggests that the data demands of SSA tend to make it less tractable for incorporating more than just one or two control variables. On the other hand, as long as two or more input-output tables exist for an economy, SDA is quite adept in analyses for which controlling for many variables is desired. Until the present paper, however, no SDA analysis has examined how a regional economy differs from its nation parent over time, which is the point of SSA.

In this paper we derive an SDA-based SSA, and we use a general, but simple, three-region multiregional I-O model to do so. That is, the approach we outline presumes access to two periods of national I-O accounts models that are already spatially decomposed into at least three regions. In this vein, our approach is quite general. We somewhat arbitrarily focused our analysis on the change in regional labor use since it parallels the main outcome in traditional SSA. We decompose regional change in labor usage into seven components. Results of a practical analysis using the decomposition, which we presented, would demonstrate the proximate contribution of each of the seven components to the change in labor use. Traditional SSA roughly accounts for just the second and seventh components. The first, third, and sixth components examine sources of interregional shifts over time. The fourth component identifies effects that emanate from regional differences in final demand expenditures. In SDA, by adding explanatory variables the effects of pre-existing ones are often diluted, similar to the case of exploratory regression analysis. The implication of this is that through the use of more explanatory variables in SDA SSA, both national mix and trend effects, may well turn out to be less crucial to regional economic growth than traditional SSA tends to have us surmise.

There is no reason that the focus of an SDA SSA has to be on the change in jobs count. We could just as easily have formulated an SDA SSA that focuses some other economic variable availed via an I-O table—like, say, the change in regional productivity, labor's share of GDP, or household consumption. Also the additive decomposition that we articulated could be expressed with more components—for example, the household consumption component of final demand could be broken out into different household classes. Further, we could have formulated a multiplicative decomposition to examine the same focal variable. A typical feature of a multiplicative approach is that it expresses the growth *rate* in some variable as the multiplication of the growth rates of its constituent parts. A pitfall of the SDA approach to SSA, of course, is its requirement of component data at the regional level.

At the outset of this paper, we mention in passing that several panels of I-O tables are now available to examine the veracity and value of new approaches like SDA SSA. In addition to the series of multiregional I-O tables for Japan, China, and Spain, some global panels of I-O tables have been constructed and analyzed in recent years. Several are briefly introduced in an issue of *Economic Systems Research* introduced by Tukker and Dietzenbacher (2013)—EORA, EXIOPOL, GTAP, IDE-JETRO, and WIOD. Of course, the decomposition proposed herein implicitly assumes spatially constant technology. Due to this assumption, it makes sense to use as a proxy for the "nation" meta-region in our decomposition countries that have similar labor skills and hence, technological capability. Thus, if one were limited to using these data sets, it would make sense to center an SDA SSA analysis on some subset of any one of these databases—i.e., all ASEAN nations, the set of countries in the EU, or perhaps the union of European nations and the G7 nations, which adds the U.S., Canada, Russia, and Japan.

In a similar vein, our framework only identifies two periods. Clearly, more periods could be analyzed using the framework we have outlined. They need only be studied serially, following the example of traditional dynamic SSA (Thirlwall 1967; Barff and Knight 1988). Indeed, many authors have already applied SDA in such a fashion. Perhaps the best example is Arto and Dietzenbacher (2014), who performed what might be termed a "dynamic" SDA to examine the effect of trade changes on the growth of global CO_2 emissions. Indeed, armed with this SSA-SDA approach and a panel of interregional I-O tables, it might be interesting to revisit the aims of Oosterhaven and Escobedo-Cardeñoso (2011) who demonstrated that regional I-O tables can be forecasted fairly well, using a set of multiregional I-O tables over a number of years.

As we see it, the set of possible applications of the general technique that we outline in this paper is potentially diverse. Indeed, the possibilities are presently bounded only by the sets of multiregional I-O tables available and the imagination of researchers. As more tables become available, the set of potential applications will diversify with them. Be imaginative and fruitful with our germ of an approach!

References

Arto I, Dietzenbacher E (2014) Drivers of the growth in global greenhouse gas emissions. Environ Sci Technol 48:5388–5394

Ashby LD (1964) The geographical redistribution of employment: an examination of the elements of change. Survey of current business, October, pp 13–20

Barff RA, Knight PL III (1988) Dynamic shift-share analysis. Growth Chang 19(2):1–10

Casler SD (1989) A theoretical context for shift and share analysis. Reg Stud 23:43–48

Creamer D (1943) Shifts of manufacturing industries, Chapter 4. In: McLaughlin GE (eds) Industrial location and national resources, December 1942. National Resources Planning Board: Washington, DC, pp 85–104. Available online in January 2016 at https://archive.org/stream/industriallocatnatre1942rich#page/85/mode/2up

Dietzenbacher E, Los B (1997) Structural decomposition techniques: sense and sensitivity. Econ Syst Res 10:307–323

Dietzenbacher E, Hoen AR, Los B (2001) Labor productivity in Western Europe (1975–1985): an intercountry, interindustry analysis. J Reg Sci 40:425–452

Dietzenbacher E, Lahr ML, Los B (2004) The decline in labor compensation's share of GDP: a structural decomposition analysis for the US, 1982–1997. In: Dietzenbacher E, Lahr ML (eds) Wassily Leontief and input-output economics. Cambridge University Press, Cambridge, pp 188–212

Dunn ES Jr (1960) A statistical and analytical technique for regional analysis. Pap Reg Sci Assoc 6:97–112

Fuchs VR (1959) Changes in U. S. manufacturing since 1929. J Reg Sci 1:1–17

Graham DJ, Spence N (1998) A productivity growth interpretation of the labour demand shift-share model. Reg Stud 32:515–525

Koopman R, Wang Z, Wei S-J (2014) Tracing value-added and double counting in gross exports. Am Econ Rev 104:459–494

Le Gallo J, Kamarianakis Y (2011) The evolution of regional productivity disparities in the European union from 1975 to 2002: a combination of shift-share and spatial econometrics. Reg Stud 45:123–139

Li H, Haynes KE (2011) Economic structure and regional disparity in China: beyond the Kuznets transition. Int Reg Sci Rev 34:157–190

Márquez MA, Ramajo J, Hewings GJD (2009) Incorporating sectoral structure into shift–share analysis. Growth Chang 40:594–618

Nazara S, Hewings GJD (2004) Spatial structure and taxonomy of decomposition in shift-share analysis. Growth Chang 35:476–490

Oosterhaven J, Escobedo-Cardeñoso F (2011) A new method to estimate input-output tables by means of structural lags, tested on Spanish regions. Pap Reg Sci 90:829–844

Ramajo J, Márquez MA (2008) Componentes Espaciales en el Modelo Shift-Share. Una Aplicación al Caso de las Regiones Peninsulares Españolas. Estadística Española 50:41–65

Rose AZ, Casler SD (1996) Input–output structural decomposition analysis: a critical appraisal. Econ Syst Res 8:33–62

Rose AZ, Chen C-Y (1991) Sources of change in energy use in the US economy, 1972–1982: a structural decomposition analysis. Resour Energ 13:1–21

Serrano M, Dietzenbacher E (2010) Responsibility and trade emission balances: an evaluation of approaches. Ecol Econ 69:2224–2232

Thirlwall AP (1967) A measure of the 'proper distribution of industry'. Oxf Econ Pap 19:46–58

Tukker A, Dietzenbacher E (2013) Global multiregional framework: an introduction and outlook. Econ Syst Res 25:1–19

Michael L. Lahr is a Research Professor at the Edward J. Bloustein School for Planning and Public Policy and the Director of Rutgers Economic Advisory Service (R/ECON™) at Rutgers, the State University of New Jersey. He obtained his Ph.D. in 1992 from the Regional Science Department at the University of Pennsylvania. Lahr's main interests are in new economic modeling methods, particularly those pertaining to interindustry analysis and regional economic development. He is Vice President of the International Input-Output Association, a past President of the Southern Regional Science Association, and a past Chairman of The North American Regional Science Council. Lahr has been has also been on the Board of Associate Editors for four journals. He has held positions with USDA's Economic Research Service, Bryn Mawr College, the Regional Science Research Institute, and Battelle Memorial Institute.

Erik Dietzenbacher is Professor in Interindustry Economics at the University of Groningen. His research interests center around input-output applications. He is also Affiliate Research Professor at the Regional Economics Applications Laboratory of the University of Illinois at Urbana-Champaign and was Guest Professor at the University of the Chinese Academy of Sciences in Beijing. He was the project coordinator of "World Input-Output Database: Construction and Applications", a large-scale collaborative research project that was funded by the EU. He is currently President of the International Input-Output Association.

Chapter 14
A Synthesis of Spatial Models for Multivariate Count Responses

Yiyi Wang, Kara Kockelman, and Amir Jamali

14.1 Introduction

Spatial data are central to regional science applications and many other disciplines. Location attributes for each observation reveal where events occur or other information (pollution levels [Goodkind et al. 2014], land values [Du and Mulley 2012], and crimes [Levine 2009]) exists, often at fine spatial resolution. There are three types of spatial data: geostatistical data, areal or lattice data, and point data.

- Geostatistical data are innate to the landscape or environment (such as soil mineral levels, rainfall, and pollutant levels) and span continuously over space. Given their continuous nature, such variables need to be collected by sampling at different locations (Deutsch and Journel 1997). The goal of geostatistical analysis is to predict values at unknown locations using sampled/observed values. For this purpose, kriging is often used: it spatially interpolates unknown values using nearby observations (Krige 1951).
- Areal or lattice data are observed at certain geographic units (e.g., vehicle registration data across counties and land use changes across parcels). These geographic units divide up the study area into small tiles (tessellations) like census tracts. The goal of areal data analysis is usually to detect and explain spatial patterns, as opposed to predicting unknown values, since there is typically

Y. Wang (✉) • A. Jamali
Civil Engineering Department, Montana State University, Bozeman, MT, 59717, USA
e-mail: yiyi.wang@montana.edu

K. Kockelman
Department of Civil, Architectural, and Environmental Engineering, University of Texas at Austin, Austin, TX, 78712, USA

R. Jackson, P. Schaeffer (eds.), *Regional Research Frontiers - Vol. 2*,
Advances in Spatial Science, DOI 10.1007/978-3-319-50590-9_14

no gap in the area of interest. Areal data are usually analyzed by spatial econometric methods (LeSage and Pace 2009; Anselin 2010).

- Point data note the location of many specific occurrences like crashes or species sightings over a period of time. "Hot Spot" analysis is often used to identify clustering patterns of these points (Lu 1998). An array of metrics can be used to portray the magnitude of clusters, like Moran's I, Geary C's Location Quotient, and the nearest neighbor index (NNI). Point data can be converted to areal data by tessellating the study area into zones and aggregating the points at each zone.

14.1.1 Motivations for Spatial Models

This chapter focuses on spatial models for analyzing areal data, in a multivariate count format (like vehicle ownership across census tracts, number of crimes across zones, and patent applications across counties). Spatial models are attractive for two reasons that are rooted in geospatial theory: spatial dependence and spatial heterogeneity.

Spatial dependence (autocorrelation) describes correlations across the same variable observed at different locations (zones). A positive spatial autocorrelation implies clustering, so values observed at nearby locations are more similar than values observed at distant locations. A negative spatial autocorrelation portrays a dispersed pattern, in which a value at one location tends to be surrounded by dissimilar values (for the same variable). Spatial heterogeneity is defined as uneven distribution of a variable over space (Vinatier et al. 2011). Spatial heterogeneity arises due to structural instability: each zone/location subscribes to a different process to generate the variable of interest. Spatial heterogeneity can be expressed in an analytical model either as heteroscedastic (non-constant) error variance or regression coefficients that vary across observational units (Anselin 2001). Simoes and Natario (2016) provide a summary of statistical tests to detect spatial heterogeneity.

Conventional econometric models do not work for data that exhibit spatial dependence and/or heterogeneity. These models assume that the error terms are distributed normally (Gaussian), retain the same variance (which violates spatial heterogeneity), and are independent across observations (which conflicts with spatial dependence). To address spatial dependence, models that recognize correlations (such as spatial autoregressive models) have been rather effective in various contexts, like crash and crime prediction (Levine et al. 1995a, b; Miaou et al. 2003; Wang and Kockelman 2013), home prices (Case et al. 2003), land use dynamics (Chakir and Parent 2009; Wang and Kockelman 2009; Wang et al. 2014), and technology innovations (LeSage and Pace 2009). To tackle spatial heterogeneity, geographically weighted regression (GWR) is regularly used through locally estimating coefficients, rendering a contextual layer of coefficient estimates that vary over space. Examples of GWR span many fields, such as ecology, wealth and epidemics (Platt 2004, Ognev-Himmelberger et al. 2009, Atkinson et al. 2003, and Nagaya et al. 2010), traffic count and crash count predictions across road networks (Zhao and Park 2004; Hadayeghi et al. 2009), and land use (Páez 2006; Wang et al. 2011).

14.1.2 Geo-Referenced Multivariate Count Data

One form of areal/lattice data is geo-referenced count data, data that take on non-negative integer values and record the number of items or events in zones of interest (e.g., number of vehicles owned across zones, crime counts across block groups, and crash counts by intersection and/or road segment). For a generic count variable, multiple levels of that variable are often observed: for example, number of vehicles by fuel economy category or number of crimes by type. These are multivariate count data. It is often of interest to gauge correlations among the different levels of a count (response) variable in addition to incorporating spatial dependence and/or heterogeneity across locations. The correlations reveal interactions among different levels of the response variable.

This chapter provides a synthesis of spatial models for analyzing count responses that have location attributes. The synthesis begins with a discussion of univariate count responses before presenting methods for multivariate settings.

14.2 Spatial Models for Univariate Count Data

Techniques for analyzing spatial count data broadly diverge depending on the type of spatial interaction one wishes to control for. As noted earlier, there are two types of spatial interactions: spatial heterogeneity and spatial dependence. GWR seeks to address spatial interactions shown as contextual variations in coefficient estimates over space (i.e., spatial heterogeneity). Hadayeghi et al. (2009) developed a GWR-Poisson model to explain traffic crashes using transportation planning factors while controlling for spatial variations across zones. For each zone, a weighted Poisson regression model was estimated using the part of the data set observed in that zone's neighborhood. Weights are assigned to all neighbors, to reflect their importance in predicting counts in the zone of interest. The weights fall as the (straight-line or network-based) distance between zones increases.

For spatial dependence, many methods exist for analyzing univariate count data. They generally fall into three categories, as follows.

14.2.1 Log-linear Spatial Models

Standard spatial models (e.g., spatial autoregressive [SAR] models or spatial error models [SEM], LeSage and Pace 2009) were developed for data generated from a Gaussian process, in which the response variable takes on a continuous form. While not inherently designed to analyze count data, these models are sometimes used for analyzing count responses that are high in magnitude (e.g., hourly traffic volumes and employment counts). To apply these models in a count response setting, the count variable is artificially transformed into a quasi-continuous variable. A count variable (e.g., species abundance or counts) is typically normalized by an exposure term so that the resulting variable represents the rate at which things happen (e.g., species abundance per square mile or an approximation of crime counts per capita). Then, the rate variable is log-transformed, to produce a new response variable. The log transformation is important because it allows for the possibility of negative predictions. Examples include Weir et al.'s (2009) study on pedestrian crashes across San Francisco census tracts and Aufhauser and Fischer's (1985) study on migration patterns.

However, the log transformation will not work when low or zero counts exist, since their logarithms are mathematically ill-defined. In addition, a Gaussian process falls short of describing discrete events (e.g., crime or traffic crashes) that have low counts (rates), making it more attractive to use a discrete random process (e.g., Poisson). Two general approaches for discrete data analysis exist: these are conditional autoregressive (CAR) Poisson models and autoregressive Poisson models. Their difference lies in where spatial autocorrelation occurs: across the error terms (as in the CAR-Poisson) or the response terms (in the autoregressive-Poisson).

14.2.2 Conditional Autoregressive (CAR) Poisson Models

A CAR-Poisson model assumes that the count variable follows a Poisson process: $y_i \sim$ Poisson (λ_i), where y_i represents the number of events observed in zone i, and λ_i denotes the expected/mean count for that zone. The expected mean relates to the explanatory variables (x_i), their coefficients, and an exposure term (E): $\lambda_i = E^{\alpha} \cdot \exp\left(x_i'\beta + \gamma_i\right)$. The nuisance term, γ_i, represents noise or uncertainty that is unexplained by the control variables and is assumed to follow a CAR specification.

CAR specifications were first used by Besag et al. (1991), and are mostly estimated using Bayesian methods. A CAR model is built from a series of conditional distributions,[1] as shown in Equation 14.1 (Cressie 1991):

$$\gamma_i \Big| \gamma_{\neq i} \sim N\left[\mu_i + \sum\nolimits_{j=1}^{n} c_{ij}\left(\gamma_j - \mu_j\right),\ \sigma_i^2\right] \tag{14.1}$$

where γ_i indicates the spatially autocorrelated variable (e.g., spatial random effects centered onzero, or a response variable—like traffic flows or household incomes), γ_{-i} denotes such variables at neighboring locations (other than location i), μ_i is the expected/mean value of γ_i (i.e., $E(\gamma_i) = \mu_i$) and assumed to be zero, σ_i^2 is the conditional variance, and c_{ij} are weights (either known or unknown) describing the proximity or closeness between locations i and j.

The CAR specification permits contiguity and distance-based weight matrices, but precludes the Kth-nearest-neighbor weighting scheme because such weights violate the symmetry condition. First-order contiguity weights are defined such that $w_{ij} = 1$ if i and j share a common border (else $w_{ij} = 0$), and $\mathbf{W}$'s diagonal elements are all zeros by construction (Cressie 1991). This type of CAR model is called a *proper* CAR model[2], and is commonly estimated using Bayesian techniques in the open-source WinBUGS software package (Spiegelhalter et al. 2003), where "BUGS" stands for Bayesian inference Using Gibbs Sampling.

---

[1]These conditional distributions lead to a multivariate normal (MVN) joint distribution of the spatially correlated variables (shown in Equation 14.2), based on the factorization theorem (Besag et al. 1991).

$$\boldsymbol{\gamma} \sim MVN_n\left[\boldsymbol{\mu}, (\mathbf{I} - \mathbf{C})^{-1}\mathbf{M}\right] \tag{14.2}$$

where the column vector $\boldsymbol{\gamma}$ is a stacked version of the n γ_i's (as is the vector $\boldsymbol{\mu}$), $\mathbf{I}$ is an identity matrix, $\mathbf{C}$ is an n by n weight matrix (defined by site contiguity or inter-observation distances), with $\mathbf{C} = [c_{ij}]$, and $\mathbf{M}$ is a diagonal matrix, with $\mathbf{M}_{\text{ii}} = \sigma_i^2$. This joint distribution is used along with the likelihood function of the data set to implement the Gibbs sampler to estimate the posterior distributions of all parameters. Note that the Equations (14.1) and (14.2) are often referred to as a Markov random field (MRF) because of the way they are derived: achieving a closed-form joint distribution by first specifying a set of conditional distributions (Banerjee et al. 2004).

The validity of the MVN distribution shown in Equation 14.2 requires that its covariance matrix, $(\mathbf{I} - \mathbf{C})^{-1}\mathbf{M}$, be symmetric and positive-definite (like any covariance matrix must), thereby necessitating certain constraints on the forms of the matrices $\mathbf{C}$ and $\mathbf{M}$. For example, one may let $\mathbf{C} = \rho\mathbf{W}$ and $\sigma_i^2 = \frac{\sigma^2}{w_{i+}}$, where ρ is referred to as the spatial autocorrelation coefficient, $\mathbf{W}$ is a row-standardized weight matrix (i.e., $\mathbf{W} = \left[w_{ij}^*\right]$ and $w_{ij}^* = \frac{w_{ij}}{w_{i+}}$), and w_{i+} is the ith row sum of $\mathbf{W}$.

[2]This is not the "intrinsic" CAR model, because the latter does not have a spatial autocorrelation coefficient, ρ, which measures the overall strength of spatial interactions. Due to the absence of the spatial autocorrelation coefficient, its joint distribution is improper or unbounded in the sample space (Gelfand and Vounatsou 2003).

14.2.3 Spatial Autoregressive Poisson Models

While the CAR-Poisson model captures spatial dependence in error terms, SAR-type models describe spatial dependence in response variables. Lagrange multiplier tests can be used to discern which type of spatial dependence prevails in a spatial data set (whether spatial dependence occurs across the error terms or the responses) (Simoes and Natario 2016). Intuitively, a spatially-lagged error term represents subtle spatial dependence due to missing variables that trend in space, whereas a spatially-lagged response term implies more direct spatial interactions in which the response observed at one zone is in part predicted by its neighbors' values in addition to its own covariates.

Cressie (1991) introduced the auto-Poisson model, in reference to models in which the mean rate, $\boldsymbol{\lambda}$, involves autocorrelated response variables, i.e., $\boldsymbol{\lambda}=\exp(\mathbf{X}\boldsymbol{\beta}+\rho\mathbf{W}\boldsymbol{y})$. More recently, Griffith (2000) and Chun (2008) developed a Poisson-based spatial filtering approach to estimate auto-Poisson models. However, these types of Poisson models permit only negative autocorrelation, an unwanted result arising from the peculiar way spatial autocorrelation enters the specification, as shown in the following equation: $\boldsymbol{\lambda} = \exp(\mathbf{X}\boldsymbol{\beta} + \rho\mathbf{W}\boldsymbol{y})$, where $\boldsymbol{\lambda}$ denotes a vector of expected mean rates, $\mathbf{X}$ is an n by k covariate matrix, $\boldsymbol{\beta}$ is a k by 1 vector of unknown coefficients, $\boldsymbol{y}$ represents a vector of observed (count) responses, $\mathbf{W}$ an n by n weight matrix, and ρ the spatial autocorrelation coefficient. In addition, the joint likelihood function under an auto-Poisson assumption requires a non-closed-form solution for the normalizing constant (in order for the joint likelihood function under the auto-Poisson specification to be proper, or integrate to 1), which impedes successful estimation (Griffith 2000).

Liesenfeld et al. (2015) developed a new method to estimate spatial models for a wide range of non-Gaussian response variables including discrete choices, count, and other limited dependent variables (e.g., truncated, censored, or self-selected). This method combined Efficient Important Sampling (EIS) and sparse matrix algorithms to achieve accurate estimation of the likelihood function associated with spatially-interacted data and can handle a large number of observations. Liesenfeld et al. (2015) provided two such demonstrations: a spatial probit model for understanding U.S. voters' decisions in the 1996 presidential election, and a spatial count model for anticipating the prevalence of start-up companies across 3,110 U.S. counties.

For count responses, the model is formulated as:

$$f\ (y\,|\lambda, X) = \prod\nolimits_{i=1}^{n} f\ (y_i\,|\lambda_i) \tag{14.3}$$

$$ln\,(\lambda)\Big|X \sim MVN\left(m, H^{-1}\right) \tag{14.4}$$

where, y is the response variable (e.g., counts), f (.) is the likelihood function (e.g., $f(y) = \frac{\lambda^y e^{-\lambda}}{y!}$ for a Poisson process), λ_i is a latent variable measuring the expected mean counts, i is an index for observation unit. The latent vector, ln(λ), follows a multivariate normal distribution centered at m with a variance-covariance matrix, H^{-1}, i.e., the inverse of a Hessian matrix, H. When a direct spatial interaction is anticipated among neighbors, the latent variable at location i is influenced by the latent variables observed at its neighbors: mathematically, $\ln(\lambda) = \rho W \ln(\lambda) + X\beta + \varepsilon$. Under this construct, the mean (m) and the Hessian matrix are defined by

$m = (I - \rho W)^{-1} X\beta$ and $H = (1/\sigma^2)\,(I_n - \rho W)'\,(I_n - \rho W)$. The rest of the parameters (ρ and W) are as defined previously.

14.3 Spatial Models for Multivariate Count Data

While univariate count models address spatial dependence for a single outcome across zones, many empirical studies are interested in gauging the interactions among multiple outcomes while controlling for spatial effects. For example, the prevalence of one disease can coincidentally affect other diseases due to shared risk factors; the growth rate of new business establishment from one industry can correlate with those of other industries in nearby areas as a result of knowledge flows and transportation accessibility; and traffic crashes often show correlations among different severity levels because of shared influence of certain infrastructure or environmental factors that are latent/unobserved in the data. To control for these interactions among more than one outcome, multivariate (MV) count models are used to simultaneously anticipate the prevalence of multiple levels of outcomes while controlling for spatial effects.

In general, four methods exist for predicting MV count data over space in the literature. Table 14.1 summarizes research studies that utilized (spatial) MV count models in light of sample size, estimation method, statistical tools used, and model specifications.

14.3.1 Multivariate Conditional Autoregressive (MCAR) Models

The conditional autoregressive (CAR) model is the most commonly used method to handle spatial count data (e.g., Jin et al. 2005; Kramer and Williamson 2013; Barua et al. 2014; Boulieri et al. 2016). Its popularity is fueled by open-source software such as WinBUGS and its twin package OpenBUGS (Spiegelhalter et al. 2003), which code and estimate the CAR specification and its extensions (e.g., a time-space CAR model and moving-average models) with hierarchical Bayesian methods.

Table 14.1 Summary table of spatial models for multivariate count data

Category	Author(s) (Year)	Sample Size	No. Response Levels	Model Specification	Estimation Method	How to control for correlations among multiple responses?	Software
Conditional Autoregressive Model (CAR)	Aguero-Valverde et al. (2016)	832 road segments	4	MCAR	Bayesian MCMC	MCAR structure	OpenBUGS 3.0
	Aguero-Valverde and Jovanis (2010)	7968 segments	6	MCAR	Bayesian MCMC	MCAR structure	OpenBUGS
	Gelfand and Vounatsou (2003)	287 locations	2	MCAR	Bayesian MCMC (Gibbs sampler)	MCAR structure	-
	Jin et al. (2005)	87 counties	2	MCAR	MCMC	MCAR structure	Coded in C
	Leyland et al. (2000)	143 zip-code areas	2	Multivariate Poisson lognormal model	Iterative general-ized least squares	Heterogen-eous error term	Software package MLwiN
	Song et al. (2006)	254 counties	4	MCAR	Bayesian MCMC	MCAR structure	-
	Wang and Kockelman (2013)	218 zones	2	MCAR	Bayesian MCMC	MCAR structure	WinBUGS

Multivariate Finite Mixture models	Alfo et al. (2009)	375 boroughs	2	Multivariate finite mixture models with spatial dependence	EM algorithm	Random terms generated from the convolution (BYM) model	Coded in MATLAB
	Karunanayake (2007)	150 grid cells	3	Multivariate Poisson finite mixture models	EM algorithm	Poisson finite model structure	Splus/R codes
Generalized Ordered-Response (GOR) Model	Castro et al. (2012)	170 inter-sections	1	GOR	Composite marginal likelihood	-	GAUSS
	Narayanamoorthy et al. (2013)	285 census tracts	4	GOR	Composite marginal likelihood	Multivariate normal distribution through a multinomial probit (MNP) component	GAUSS
Spatio-Temporal Models	Aldor-Noiman et al. (2013)	188 census tracts	4	Integer-valued first-order autoregressive time-series model with CAR structures	Bayesian MCMC	A Dirichlet prior placed on the rate parameters of the Poisson processes	-
	Boulieri et al. (2016)	7932 electoral wards	2	Poisson CAR model with a BYM structure and a random walk	Bayesian MCMC	Multivariate spatially structured and unstructured effects	OpenBUGS
	Schmidt and Rodriguez (2010)	160 sites	4	Multivariate Poisson lognormal mixture model with a linear model of coregionalization (LMC)	Bayesian MCMC	Multivariate normal distribution of the error terms (which permits negative covariances)	OX

A multivariate CAR structure builds upon the univariate CAR-Poisson structure noted earlier and was enhanced by studies in genome analysis (Gelfand and Vounatsou 2003), disease mapping (Jin et al. 2005), traffic safety (Wang and Kockelman 2013; Aguero-Valverde et al. 2016), alternative-fuel vehicles (Chen et al. 2013; Bansal et al. 2015), and location decisions of new business establishment (Wang and Kockelman 2013).

The CAR structure defines the spatial random term for response type k observed in zone i (ϕ_{ik}) by a multivariate normal distribution. For an example involving two levels of responses, $\boldsymbol{\phi}_2 \sim N(\mathbf{0}, [(\mathbf{D} - \alpha_2 \mathbf{W})\tau_2]^{-1})$ and $\boldsymbol{\phi}_1 | \boldsymbol{\phi}_2 \;\; N(\mathbf{A}\boldsymbol{\phi}_2, [(\mathbf{D} - \alpha_1 \mathbf{W})\tau_1]^{-1})$, where τ_1 andτ_2 scale up or down the variance-covariance matrices; α_1 and α_2 measure the strength of spatial dependence; $\mathbf{W}$ is the spatial weight matrix (defined by contiguity or distance, though the former is more common in empirical studies, thanks to the computational benefits of sparse matrices); and $\mathbf{D}$ is a diagonal matrix with the ith diagonal element denoting the ith row sum of the weight matrix $\mathbf{W}$. More details are deferred to Wang and Kockelman (2013) for a two-level response setting and Bansal et al. (2015), Gelfand and Vounatsou (2003), and Aguero-Valverde et al. (2016) for response variables involving three or more levels.

14.3.2 Finite Mixture Models with Spatial Dependence

A standard finite mixture model provides a flexible alternative to analyze heterogeneous data and is typically estimated by the expectation-maximization (EM) algorithm (Gupta and Chen 2010). In a finite mixture model, the probability density function for a population (data) is expressed by a weighted average of the distribution functions of its sub-populations:

$$p(y|\Theta) = w_1 f_1(y|\theta_1) + w_2 f_2(y|\theta_2) + \cdots + w_K f_K(y|\theta_K) \tag{14.5}$$

where $\Theta = (\theta_1, \theta_2, \cdots, \theta_K; w_1, w_2, \cdots, w_K)$ represents the parameter space; the weights are positive and sum to numeral one; and $f(.)$ represents a distribution function (e.g., Poisson distribution with a latent parameter, λ_k, to measure the mean/expected level for a sub-population, if y is count). The model captures heterogeneity by compartmentalizing the probability density function of the population into discrete components associated with the sub-populations (Park 2010).

For spatial data, these discrete components can serve as proxies for the geographical clusters that exhibit unique trends or coefficients, hence controlling for area-specific heterogeneity (Alfo et al. 2009). Alfo et al. (2009) extended a standard finite mixture to control for spatial dependence within each cluster using the convolution method (also known as the Besag-York-Mollie [BYM] model, Besag et al. 1991) and the correlations among two levels of outcomes (e.g., two diseases).

Specifically, the log-transformed mean is decomposed into three parts:

$$\log(\lambda_{k1}) = \alpha_1 + \mu_k + \beta_{k1} \tag{14.6}$$

$$\log(\lambda_{k2}) = \alpha_2 + \mu_k/\delta + \beta_{k2} \tag{14.7}$$

where, α_1 and α_2 are constant terms representing the base-line risks associated with each disease, μ_k represents the shared factors that influence both outcomes, and β_{k1} and β_{k2} represent factors specific to each outcome. In addition to area-specific heterogeneity (a fortuitous property of all finite mixture-type models), this model specification also allows for spatial dependence across clusters by imposing a CAR structure on the three random terms, μ_k, β_{k1}, and β_{k2}.

The model was applied to estimate the prevalence of two heart diseases across 375 boroughs in Italy's Lazio region (Alfo et al. 2009), among other applications in health geography (see, e.g., Anderson et al. 2014). While the finite mixture models can define clusters in a meaningful way, the models can incur excessive computation time and are considered a special type of the generalized MCAR models (Alfo et al. 2009).

14.3.3 Generalized Ordered-Response Models

Some researchers have modeled spatial count data from an *ordered response* perspective that is rooted in utility-maximization choice theory. For example, in the context of intersection pedestrian crashes, Castro et al. (2012) utilized a continuous latent variable to proxy for traffic crash propensity and defined cut-off values to divide the latent variable into mutually exclusive intervals, with each interval representing a certain level of crash frequency. The model was cast in an ordered probit setting and estimated by a composite marginal likelihood approach.

Bhat et al. (2014) enhanced the model by permitting multivariate correlations through a multinomial probit (MNP) kernel. A MNP model is traditionally used in consumer choice or decision science to anticipate the influences of external variables on a person's choices (e.g., voting decision, vehicle purchase choice, etc.). In the context of multivariate count data modeling, each choice alternative can be used to represent each level of outcome. This method takes advantage of the quasi-concave property of the utility function and associated computational benefits. The model was estimated using the maximum composite marginal likelihood (MACML) approach (Bhat 2011).

14.3.4 Spatiotemporal Models

Aldor-Noiman et al. (2013) accounted for spatial and temporal dependencies in modeling weekly counts of different violent crimes across 188 Washington D.C. census tracts. Four crime types were analyzed simultaneously: rape, robbery, arson, and aggravated assault. The data present two challenges: low counts and irregular spatial structure. In the study area, two disjointed zones have crime rates that are correlated and nearby zones have opposite crime rates (due to heterogeneous demographics and natural boundary), diverging from a regular spatial data with clear spatial clustering. An integer-valued first-order autoregressive process, INAR(1), was used to capture temporal correlations among weekly crime rates. The use of INAR(1) is innovative because it incorporates two latent factors: a random term for seasonal effects and a zone-specific rate function that carries spatial dependence through a Dirichlet prior. A nonparametric Bayesian approach was used to estimate the multivariate Poisson-INAR(1) model, coupled with multiple shrinkage to handle the large sample size. "Bayesian nonparameteric methods have previously been studied as tools for data-driven clustering analysis" (Aldor-Noiman et al. 2013, p. 4) and appear to be as an effective way to analyze multiple correlated low-count time series (e.g., wild fires and earthquakes). The Dirichlet process also offers advantages by presenting a sparse neighborhood structure, similar to how a sparse spatial weight matrix functions in a Bayesian parametric setting.

14.4 Conclusions

This chapter describes the various spatial models that have been used to analyze univariate and multivariate count responses with location attributes. Two types of spatial effects are generally considered: spatial dependence (i.e., interactions among neighbors directly through spatially correlated response terms or indirectly through spatially lagged nuisance terms) and spatial heterogeneity (to describe contextual differences via spatially variable coefficient values).

For univariate count data, many spatial models exist, including a CAR model to explain spatial dependence in the error terms, a Poisson autoregressive model to convey more direct influence among neighbors through the response terms, and a GWR-Poisson model to allow coefficients that vary across locations. Goodchild and Haining (2003) suggested that the CAR model best applies to regions having more "local" spatial effects, like first-order-neighbor influence, whereas other spatial stochastic processes (which include the SAR and spatial error models [SEMs]) are more suitable for situations with higher-order dependencies, and thus exhibit more "global" spatial effects or relationships/interactions.

For multivariate count data, spatial effects enter the models chiefly through CAR-type interactions across error terms. The multivariate CAR structure is the most common approach to analyze such data due in part to the wide usage of open-source statistical software. However, such models only describe spatial interactions across the error terms and fall short when a more direct representation of spatial interaction is desired. By comparison, generalized ordered response (GOR) models (Bhat 2011), the spatial autoregressive Poisson model (Liesenfeld et al. 2015), and the Poisson mixture models (e.g., Schmidt and Rodriguez 2010) offer more flexible specifications: e.g., the spatial autoregressive Poisson models allow for direct spatial interactions of a variety of limited dependent variables, and the GOR models and the Poisson mixture models permit both negative and positive correlations among response levels. Future research should consider testing among these methods with respect to prediction accuracy, transferability, and computation. Efforts could also be spent to explore new ways to expand the computation of multivariate count models as large-scale spatial data (e.g., GPS traces and naturalistic driving data) become more regularly recorded and used in geography, transportation, and regional science.

The future of spatial multivariate count modeling presents both challenges and opportunities. The foremost challenge is small sample size as seen in the moderate number of observation units used in many of the reviewed studies. With the advent of crowdsourcing and voluntary geographic information, comes the need for analytical tools that can handle thousands of data points made over a large geography (e.g., pavement cracks observed across a road network, public opinions on designs or prototypes of a commercial product [Brabham 2008], and GPS traces of trips made by millions of households across a region) while portraying complex spatial (and temporal) interactions. The most common tool used so far is OpenBUGS, an open-source software that implements a number of complex spatial and time-series models through Bayesian MCMC methods (e.g., Gibbs sampling and Metropolis-Hastings algorithms). It is a variation of WinBUGS, which can also handle spatial models but is restricted to only one sampling method (Gibbs sampling).

Another challenge relates to computing issues (e.g., long run time and convergence) that complex models frequently encounter. While models involving moderate sample size (e.g., hundreds of data points) can be estimated within minutes, models with large sample size (e.g., more than thousands of data points) require excess run times, see, e.g., Aguero-Valverde and Jovanis (2010) reported that two days elapsed for their multivariate CAR model to converge after completing two chains, each with 100,000 Bayesian draws (for each parameter); and Boulieri et al. 2017 spent 20–27 hours to complete the 50,000 Bayesian draws (for each parameter) before reaching convergence for their Poisson Log-normal CAR model with a BYM structure. Both models were run in OpenBUGS. Run time is chiefly influenced by how fast the parameter draws converge to a stable value (if using Bayesian method) or how fast the algorithms locate the optimal solution of the likelihood function (if using maximum likelihood estimation or expected moment method). To improve computation efficiency, an analyst can consider reducing the complexity of spatial weight matrices (e.g., through sparse matrix algorithm [Finley et al. 2013]) and enhance convergence property, e.g., tweaking the acceptance rate of the M-H (so that chains converge at a faster rate) or improving parameter identification

(Waller et al. 1997) by using an appropriate value for the precision parameters associated with spatial (and heterogeneity) error terms or assigning hyperpriors for these precision parameters (Eberly and Carlin (2000).

In terms of emerging opportunities, a potentially transformative one is seen in extending advanced spatial models in settings that use geo-referenced, real-time input data to make forecasts about current or near-future values (i.e., nowcasting [e.g., Lampos et al. 2015, Preis and Moat 2014]). Recent years have seen a rapid growth of real-time data with location attributes, from Google's influenza reports (which exploit Internet users' search queries), through pedestrian or cyclist route and volume data collected from smart-phone applications (Smith 2015), to vehicle and driver information streamed from connected and instrumented vehicles. Coupled with nowcasting technology, these data offer critical information for developing a real-time advisory system, such as anticipating a flu trend and offering insight for medical surveillance, or anticipating crash risk of pedestrians (or cyclists) and forewarning road users of collision risk as they navigate the network. Spatial models can enhance the regression techniques used in the nowcasting literature by controlling for spatial dependence and other interactions typically found in geo-referenced data.

References

Aldor-Noiman S, Brown LD, Fox EB, Stine RA (2013) Spatio-temporal low count processes with application to violent crime events. Cornell University Library. Accessed at URL: http://arxiv.org/pdf/1304.5642.pdf

Alfo M, Nieddu L, Vicari D (2009) Finite mixture models for mapping spatially dependent disease counts. Biom J 51(1):84–97

Anderson C, Lee D, Dean N (2014) Identifying clusters in Bayesian disease mapping. Biostatsitics 15:457–469

Anselin, L. (2001) Chapter 14. Spatial econometrics. A companion to theoretical econometrics. Blackwell Publishing Ltd. http://web.pdx.edu/~crkl/WISE/SEAUG/papers/anselin01_CTE14.pdf

Anselin L (2010) Thirty years of spatial econometrics. Pap Reg Sci 89:3–25

Atkinson P, German S, Sear D, Clark M (2003) Exploring the relations between riverbank erosion and geomorphological controls using geographically weighted logistic regression. Geogr Anal 35(1):58–82

Aufhauser E, Fischer MM (1985) Log-linear modeling and spatial analysis. Environ Plan A 17(7):931–951

Aguero-Valverde J, Jovanis PP (2010) Spatial correlation in multilevel crash frequency models effects of different neighboring structures. Transp Res Rec J Transp Res Board 2165:21–32. doi:10.3141/2165-03

Aguero-Valverde J, Kun-Feng (Ken) W, Eric TD (2016) A multivariate spatial crash frequency model for identifying sites with promise based on crash types. Accid Anal Prev 87:8–16

Banerjee S, Carlin BP, Gelfand AE (2004) Hierarchical modeling and analysis for spatial data. Chapman and Hall–CRC, Boca Raton

Bansal P, Kockelman K, Wang Y (2015) Hybrid electric vehicle ownership and fuel economy across texas: application of spatial models. Transportation Research Record No. 2495: 53–64

Barua S, El-Basyouny K, Islam MT (2014) A full Bayesian multivariate count data model of collision severity with spatial correlation. Anal Methods Accid Res 3-4:28–43

Besag J, York J, Mollié A (1991) Bayesian image restoration, with two applications in spatial statistics (with discussion). Ann Inst Stat Math 43:1–59

Bhat CR (2011) The maximum approximate composite marginal likelihood (MACML) estimation of multinomial probit-based unordered response choice models. Transp Res B 45:923–939

Bhat CR, Born K, Sidharthan R, Bhat PC (2014) A count data model with endogenous covariates: formulation and application to roadway crash frequency at intersections. Anal Methods Accid Res 1:53–71

Boulieri A, Liverani S, de Hoogh K, Blangiardo M (2017) A space–time multivariate Bayesian model to analyze road traffic accidents by severity. J Royal Stat Soc A 180(1):119–139

Case B, Clapp J, Dubin R, Rodriguez M (2003) Modeling Spatial and temporal house price patterns: a comparison of four models. J Real Estate Financ Econ 29(2):167–191

Castro M, Paleti R, Bhat CR (2012) A latent variable representation of count data models to accommodate spatial and temporal dependence: application to predicting crash frequency at intersections. Transp Res B 46:253–272

Chakir R, Parent O (2009) Determinants of land use changes: a spatial multinomial probit approach. Pap Reg Sci 88(2):327–344

Chen D, Wang Y, Kockelman K (2013) Where are the electrical vehicles? A spatial model for vehicle-choice count data. J Transp Geogr 43:181–188

Chun Y (2008) Modeling network autocorrelation within migration flows by eigenvector spatial filtering. J Geogr Syst 10(4):317–344

Cressie NA (1991) Statistics for spatial data. Wiley, New York

Deutsch CV, Journel AG (1997) GSLIB: geostatistical software library and user's guide (applied geostatistics series), 2nd edn. Oxford University Press, New York

Du H, Mulley C (2012) Understanding spatial variations in the impact of accessibility on land value using geographically weighted regression. J Transp Land Use 5(2):46–59

Eberly LE, Carlin BP (2000) Identifiability and convergence issues for Markov chain Monte Carlo fitting of spatial models. Stat Med 19:2279–2294

Finley AO, Banergee S, Gelfand A (2013) spBayes for large univariate and multivariate point-referenced spatio-temporal data models. Working paper available at https://arxiv.org/pdf/1310.8192.pdf

Gelfand A, Vounatsou P (2003) Proper multivariate conditional autoregressive models for spatial data analysis. Biostatistics 4(1):11–25

Goodchild MF, Haining RP (2003) GIS and spatial data analysis: converging perspectives. Papers Reg Sci 83:363

Goodkind AL, Coggins JS, Marshall JD (2014) A spatial model of air pollution: the impact of the concentration-response function. J Assoc Environ Resour Econ 1(4):451–479

Griffith D (2000) A linear regression solution to the spatial autocorrelation problem. J Geogr Syst 2:141–156

Gupta MR, Chen Y (2010) Theory and use of the EM algorithm. doi:10.1561/2000000034

Hadayeghi A, Shalaby A, Persaud B (2009) Development of planning level 2 transportation safety tools using geographically weighted poisson regression. Accid Anal Prev 42(2):676–688

Jin X, Carlin BP, Banerjee S (2005) Generalized hierarchical multivariate CAR models for areal data. Biometrics 61(4):950–961

Karunanayake CP (2007) Multivariate poisson hidden Markov models for analysis of spatial counts. Doctor of Philosophy thesis, Department of Mathematics and Statistics, University of Saskatchewan, Saskatoon, SK, Canada

Kramer MR, Williamson R (2013) Multivariate Bayesian spatial model of preterm birth and cardiovascular disease among Georgia women: evidence for life course social determinants of health. Spat Spatiotemporal Epidemiol 6:25–35

Krige DG (1951) A statistical approach to some basic mine valuation problems on the Witwatersrand. J Chem Metall Min Soc S Afr 52(6):119–139

Lampos V, Andrew C, Miller SC, Stefansen C (2015) Advances in nowcasting influenza-like illness rates using search query logs. Scientific reports 5, Article number: 12760. Available at http://www.nature.com/articles/srep12760

LeSage J, Pace K (2009) Introduction to spatial econometrics. Chapman & Hall/CRC/Taylor & Francis Group, Boca Raton, FL

Levine L (2009) Introduction to the special issue on Bayesian journey to crime modeling. J Investig Psychol Offender Profiling 6(3):167–185

Levine N, Kim K, Nitz L (1995a) Spatial analysis of Honolulu motor vehicle crashes: I. Spatial patterns. Accid Anal Prev 27(5):663–674

Levine N, Kim K, Nitz L (1995b) Spatial analysis of Honolulu motor vehicle crashes: II. Zonal generators. Accid Anal Prev 27(5):675–685

Leyland A, Langford I, Rasbash J, Goldstein H (2000) Multivariate spatial models for event data. Stat Med 19:2469–2478

Liesenfeld R, Richard JF, Vogler J (2015) Likelihood evaluation of high-dimensional spatial latent Gaussian models with Non-Gaussian response variables. Available at SSRN: SSRN-id2196041 2

Lu Y (1998) Spatial cluster analysis for point data: location quotients versus kernel density. Department of Geography, State University of New York at Buffalo. http://dusk.geo.orst.edu/ucgis/web/oregon/papers/lu.htm

Miaou S-P, Song J, Mallick B (2003) Roadway traffic crash mapping: a space-time modeling approach. J Transp Stat 6(1):33–58

Nakaya T, Fotheringham S, Brunsdon C, Charlton M (2010) Geographically weighted poisson regression for disease association mapping. Stat Med 24(17):2695–2717

Narayanamoorthy S, Paleti R, Bhat CR (2013) On accommodating spatial dependence in bicycle and pedestrian injury counts by severity level. Transp Res B 55:245–264

Ognev-Himmelberger Y, Pearsall H, Rakshit R (2009) Concrete evidence and geographically weighted regression: a regional analysis of wealth and the land cover in Massachusetts. Appl Geogr 29(4):478–487

Páez A (2006) Exploring contextual variations in land use and transport analysis using a 35 probit model with geographical weights. J Transp Geogr 14:167–176

Park BJ (2010) Application of finite mixture models for vehicle crash data analysis. Texas A&M University Dissertation. URL: http://oaktrust.library.tamu.edu/bitstream/handle/1969.1/ETD-TAMU-2010-05-7667/PARK-DISSERTATION.pdf?sequence=2. Accessed 30 May 2016

Platt R (2004) Global and local analysis of fragmentation in a mountain region of Colorado. Agric Ecosyst Environ 101:207–218

Preis T, Moat HS (2014) Adaptive nowcasting of influenza outbreaks using google searches. Royal Society Open Science article. doi:10.1098/rsos.140095.

Schmidt AM, Rodriguez MA (2010) Modelling multivariate counts varying continuously in space. Book chapter in Bayesian Statistics, 9. ISBN: 9780199694587

Simoes P, Natario I (2016) Spatial econometric approaches for count data: an overview and new directions. IntJ Soc Behav Educ Econ Bus Ind Eng 10(1):348–356

Smith A (2015) Crowdsourcing pedestrian and cyclist activity data. US Department of Transportation Federal Highway Administration Report DTFHGI-11-H-00024. Available at http://www.pedbikeinfo.org/cms/downloads/PBIC_WhitePaper_Crowdsourcing.pdf

Song JJ, Ghosh M, Miaou S, Mallick B (2006) Bayesian multivariate spatial models for roadway traffic crash mapping. J Multivar Anal 97(1):246–273

Spiegelhalter D, Thomas A, Best N, Lunn D (2003) WinBUGS user manual version 1.4. URL: http://voteview.org/manual14.pdf

Vinatier F, Tixier P, Duyck PF, Lescourret F (2011) Factors and mechanisms explaining spatial heterogeneity: a review of methods for insect populations. Methods Ecol Evol 2(1):11–22

Waller LA, Carlin BP, Xia H, Gelfand AE (1997) Hierarchical spatio-temporal mapping of disease rates. J Am Stat Assoc 92(438):607–617

Wang X, Kockelman KM (2009) Application of the dynamic spatial ordered probit model: patterns of land development change in Austin, Texas. Pap Reg Sci 88(2):345–366

Wang Y, Kockelman K (2013) A Poisson-lognormal conditional autoregressive model for multivariate spatial analysis of pedestrian crash counts across neighborhoods. Accid Anal Prev 60:71–84

Wang Y, Kockelman K, Damien P (2014) A spatial autoregressive multinomial probit model for anticipating land use change in Austin, Texas. Ann Reg Sci 52:251–278

Wang Y, Kockelman K, Wang X (2011) Anticipating land use change using geographically weighted regression models for discrete response. Transportation Research Record No. 2245:111–123

Weir M, Weintraub J, Humphreys E, Seto E, Bhatia R (2009) An area-level model of vehicle-pedestrian injury collisions with implications for land use and transportation planning. Accid Anal Prev 41:137–145

Zhao F, Park N (2004) Using geographically weighted regression models to estimate annual average daily traffic. Transp Res Rec 1879:99–107

Yiyi Wang is an assistant professor of transportation engineering in the Civil Engineering Department at Montana State University (MSU)—Bozeman. Her research primarily focuses on the demand and safety aspects of active transportation (walking and biking), traffic safety modeling, and multi-modal transportation. She is a faculty research associate of MSU's Western Transportation Institute. She also serves as the co-advisor of the Institute of Transportation Engineers Student Chapter at MSU. She earned her Ph.D. in Civil Engineering in 2013 from the University of Texas at Austin.

Kara Kockelman is UT Austin's Schoch Professor of Civil, Architectural and Environmental Engineering. She holds a PhD, MS, and BS in civil engineering, a Masters of City Planning, and a minor in economics from the University of California at Berkeley. She has received a Google Research Award, NSF CAREER Award, and MIT's *Technology Review* Top 100 Innovators Award. She is the author of over 130 archival journal articles and serves on several committees of the Transportation Research Board. Dr. Kockelman's research with her students emphasizes the impacts of connected and automated vehicles, statistical modeling of urban systems (including models of travel behavior, trade, and location choice), energy and climate issues (vis-à-vis transport and land use decisions), the economic impacts of transport policy, and crash occurrence and consequences.

Amir Jamali is a PhD student in transportation engineering in the Civil Engineering Department at Montana State University (MSU), Bozeman. His research interests include Intelligent Transportation Systems (ITS), optimization modeling, and traffic safety modeling. He has earned his Master's degree in 2014 from the Sharif University of Technology in Tehran, Iran.

Chapter 15
Modeling of Infectious Diseases: A Core Research Topic for the Next Hundred Years

I Gede Nyoman Mindra Jaya, Henk Folmer, Budi Nurani Ruchjana, Farah Kristiani, and Yudhie Andriyana

15.1 Introduction

Incidence of disease is an under-researched topic in regional science. This is unfortunate because it frequently has far-reaching welfare impacts at household, regional, national, and even international levels. For the individual, health problems may range from minor nuisance to death. However, not only the victims but also their family members are affected if they fall ill (e.g., because of an increase in their household tasks or loss of income). Other, mainly financial, implications are related to seeing a doctor or buying medicine. Incidence of disease may also lead to loss of leisure or school days. Another nuisance is restriction of the movement of people to prevent the spread of a disease.

I.G.N.M. Jaya (✉)
Statistics Department, Universitas Padjadjaran, Kabupaten Sumedang, Indonesia
e-mail: mindra@unpad.ac.id

H. Folmer
Faculty of Spatial Science, University of Groningen, Groningen, The Netherlands

B.N. Ruchjana
Mathematics Department, Universitas Padjadjaran, Kabupaten Sumedang, Indonesia

F. Kristiani
Mathematics Department, Parahyangan Catholic University, Kota Bandung, Indonesia

Y. Andriyana
Statistics Department, Universitas Padjadjaran, Kabupaten Sumedang, Indonesia

R. Jackson, P. Schaeffer (eds.), *Regional Research Frontiers - Vol. 2*,
Advances in Spatial Science, DOI 10.1007/978-3-319-50590-9_15

Regional impacts of disease incidence consist in the first place of the impacts on the households that are directly or indirectly affected. However, in addition, there are costs caused by precautionary actions and production losses. In the case of epidemics, such as the Ebola virus disease, a regional system may be paralyzed. Given its welfare impacts and soaring incidence, inter alia, because of climate change, increasing population density, higher mobility, and increasing immunity to several common medicines, the incidence and spread of diseases should become regular research topics in regional science. For recent studies in regional science devoted to the topics, we refer to Ando and Baylis (2013) and Congdon (2013).

Methodological reasons also explain why regional scientists should pay (more) attention to the analysis of the incidence of diseases and its consequences. Although both regional science and epidemiology analyze the spatial distributions of their research topics and apply spatial analytical techniques, interesting methodological differences between them open possibilities for cross-fertilization. Whereas the units of analysis in regional science usually are administrative entities, such as the US states or counties with "large" populations, the spatial units in epidemiology are "small," such as neighborhoods, as required by the effective application of prevention or control measures. Given that the interest in regional science in small region phenomena, such as crime or the development of housing prices at the neighborhood level, is growing, the methods applied in epidemiology may turn out to be applicable in regional science as well. On the other hand, spatial spillover, which is a core issue in regional science for which a large variety of econometric approaches has been developed, has played a less significant role in epidemiology. Considering that infectious diseases tend to spatially spill over, epidemiology may benefit from the spatial spillover models and econometric approaches in regional science.

An important step in the analysis of regional impacts of a disease is the prediction of its incidence. The main objective of this study is to present an overview of the most common statistical methods to predict incidence of *infectious* diseases, to outline their pros and cons and the conditions under which they can be applied. The paper is restricted to infectious diseases. Typical for this type of diseases is that they are transmitted in space (see Sect 15.2). The key concepts in the analysis and prediction of the incidence of an infectious disease are the standardized mortality/morbidity ratio (SMR) and its standard error. In the paper, we discuss three types of approaches that have been used to estimate the key parameters of infectious disease incidence: maximum likelihood (ML), Bayesian methods, and nonparametric methods.

The paper is organized as follows: In Sect. 15.2, we discuss the types of infectious diseases and the basic model used to describe their occurrence. In Sect. 15.3, we discuss the main estimators that have been developed and applied to model the incidence of infectious diseases, i.e., ML, Bayesian smoothing, nonparametric methods, and econometric methods). In Sect. 15.4, we summarize the main findings and present conclusions, including a research agenda.

15.2 Basic Characteristics of Infectious Diseases

Infectious or transmissible diseases are caused by pathogenic microorganisms and transmitted from person to person by direct or indirect contact. Bacteria, viruses, or fungus are examples of the pathogenic agents.

Based on incidence, four types of infectious diseases are usually distinguished. A disease that occurs occasionally in a population is classified as *sporadic*; if it occurs constantly, it is *endemic*; if a large number of victims are infected in a short period, it is *epidemic*; and if it occurs worldwide in a short period, it is *pandemic*.

Infectious diseases have three transmission mechanisms: *contact*, *vehicle*, and *vector transmission*. In the first mechanism, the transmission is by direct person-to-person contact or indirect by contact with nonliving objects (such as contaminated soils) or by mucus droplets in coughing, sneezing, laughing, or talking. In the second mechanism, media, such as air (airborne), food (food-borne), or water (waterborne), are the transmitting agents. Finally, a vector is a mechanism that transports infectious agents from an infected person or animal to susceptible individuals. Vectors consist of two types: biological and mechanical. In the case of a biological vector, the agent reproduces in the vector's body that carries it to the susceptible person. Examples of biological vectors are mosquitoes, ticks, and bugs. A mechanical vector picks up and transports the agent outside of its body. The vector itself is not infected by the agent. An example is a housefly. Vector transmission is the most common transmission mechanism. For more details about transmission and its mechanisms, we refer to, e.g., Chen et al. (2015).

15.3 Infectious Disease Modeling

The basic concept in modeling the relative risk of an infectious disease is the SMR. It is used to identify high-risk regions. It is defined as follows: assume y_i and e_i are the observed and expected number of cases in region i, $(i = 1, 2, 3, \ldots, N)$, respectively. The SMR is then defined as follows:

$$SMR_i = \frac{y_i}{e_i}, \tag{15.1}$$

where e_i defined as

$$e_i = N_i \times \frac{\sum_{i=1}^{n} y_i}{\sum_{i=1}^{n} N_i}, \tag{15.2}$$

and N_i is the size of the population at risk in region i. A larger than one (15.1) SMR means that the region concerned has a larger actual incidence than its expectation; such region is classified as a high-risk region. By contrast, a region with a smaller than one (15.1) SMR is a low-risk region (Tango 2010).

15.3.1 ML

The traditional estimator of relative risk is ML (Shaddick and Zidek 2016). For count data and y_i, a "small" non-negative, discrete number, the Poisson distribution is typically chosen to model infectious disease incidence. With mean and variance $e_i\theta_i$ respectively, where θ_i is the relative risk parameter in region i, the following is obtained:

$$y_i \middle| e_i\theta_i \sim \text{Poisson}\left(e_i\theta_i\right). \tag{15.3}$$

The simplest model assumes no covariate and random term in the model. The ML estimator of θ_i is

$$\widehat{\theta}_i^{ML} = \frac{y_i}{e_i}, \tag{15.4}$$

which is unbiased. The variance is

$$\widehat{V\left(\widehat{\theta}_i^{ML}\right)} = \frac{\widehat{\theta}_i^{ML}}{e_i}. \tag{15.5}$$

For small e_i, (15.4) and (15.5) are "large" which leads to imprecise estimation of relative risk. For example, two similar regions, A and B, have the same population at risk, that is, they have the same expected number of cases, e_i. Suppose that e_i is 0.1 and that in region A one case is found and in B, zero. Hence, $\widehat{\theta}_i^{ML}$ in region A is 10 and in region B, zero. Region A has extreme $\widehat{\theta}_i^{ML}$ compared with region B, while the number of cases differs by 1 only. It follows that the ML-estimated relative risk may be very unstable and lead to wrong conclusions (Pringle 1996). Consequently, more appropriate methods for disease modeling and mapping are required. One class of such methods is smoothing. Smoothing techniques exploit information from neighboring regions to adjust the estimate for a given region. The basic principle is *shrinkage*. That is, ML estimates with small expected rates or high variances will be "shrunk" toward the overall mean, whereas those with small variances will essentially remain unchanged. Smoothing thus decreases the mean squared error (Anselin et al. 2006). Bayesian and nonparametric techniques are two popular smoothing methods used in disease modeling and mapping.

15.3.2 Bayesian Smoothing

Bayesian smoothing methods are statistical approaches to update unknown parameters using information from observations. As a first step, prior information on the parameter of interest is specified in terms of a probability distribution. Next, empirical evidence (data) is obtained and combined with the prior information using Bayes' theorem, which leads to a posterior probability distribution of the parameters. The posterior becomes the basis for statistical inference (Congdon 2010). Specifically, the observed data $y = (y_1, \ldots, y_n)^T$ is assumed to be generated from a probability distribution $f(y_i | \theta_i)$ with unknown parameters $\theta = (\theta_1, \ldots, \theta_n)^T$. The unknown parameters θ, in turn, are assumed to be random variables with prior $f(\theta_i | \boldsymbol{\gamma})$ with unknown hyperparameter $\boldsymbol{\gamma} = (\gamma_1, \ldots, \gamma_k)^T$. The posterior density of θ_i, given the data y_i, the conditional density $f(y_i | \theta_i)$, and the conditional density $f(\theta_i | \boldsymbol{\gamma})$, is

$$f(\theta_i | y_i, \boldsymbol{\gamma}) = \frac{f(y_i | \theta_i) \times f(\theta_i | \boldsymbol{\gamma})}{f(y_i | \boldsymbol{\gamma})}, \tag{15.6}$$

where $f(y_i | \boldsymbol{\gamma})$ is the marginal likelihood of the data given hyperparameter $\boldsymbol{\gamma}$. To ensure that the posterior distribution, $f(\theta_i | y_i)$, is a proper density, the marginal likelihood, $f(y_i | \boldsymbol{\gamma})$, is taken as a normalizing constant, which is found by integrating the likelihood, $f(y_i | \theta_i)$, over the joint prior density:

$$f(y_i | \boldsymbol{\gamma}) = \int f(y_i | \theta_i) \times f(\theta_i | \boldsymbol{\gamma}) \, d\theta_i. \tag{15.7}$$

Based on the above mentioned description, (15.6) can be written as follows:

$$f(\theta_i | y_i, \boldsymbol{\gamma}) \propto f(y_i | \theta_i) \times f(\theta_i | \boldsymbol{\gamma}). \tag{15.8}$$

The estimated posterior density $f(\theta_i | y_i, \widehat{\boldsymbol{\gamma}})$ is used to make inferences about θ_i, where $\widehat{\boldsymbol{\gamma}}$ is an estimate of $\boldsymbol{\gamma}$.

Bayesian approaches are composed of two classes: empirical Bayes (EB) and full Bayes (FB). Each is made up of several types. In the case of EB, parameters $\boldsymbol{\gamma}$ are replaced by point estimates of hyperparameter based on the marginal distribution of y_i. In the case of FB, a prior distribution $f(\gamma_1), \ldots, f(\gamma_k)$, is specified for the hyperparameter $\boldsymbol{\gamma}$ (Hog et al. 2005).

A typical example of each case is presented below.

15.3.2.1 Empirical Bayes Poisson-Lognormal Model[1]

The empirical Bayes Poisson-lognormal (EBPLN) model was introduced by Clayton and Kaldor (1987). It can be summarized as follows: The prior distribution of the relative risk, θ, is assumed to have a multivariate lognormal distribution. That

[1]Other EB models are the Poisson-Gamma model and the linear empirical Bayes model. See, e.g., Clayton and Kaldor (1987) and Lawson et al. (2000) for details.

is, the log of the relative risk, $\boldsymbol{\zeta} = \log(\theta)$; $\boldsymbol{\zeta} = (\zeta_1, .., \zeta_n)^T$, is assumed to follow a multivariate normal distribution with mean $\boldsymbol{\mu}$ and covariance matrix $\boldsymbol{\Sigma}$. Hence, the density function of $\boldsymbol{\zeta}$ is as follows:

$$f(\boldsymbol{\zeta}|\boldsymbol{\mu},\boldsymbol{\Sigma}) = (2\pi)^{-\frac{1}{2n}}(\theta_1 \dots \theta_n)^{-1}|\boldsymbol{\Sigma}|^{-\frac{1}{2}}\exp\left\{-\frac{1}{2}(log\boldsymbol{\theta} - \boldsymbol{\mu})'\boldsymbol{\Sigma}^{-1}(log\boldsymbol{\theta} - \boldsymbol{\mu})\right\}. \tag{15.9}$$

The EB estimator is obtained from the expectation of the relative risk θ given y, $E(\theta|y)$. However, the posterior distribution of the Poisson-lognormal is not a closed form, that is, it has no analytical solution for $E(\theta|y)$. As a way out, Clayton and Kaldor (1987) proposed a quadratic approximation by substituting θ_i for $\exp(\zeta_i)$ to construct the Poisson likelihood $\boldsymbol{\zeta}$ given y. The likelihood thus is

$$L(\boldsymbol{\zeta}|\mathbf{y}) = \prod_{i=1}^{n} f(y_i|\zeta_i) = \prod_{i=1}^{n}\left(\frac{\exp(-e_i\exp(\zeta_i))(e_i\exp(\zeta_i))^{y_i}}{y_i!}\right). \tag{15.10}$$

The EB estimator using the quadratic approximation requires the estimate of the vector of parameters $\boldsymbol{\zeta}$. Clayton and Kaldor (1987) proposed ML to estimate $\boldsymbol{\zeta}$. The ML estimator of $\tilde{\zeta}_i = \log\left(\frac{y_i}{e_i}\right)$. However, this solution does not hold for $y_i = 0$. Therefore, Clayton and Kaldor (1987) suggested to add the constant 0.50 to y_i, that is,

$$\tilde{\zeta}_i = \log\left(\frac{y_i + 0.5}{e_i}\right). \tag{15.11}$$

Equation (15.11) is an explicit solution of the EB estimate of $\boldsymbol{\zeta}$ based on quadratic approximation. However, the solution is not based on the expectation of the posterior distribution of the Poisson-lognormal model, $f(\boldsymbol{\zeta}|y,\boldsymbol{\mu},\boldsymbol{\Sigma})$. With the quadratic approximation of the likelihood function over the lognormal prior, the posterior distribution of $\boldsymbol{\zeta}$ given the data y is

$$f(\boldsymbol{\zeta}|\mathbf{y},\boldsymbol{\mu},\boldsymbol{\Sigma}) \propto f(\mathbf{y}|\boldsymbol{\zeta})\, f(\boldsymbol{\zeta}|\boldsymbol{\mu},\boldsymbol{\Sigma}), \tag{15.12}$$

which follows a multivariate normal with mean $\boldsymbol{\mu}$ and variance $\boldsymbol{\Sigma}$ (Leonard 1975; see Clayton and Kaldor 1987, for details). Estimating $\boldsymbol{\mu}$ and $\boldsymbol{\Sigma}$ is thus necessary to obtain an explicit solution for $\boldsymbol{\zeta}$ based on $f(\boldsymbol{\zeta}|y,\boldsymbol{\mu},\boldsymbol{\Sigma})$. The EM algorithm can be used for this purpose. In the simplest case, the ζ_i are taken as $i.i.d\, N(\mu,\sigma^2)$. Given that the distribution of the ζ_i has two parameters, μ and σ^2, the EBPLN, $\widehat{\zeta}_i^{\text{EBPLN}}$, becomes (Meza 2003):

$$\widehat{\zeta}_i^{EBPLN} = \frac{\widehat{\mu} + \widehat{\sigma}^2(y_i + 0.5)\tilde{\zeta}_i - 0.5\widehat{\sigma}^2}{1 + \widehat{\sigma}^2(y_i + 0.5)}, \tag{15.13}$$

with

$$\widehat{\mu} = \frac{1}{n}\sum_{i=1}^{n}\widehat{\zeta}_i^{EBPLN}, \tag{15.14}$$

$$\widehat{\sigma}^2 = \frac{1}{n}\left(\widehat{\sigma}^2\sum_{i=1}^{n}\left[1+\widehat{\sigma}^2\left(y_i+0.5\right)\right]^{-1} + \sum_{i=1}^{n}\left(\widehat{\zeta}_i^{EBPLN} - \widehat{\mu}\right)^2\right). \tag{15.15}$$

The EBPLN estimator of the relative risk is $\widehat{\theta}_i^{\mathrm{EBPLN}} = \exp\left(\widehat{\zeta}_i^{\mathrm{EBPLN}}\right)$.

The EM algorithm to (iteratively) obtain the estimates of μ and σ^2 using Equations (15.13), (15.14), and (15.15) is as follows:

(1) Obtain the initial values of $\left\{\widehat{\zeta}_i, \widehat{\mu}, \widehat{\sigma}^2\right\}$:

   (a) $\widehat{\zeta}_i = \log\left(\frac{y_i+0.5}{e_i}\right)$
   (b) $\widehat{\mu} = \frac{1}{n}\sum_{i=1}^{n}\widehat{\zeta}$
   (c) $\widehat{\sigma}^2 = \frac{1}{n}\sum_{i=1}^{n}\left(\widehat{\zeta} - \widehat{\mu}\right)^2$

(2) *Expectation* (E) Step: Estimate the relative risk using Equation (15.13).
(3) *Maximization* (M) Step: Update the parameter estimates $\widehat{\mu}$ and $\widehat{\sigma}^2$ using Equations (15.14) and (15.15).
(4) Repeat Steps 2–3 until a predetermined precision is obtained, e.g., $\left|\widehat{\zeta}_i^{\mathrm{EBLN}(t+1)} - \widehat{\zeta}_i^{\mathrm{EBLN}(t)}\right| \leq 1e - k$, with k a positive integer.

15.3.2.2 Full Bayesian Poisson-Lognormal Model[2]

Full Bayesian (FB) estimation is more widely used in Bayesian disease modeling than EB because it is more flexible in defining the prior hyperparameter $\boldsymbol{\gamma} = (\gamma_1, \ldots, \gamma_k)^T$, and because it can provide a measure of uncertainty of the estimates of relative risks (Maiti 1998). The quality of the FB estimates depends on the accuracy in determining a hyperprior distribution.

In FB, the posterior parameters can be estimated using Markov chain Monte Carlo (MCMC) simulation, such as the Gibbs sampler and Metropolis-Hastings (M-H) or integrated nested Laplace approximation (INLA). The procedure is as follows: As in the case of EBPLN, FBPLN assumes the log relative risk, ζ_i, to follow a normal distribution, that is, $\zeta_i \sim \mathrm{i.i.d}\ \mathrm{Normal}(\mu, \sigma^2)$.

[2] Another FB model is the Poisson Gamma model. See, e.g. Lawson (2006) for an overview.

The basic FBPLN model may be written as follows (Meza 2003):

(i) $y_i \Big| \theta_i \overset{iid}{\sim} \text{Poisson}\,(e_i\theta_i)$

(ii) $\zeta_i = \log(\theta_i) \Big| \mu, \sigma^2 \overset{iid}{\sim} N(\mu, \sigma^2)$

(iii) $f(\mu, \sigma^2) \propto f(\mu) f(\sigma^2)$ with
$f(\mu) \propto 1; \sigma^{-2} \sim \text{Gamma}\,(a, b)\,; a \geq 0, b > 0$

Commonly, the prior parameters (a, b) are assumed to be known. Obtaining the posterior distribution of $\theta_i|y_i$ involves high-dimensional integrals that are difficult to sample directly from. However, sampling from the full conditional distribution of each parameter is often easy. The Gibbs sampler can be used to estimate the posterior distribute on (Maiti 1998). The full conditional distribution to implement Gibbs sampling can be written as follows:

(i) $f(\theta_i|\mu, \sigma^2, y_i) \propto \theta_i^{y_i-1} \exp\left[-e_i\theta_i - \frac{1}{2\sigma^2}(\zeta_i - \mu)^2\right]$

(ii) $[\mu|\theta_i, \sigma^2, y_i] \sim N\left(\frac{1}{n}\sum_i \zeta_i, \frac{\sigma^2}{m}\right)$

(iii) $[\sigma^2|\theta_i, \mu, y_i] \sim G\left(\frac{n}{2} + a, \frac{1}{2}\sum_i (\zeta_i - \mu)^2 + b\right)$

MCMC samples can be directly generated from (ii) and (iii) using the M-H algorithm. Several software programs can be used to estimate the FBPLN. The WinBUGS software program is generally used.

For computational purposes, ζ_i is decomposed into two components, β_0 and u_i. β_0 is the overall level of the log relative risk, whereas u_i is the residual.

$$\log(\theta_i) = \beta_0 + u_i, \tag{15.16}$$

$$u_i \sim i.i.d\ Normal\left(0, \sigma_u^2\right).$$

The parameters β_0 and u_i have a hyperprior distribution as follows:

$$\beta_0 \sim i.i.d\ Normal\left(0, \sigma_{\beta_0}^2\right),$$

$$1/\sigma_u^2 \sim Gamma\,(a, b).$$

Using noninformative prior, the value of $\sigma_{\beta_0}^2$ is usually replaced by a large number, for example, $\sigma_{\beta_0}^2 = 10^5$ and for $a = 0.5$ and $b = 0.0005$ (Tango 2010).

15.4 The Besag, York, and Mollie (BYM) FB Model

ML and the traditional Bayesian approaches do not accommodate spatial trend, covariates, and spatially uncorrelated and spatially correlated heterogeneity. The FBPLN model can be extended to include those components. To consider spatially correlated heterogeneity, Clayton and Kaldor (1987) proposed the conditional autoregressive (CAR) model for the log relative risk. The CAR model is defined as follows:

$$E\left(\zeta_i|\zeta_{j(j\neq i)}\right) = \mu_i + \rho\sum_j w_{ij}\left(\zeta_j - \mu_j\right)$$

$$Var\left(\zeta_i|\zeta_{j(j\neq i)}\right) = \sigma^2, \tag{15.17}$$

where $w_{ij}$is an element of the spatial weights matrix **W.** To simplify computations, μ_i is assumed to be equal to μ.

The "complete" FBLN model to estimate the relative risk was developed by Besag et al. (1991), denoted BYM. Considering its "completeness", it has become a popular model in Bayesian disease modeling and mapping, especially of infectious diseases. The BYM model reads as follows (Lawson et al. 2000):

$$\log\left(\theta_i\right) = t_i + u_i + v_i, \tag{15.18}$$

where t_i denotes the spatial trend and covariates, u_i denotes the spatially uncorrelated heterogeneity, and v_i denotes the spatially correlated heterogeneity (Lawson et al. 2000). A typical spatial trend regression model reads as follows:

$$t_i = \sum_{h=1}^{H}\left(a_h x_i^h + b_h y_i^h\right) + \sum_{k=1}^{K} c_k z_k, \tag{15.19}$$

where $\{(x_i, y_i)\}$ are the centroids of the i-th region, H is the degree of the trend (e.g., $h = 1$: linear trend; $h = 2$:quadratic trend), K is the number of covariates, and z is the vector of covariates.

In the case of count data, over-dispersion frequently occurs, that is, the variance observed is greater than the mean. Over-dispersion has two types: spatially uncorrelated and spatially correlated heterogeneity (Lawson 2006). Spatially uncorrelated heterogeneity occurs because of observations with small or zero cases, differences in the number of subpopulation, and omitted environmental or ecological factors, such as pollution, rainfall, humidity, temperature, and radiation. Spatially uncorrelated heterogeneity is accommodated by defining a non-informative prior[3] for u_i, usually

[3]A noninformative prior is used to denote lack of information about the parameter of interest (Lawson 2013).

the normal distribution (Lawson et al. 2003):

$$u_i \sim i.i.d \text{ Normal}\left(0, \sigma_u^2\right). \tag{15.20}$$

Spatially correlated heterogeneity, v_i, occurs because of spatial clustering or spatial autocorrelation (Lawson 2006). It can be considered using information relating to adjacent regions, based on the assumption that adjacent regions with similar spatial characteristics have similar relative risks.

A conditional autoregressive (CAR) prior is usually used to capture spatially correlated heterogeneity. Besag et al. (1991) proposed the following CAR prior:

$$v_i \middle| v_{j \neq i} \sim \text{Normal}\left(\frac{\sum_j w_{ij} v_j}{\sum_j w_{ij}}, \frac{\sigma_v^2}{\sum_j w_{ij}}\right), \tag{15.21}$$

where w_{ij} denotes spatial dependence between regions i and j.

A limitation of the Besag prior is that it is only appropriate for strong spatial autocorrelation. If weak spatial autocorrelation exists, the CAR prior produces random effects that are overly smooth (Lee 2013). To overcome this limitation, spatially uncorrelated heterogeneity u_i should be used. To accommodate varying strengths of spatial autocorrelation, Leroux et al. (1999) and Stern and Cressie (1999) proposed alternative CAR priors. The Leroux et al. (1999) CAR prior reads as follows:

$$v_i \middle| v_{j \neq i} \sim N\left(\frac{\rho \sum_j w_{ij} v_j}{\rho \sum_j w_{ij} + 1 - \rho}, \frac{\sigma_v^2}{\rho \sum_j w_{ij} + 1 - \rho}\right), \tag{15.22}$$

The Stern and Cressie (1999) CAR prior is as follows:

$$v_i \middle| v_{j \neq i} \sim N\left(\frac{\rho \sum_j w_{ij} v_j}{\rho \sum_j w_{ij}}, \frac{\sigma_v^2}{\rho \sum_j w_{ij}}\right). \tag{15.23}$$

In both cases, ρ is the spatial autocorrelation parameter. Using the Leroux or Stern and Cressie prior renders spatially uncorrelated heterogeneity $u_i$redundant.

The FBPLN model, including spatial effects, may be written as follows (Rao 2003):

(i) $y_i | \theta_i \sim \text{Poisson}(e_i \theta_i)$

(ii) $\zeta_i \middle| \zeta_{j(j \neq i)}, \rho, \sigma^2 \sim N\left(\mu + \rho \sum_{il} w_{il} (\zeta_l - \mu), \sigma^2\right)$

(iii) $f\left(\mu, \sigma^2, \rho\right) \propto f(\mu) f\left(\sigma^2\right) f(\rho)$ with
$f(\mu) \propto 1; \sigma^{-2} \sim \text{Gamma}(a, b); a \geq 0, b > 0, \rho \sim U(0, \rho_0)$

where ρ_0 denotes the maximum value of ρ in the CAR model and $\boldsymbol{W} = (w_{il})$ is the "adjacency" matrix. Maiti (1998) proposed Gibbs sampling combined with the M-H algorithm to estimate the model.

The BYM model can be summarized as follows:

$$\eta_i = \beta_0 + \boldsymbol{X}_i^T \boldsymbol{\beta} + u_i + v_i, \tag{15.24}$$

where $\eta_i = \log(\theta_i)$, $\beta_0$is the overall relative risk, $\boldsymbol{X}_i^T = (X_{i1}, .., X_{iK})$ is a vector covariates, $\boldsymbol{\beta} = (\beta_1, .., \beta_K)^T$ is a vector regression coefficients, and u_i and $v_i$denote are spatially uncorrelated and spatially correlated heterogeneity, respectively. The following hyperparameter distributions of β_0, u_i and v_i are usually applied:

$$\beta_0, \beta_1, .., \beta_k \sim i.i.d\text{Normal}\left(0, \sigma_\beta^2\right),$$

$$1/\sigma_u^2 \sim \text{Gamma}(a, b),$$

$$1/\sigma_v^2 \sim \text{Gamma}(a, b).$$

As a non-informative prior, large values for σ_β^2 are usually taken, for example, $\sigma_\beta^2 = 10^5$ and for $a = 0.5$ and $b = 0.0005$ (Tango 2010).

The above-mentioned model only accounts for the spatial pattern of diseases but does not incorporate temporal variation. A model that includes temporal variation is a spatio-temporal model. Spatio-temporal modeling has been widely applied to analyze the spatial distribution of disease incidence and its trend, notably to detect hotspots (Lawson 2014). The most common approach is based on the assumption that a log-linear relationship exists between the relative risk and the calendar time within regions, that is, that the time trend varies from region to region (Lawson 2014). Thus

$$\mathrm{y_{it}} \,\middle|\, \mathrm{e_{it}\theta_{it}} \sim \text{Poisson}\left(\mathrm{e_{it}\theta_{it}}\right),$$

$$\eta_{\mathrm{it}} = \beta_0 + \mathrm{X_{it}^T}\beta + \mathrm{u_i} + \mathrm{v_i} + \omega_{\mathrm{t}} + \psi_{\mathrm{t}} + \phi_{\mathrm{it}}, \tag{15.25}$$

where $\eta_{it} = \log(\theta_{it}) u_i$ and v_i denote spatially uncorrelated and spatially correlated heterogeneity, respectively; ω_j and ψ_t denote temporally uncorrelated and temporally-correlated heterogeneity, and ϕ_{ij} is a spatio-temporal interaction effect. This model varies based on the structure of the space-time structure. Model (15.25) is commonly estimated using Bayesian techniques.

15.4.1 Nonparametric Estimation

The most popular nonparametric smoothing technique is the Nadaraya-Watson kernel smoother. It is defined as the weighted average of the ML estimates of the other regions (Lawson et al. 2000):

$$\theta_i^{NP} = \sum_{j \neq i}^{n} \omega_j \theta_j^{ML}, \tag{15.26}$$

with ω_j weights for values of neighboring regions defined as follows:

$$\omega_j = \frac{K\left(\left(\theta_i^{ML} - \theta_j^{ML}\right)/h\right)}{\sum_i^n K\left(\left(\theta_i^{ML} - \theta_j^{ML}\right)/h\right)}, \tag{15.27}$$

where $K(.)$ is a zero mean, radially symmetric probability density function, usually the standard Gaussian distribution:

$$K(z) = (2\pi)^{-1/2} \exp\left(-\frac{z^2}{2}\right), \tag{15.28}$$

with h the bandwidth based on the minimum value of the least squares cross-validation criteria (Simonoff 1999):

$$CV(h) = \frac{1}{n} \sum_{i=1}^{n} \left(\widehat{\theta}_i^{NP} - \overline{\hat{\theta}_{(-i)}^{ML}}\right)^2, \tag{15.29}$$

Where $\overline{\hat{\theta}_{(-i)}^{ML}}$ denotes the average relative risk estimate using ML without the ith observation.

For an application to relative risk estimation, see Kesall and Diggle (1998). The nonparametric model can be extended to include time variation and spatial dependence as follows:

$$\log\left(\lambda_{it} | y_{it}\right) = \log\left(n_{it}\right) + \log(m) + S_0(t) + \alpha_i + S_i(t), \tag{15.30}$$

where λ_{it} is a mean of Poisson distribution, n_{it} is the population count for the region i in year t; m is the overall mean of the relative risk; $S_0(t)$ is the fixed global of the relative risk trend; α_i is the random spatial effect, which may be spatially correlated; and $S_i(t)$ is the random temporal effect for the region i (MacNab and Dean 2002).

15.4.2 Spatial Econometric Models

The models discussed in the previous sections (explicitly) do not consider spatial dependence even though spatial spillovers are typical for infectious diseases. Particularly, the response variable in one region usually depends on the values of the response variable in neighboring regions (Lawson 2014; Chen et al. 2015), as in the case of dengue fever. Similarly, the status of covariates (e.g., vegetation or water quality) in one region may affect the response variable not only in that region but also in neighboring regions. Finally, spatial dependence may occur among the error terms.

One of the reasons that spatial econometric models have received little attention in epidemiology is that these models have been developed for continuous data rather than count data, especially with respect to the dependent variable. Following Lambert et al. (2010) and Bivand et al. (2014), we specify the spatially lagged (SL) mixed Poisson regression model of relative risk for count data with spatially lagged dependent variable, spatially uncorrelated (u_i) and spatially correlated (v_i) heterogeneity as components of the error term (ε_i), as follows:

$$\boldsymbol{\eta} = \rho_{Lag}\boldsymbol{W}\boldsymbol{\eta} + \beta_0\mathbf{1}_n + \boldsymbol{X}\boldsymbol{\beta} + \boldsymbol{\varepsilon} \tag{15.31}$$

where $\boldsymbol{\eta} = (\eta_1, .., \eta_n)^T$ with $\eta_i = \log(\theta_i)$, β_0 is the overall relative risk, $\mathbf{1}_n$ is a unit vector of length n, $\boldsymbol{X}$ is a matrix of covariates of size (nxK), $\boldsymbol{\beta} = (\beta_1, .., \beta_k)^T$ is a vector of coefficients, and $\boldsymbol{W}$ is a symmetric adjacency matrix with zero diagonal elements, ρ_{Lag} is the spatial lag parameter that measures infectious disease spillover among regions.

A more general model with wider applicability is the spatial Durbin-Poisson (SD-Poisson) model that allows for spatial spillovers of the covariates in addition to a spatially lagged dependent variable. The SD-Poisson model reads as follows:

$$\boldsymbol{\eta} = \rho_{Lag}\boldsymbol{W}\boldsymbol{\eta} + \beta_0\mathbf{1}_n + \boldsymbol{X}\boldsymbol{\beta} + \boldsymbol{W}\boldsymbol{X}\boldsymbol{\delta} + \boldsymbol{\varepsilon}, \tag{15.32}$$

where $\boldsymbol{\delta} = (\delta_1, \ldots, \delta_K)^T$ denotes a vector of coefficients for the spatially lagged covariates $\boldsymbol{WX}$ (Bivand et al. 2014).

From models (15.31) and (15.32), the direct and indirect (spillover) effects can be calculated. To estimate the SL-Poisson model, Lambert et al. (2010) proposed two step limited information maximum likelihood, and Bivand et al. (2014) developed a Bayesian estimator using INLA.

15.5 Summary and Research Recommendations

Incidences of infectious diseases have been soaring. According to the World Health Organization (2005), climate change, extreme weather, and environmental factors, such as lack of access to clean water and poor sanitation facilities, have contributed to the outbreaks. Socioeconomic conditions, including income, employment, education, and health behavior, are also important factors that influence the transmission of infectious diseases. Increasing urbanization, higher population density, higher mobility, and increasing resistance to several common medicines accelerate the transmission from one location to another because of more contacts between infected and susceptible people (Fong 2013).

Infectious diseases often have serious direct and indirect effects at the individual, household, and regional levels ranging from increased morbidity and mortality to the paralysis of an entire region or even a country. Early identification of an endemic is an important first step to prevent its transmission and to reduce its effects. Implementation of such early warning systems (EWSs), including roadmaps to prevent or restrict the spread of an infectious disease, is still in its infancy in most (developing) countries (Lowe et al. 2011). Therefore, the development and implementation of EWSs based on information about when and where outbreaks will occur and what factors influence transmission is a high-priority research topic. A related research topic is how to use EWS information in taking appropriate and efficient actions to manage transmission and to prevent epidemics. The development and implementation of an EWS requires intensive interaction between natural and social regional scientists.

An important component of an EWS is the identification of high-risk regions and spatial clustering. For this purpose, predictive models are required (Chen et al. 2015). In this paper, an overview of the most common approaches in disease incidence modeling has been presented. Four types of approaches have been discussed, namely, ML, Bayesian smoothing, nonparametric smoothing, and spatial econometric methods. An important conclusion that emerges from the overview presented in Sect. 15.3 is that the first three types of models do not adequately account for the basic characteristic of infectious diseases, i.e., spatial spillover. Admittedly, several of the approaches that have been commonly applied in infectious disease modeling account for similarities among spatial units, notably climate and environmental conditions, which significantly affect habitat suitability and distribution of vectors. However, this is not the same as accounting for spatial spillover. Spatial spillover means that the sheer presence of an infectious disease in one region, at present or in the past, increases the likelihood of occurrence in neighboring regions. Another type of spatial dependence relates to the covariates in that covariates in one region affect the response variable not only in that region but also in neighboring regions.

A major research topic for the immediate future is the development of models that can explain and predict the spatio-temporal distribution of infectious diseases. For that purpose, epidemiological and spatio-temporal econometric models could be combined. The basic structure of such a model that links the log of the relative risk to its predictors is as follows:

$$\eta_{it} = \beta_0 + \rho_1 \sum_{j=1}^{n} w_{ij}\eta_{jt} + \rho_2 \sum_{j=1}^{n} w_{ij}\eta_{jt-1} + \rho_3\eta_{it-1} + \sum_{k=1}^{K} \beta_{1k}X_{kit} + \sum_{k=1}^{K} \beta_{2k}X_{kit-1}$$
$$+ \sum_{k=1}^{K} \beta_{3k} \sum_{j=1}^{n} w_{ij}X_{kjt} + \sum_{k=1}^{K} \beta_{4k} \sum_{j=1}^{n} w_{ij}X_{kjt-1} + u_i + v_i + \omega_t + \psi_t + \phi_{it}, \tag{15.33}$$

where $\eta_{it} = \log(\theta_{it})$; ρ_1 and ρ_2 denote the spatial lag coefficients of the log relative risk without and with time lag, respectively; ρ_3 denotes a temporal lag coefficient of the log relative risk; β_{1k} and β_{2k} denote the regression coefficients with and without temporal lag of the k_{th} covariates, respectively; β_{3k} and β_{4k} denote the spatial lag coefficients of the covariates with and without temporal lag, respectively; u_i and v_i denote spatially uncorrelated and spatially correlated heterogeneity, respectively; ω_j and ψ_t denote temporally uncorrelated and temporally correlated heterogeneity and ϕ_{it} is a spatio-temporal interaction effect. Correlated heterogeneity is variability that occurs because of spatial or temporal dependence; uncorrelated heterogeneity is variability that occurs because of random spatial or temporal variation (Lawson 2006; Bernardinelli et al. 1995).

Model (15.33) is a complex model with a discrete (Poisson distributed) dependent variable, involves many covariates, and is influenced by location and time heterogeneity. Spatial panel econometrics comes to mind to estimate model (15.33). However, spatial panel econometrics has been developed for continuous response variables, while epidemiological data are commonly measured in count format. Therefore, models such as (15.33) cannot be estimated by conventional approaches. The development of appropriate estimators of such models is an important topic for further research. We expect that Bayesian statistics will be increasingly used in epidemiology and regional science models of count data (see also Congdon 2013). For complex models, such as the spatio-temporal varying coefficient model, the calculation of the likelihood function, along with the problem of identifiability of the parameters, is very difficult. The Bayesian method can solve this problem (Martinez and Achcar 2014).

We also expect the random effect generalized linear mixed model and Bayesian inference with INLA to become popular in infectious disease modeling. INLA is a relatively new approach to Bayesian statistical inference for latent Gaussian Markov random fields. The main advantage of the INLA approach over MCMC is that it can compute significantly faster (Rue et al. 2007).

References

Ando AW, Baylis K (2013) Spatial environmental and natural resource economics. In: Fischer MM, Nijkamp P (eds) Handbook of regional science. Springer, New York, pp 1029–1048

Anselin L, Lozano N, Koschinsky J (2006) Rate transformations and smoothing. University of Illinois, Urbana

Bernardinelli L et al (1995) Bayesian analysis of space-time variation in disease risk. Stat Med 14:2433–2443

Besag J, York J, Mollié A (1991) Bayesian image restoration with two applications in spatial statistics. Ann Inst Stat Math 43:1–59

Bivand RS, Gómez-Rubio V, Rue H (2014) Approximate bayesian inference for spatial econometrics models. Spatial Statistics 9:146–165

Chen D, Moulin B, Wu J (2015) Sntroduction to analyzing and modeling spatial and temporal dynamics of infectious diseases. In: Chen D, Moulin B, Wu J (eds) Analyzing and modeling spatial and temporal dynamics of infectious diseases. Wiley, Hoboken, NJ, pp 3–17

Clayton D, Kaldor J (1987) Empirical bayes estimates of age-standardized relative risks for use in disease mapping. Biometrics 43(3):671–681

Congdon P (2010) Bayesian hierarchical method. Tylor & Francis Group, New York

Congdon P (2013) Bayesian spatial statistical modeling. In: Fischer MM, Nijkamp P (eds) Handbook of regional science. Springer, New York, pp 1419–1434

Fong I (2013) Emerging infectious diseases of the 21st century, challenges in infectious diseases. Springer, Toronto

Hog RV, McKean JW, Craig AT (2005) Introduction to mathematical statistics. Pearson Prentice Hall, Upper Saddle River, NJ

Kesall JE, Diggle PJ (1998) Spatial variation in risk of disease: a nonparametric binary regression approach. Appl Stat 47(2):559–573

Lambert DM, Brown JP, Florax RJ (2010) A two-step estimator for a spatial lag model of counts: theory, small sample performance and an application. Reg Sci Urban Econ 40(4):241–252

Lawson AB (2006) Statistical methods methods in spatial epidemiology. Wiley, Chichester

Lawson AB (2013) Bayesian disease mapping, hierarchical modeling in spatial epidemiology, 2nd edn. CRC Press/Taylor & Francis Group, Boca Raton, FL

Lawson AB (2014) Hierarchical modeling in spatial WIREs. Comput Stat. doi:10.1002/wics.1315

Lawson AB, Biggeri B et al (2000) Disease mapping models: an empirical evaluation. Stat Med 19:2217–2241

Lawson AB, Browne WJ, Rodeiro CL (2003) Disease mapping with WinBUGS and MLwiN. Wiley, Chichester

Lee D (2013) CARBayes: an R package for bayesian spatial modeling with conditional autoregressive priors. J Stat Softw 55(13):1–24

Leonard T (1975) Bayesian estimation methods for two-way contingency tables. J R Stat Soc ser B 37:23–37

Leroux B, Lei X, Breslow N (1999) Estimation of disease rates in small areas: a new mixed model for spatial dependence. In: Halloran ME, Berry D (eds) Statistical models in epidemiology, the environment, and clinical trials. Springer, New York, pp 135–178

Lowe R, Bailey TC, Stephenson DB et al (2011) Spatiotemporal modeling of climate-sensitive disease risk: towards an early warning system for dengue in Brazil. Comput Geosci 37:371–381

MacNab YC, Dean C (2002) Spatiotemporal modeling of rates for the construction of disease maps. Stat Med 21:347–358

Maiti T (1998) Hierarchical bayes estimation of mortality rates disease mapping. J Stat Plan Inference 69(2):339–348

Martinez EZ, Achcar AJ (2014) Trends in epidemiology in the 21st century: time to adopt Bayesian methods. Cad Saúde Pública 30(4):703–714

Meza JL (2003) Empirical bayes estimation smoothing of relative risks in disease mapping. J Stat Plan Inference 11:43–62

Pringle D (1996) Mapping disease risk estimates based on small number :an assessment of empirical bayes techniques. Econ Soc Rev 27(4):341–363
Rao J (2003) Small area estimation. Wiley, Ottawa
Rue H, Martino S, Chopin N (2007) Approximate bayesian inference for latent gaussian models using integrated nested laplace approximations. Statistics Report No 1. Norwegian University of Science and Technology
Shaddick G, Zidek JV (2016) Spatiotemporal methods in environmental epidemiology. CRC Press/Taylor & Francis Group, New York
Simonoff JS (1999) Smoothing methods in statistics. Springer, New York
Stern H, Cressie N (1999) Inference for extremes in disease mapping. In: Lawson AB, Biggeri A, Bohning D et al (eds) Disease mapping and risk assessment for public health. Wiley, New York, pp 63–84
Tango T (2010) Statistical methods for disease clustering theory and methods. Springer, London
WHO (2005) Using climate to predict infectious disease epidemics. WHO, Geneva

I Gede Nyoman Mindra Jaya is a lecturer, Department of Statistics, Universitas Padjadjaran, Bandung, Indonesia. His research interests include research methodology, spatial and spatiotemporal econometrics, spatial and spatiotemporal disease mapping, and Bayesian modeling. He is a Ph.D. student, in Faculty of Spatial Science, University of Groningen, The Netherlands since 2013.

Henk Folmer is a professor, Department of Economic Geography, Faculty of Spatial Sciences, Groningen University, The Netherlands, and professor and academic dean, College of Economics and Management. Northwest Agriculture and Forestry University, China. His research interests include research methodology, (spatial) econometrics, environmental and resource economics, life satisfaction and subjective wellbeing. He earned the Ph.D. in economics from the University of Groningen in 1984. He holds an honorary doctorate from the University of Gothenburg, Sweden, and received the Outstanding Foreign Expert Award for Economic and Social Development of the Province of Shaanxi, China, in 2014.

Budi Nurani Ruchjana is a professor, Department of Mathematics, Faculty of Mathematics and Natural Sciences, Universitas Padjadjaran, Indonesia. Her research interest are stochastic process, time series analysis, geostatistics, spatiotemporal modeling and its applications. She earned the Ph.D. in Mathematics and Natural Sciences from Institut Teknologi Bandung at 2002 with a concentration in applied statistics. She is a Dean of Faculty of Mathematics and Natural Sciences, Universitas Padjadjaran for period 2012–2016. She is also a President of the Indonesian Mathematical Society (IndoMS) period 2014–2016 and a member of Commission Developing Country International Mathematical Union (CDC IMU) for period 2015–2018.

Farah Kristiani is a lecturer, Department of Mathematics, Parahyangan Catholic University, Bandung, Indonesia. Her primary research interests are applied statistics in dengue disease mapping; Bayesian modeling; and actuarial science in life insurance. She is a Ph.D. student, in Mathematics Department from Sultan Idris Education University, Malaysia since 2013.

Yudhie Andriyana is a lecturer at Statistics Department, Universitas Padjadjaran, Indonesia. He is currently assigned as the head of Master Program in Statistics, Faculty of Mathematics and Natual Sciences Universitas Padjadjaran. His research interest is Nonparametric Regression, especially working on quantile objective function in varying-coefficient models. He earned his Ph.D in Statistics from KU Leuven, Belgium, in 2015.

Part III
Open Source and Open Science

Chapter 16
Object Orientation, Open Regional Science, and Cumulative Knowledge Building

Randall Jackson, Sergio Rey, and Péter Járosi

16.1 The Future of Regional Science Modeling

Integrating human and physical systems is a daunting challenge that spans a great many problem domains, including social and economic production systems, residential behaviors, environmental exchange, and resource and land use. Because so much current research continues to be focused within rather than across these areas, our cumulative knowledge in many respects is little more than a simple summation of various disciplinary and sub-disciplinary learning curves, rather than a truly integrated, synergistic base of understanding. Indeed, a complete understanding of any subdomain may not even be possible in the absence of domain integration. Even *within* some subdomains, there may be very few instances of truly cumulative science, where one scholar's work adopts another's directly as the foundation for a new and tightly integrated cumulative model. If it were possible to speed the diffusion of modeling innovations and research findings within and among subdomains, the cumulative frontiers of knowledge could be expected also to advance apace.

We believe that the future of research in regional science, and indeed in all social science modeling, will be based on a research infrastructure that leverages the power of networked individuals focusing their collective intellect on problem

R. Jackson (✉)
Regional Research Institute, West Virginia University, Morgantown, WV, USA
e-mail: rwjackson@mail.wvu.edu

S. Rey
Arizona State University, Phoenix, AZ, USA
e-mail: srey@asu.edu

P. Járosi
West Virginia University, Morgantown, WV, USA
e-mail: peter.jarosi@mail.wvu.edu

R. Jackson, P. Schaeffer (eds.), *Regional Research Frontiers - Vol. 2*,
Advances in Spatial Science, DOI 10.1007/978-3-319-50590-9_16

solving in a community effort as we move science from the domain of individual ivory towers and research silos to a fully integrated common workplace. The research environment we envision stands to *accelerate research integration and cumulative knowledge-building* within and across human and physical systems problem domains.

16.2 OS^2: Open Science and Open Source

Open science and open source are strongly related but not identical concepts. Open science refers to a scientific field that moves forward as a collective and is open to all participants. Open source refers to equal public access to and development of problem domain content, primarily the computer code that supports modeling and solution algorithms applied within a given problem domain. We refer to this powerful combination of open science and open source development as OS^2.

16.2.1 Open Science

The rise of the open science movement is a recent phenomenon, and as such, regional modeling has been slow to engage (Rey 2014). A key tenet of open science is that for the traditional error-detection and self-correction mechanisms to be fully effective, all aspects of the scientific process need to be open. In theory, open access to the data, models, and workflow that underly a scientific study should allow other researchers to reproduce its findings. Reproducibility removes the veil from scientific findings and eliminates the need for blind faith in science and the scientist.

Reproducibility is vital to the integrity of the scientific process and assumes a central position in the open science movement, yet open science is about much more than enhancing reproducibility. New forms of open collaboration and open publishing hold the potential to advance the pace of scientific discovery and to ensure the provenance of scientific knowledge. While collaboration has always been central to scientific progress, the scale of collaboration afforded by new technologies is now on the brink of a radical transformation. Advances in high performance computing (HPC) in the form of distributed systems provides unprecedented opportunities for addressing scientific problems once viewed as beyond reach. However, realization of this potential will require collaboration among domain scientists and with computer scientists with HPC expertise. That collaboration, in turn, will require open computing frameworks with well-developed application programming interfaces (API). Scaling existing regional modeling software to take advantage of advances in modern HPC architectures is one area where this form of collaboration will have high payoff.

In many ways, the lineage of these "new" open science practices can be traced to the open source movement. Community innovation networks are already commonplace in open source software development, where legions of developers often contribute to evolutionary community resource infrastructures such as XWindows, the Linux operating system, and the Python language and its numerous graphical and numerical processing libraries. Indeed, the suggestion that this kind of approach should be adopted in social sciences dates back at least two decades to Jackson's "Object-Oriented Modeling in Regional Science: An Advocacy View" (1994); a call to action that failed to gain momentum for two main reasons. First, object orientation, essential to the success of the proposed approach, was still in the early stages of development and was not stably supported in widely used and freely accessible computer software. This has changed dramatically in recent years, especially notable in the popular and widely used open source Python programming language. Second, the notion of collaborative innovation networks (Gloor 2002, 2006; Gloor et al. 2004) and associated support infrastructures had not yet been formally recognized or well established.

Common workplaces such as GitHub.com, which provides controlled access, version control, and other mechanisms, such as code repositories and community forums that rationalize the development process are now much more common, more effective, and well supported. The development and convergence of these tools, along with a winnowing of methods for modeling national and regional economic systems makes this a perfect time to *move from silo-based research efforts to a mode of collective open science knowledge building*.

16.2.2 Open Source

Our choice of open source software and development practices in implementation of the modeling framework also reflects the philosophy of open science that informs our project. Recent developments in the Python programming language make it an ideal platform for the development of these models. Python is an object-oriented scripting language that facilitates rapid prototyping of software. Because the structure of Python's numerical functions and algorithms (e.g., in NumPy and SciPy) will be readily recognizable by those who program using traditional econometric modeling software (e.g., GAUSS and MatLab), leveraging legacy code written in those languages and porting to an object-oriented design becomes feasible.

The Python scientific community also has been at the forefront of the recent drive for reproducible research. Tools such as the Jupyter Notebook (http://jupyter.org) allow modelers to combine live code with narrative text, equations, visualizations and other media to encode a complete and reproducible record of computation. These notebooks can be made available to other researchers via GitHub repositories, to facilitate open collaboration.

By relying on public GitHub repositories, collaboration on regional modeling projects not only becomes more efficient, but also may achieve currently unparalleled scalability. Any interested regional modeler can now "fork" the project to begin their own exploration of the underlying code base. That exploration can take place without the modeler having to first receive permission for copying the project. Thus, the entry costs for engaging with the modeling project fall dramatically.

Not only does OS^2 allow for an expansion of the modeling community, but it does so in a highly efficient way. Individual efforts undertaken as part of the community receive rapid feedback, often virtually at the moment of the newly shared contribution. This can include the user tests, bug reports, new feature requests, etc. In this way, the research work flow can become a nearly continuous iterative process among any collaborators, anywhere.

Wallach (2016) has argued that research at the frontier of the social sciences is no longer a choice between computer science *or* social science but must be a synergy of the two moving forward. We see OS^2 as an integrating framework that addresses this call by fusing the practice of regional modeling together with modern principles of computer science.

16.3 Object Orientation

Object orientation is an abstraction mechanism that is used to focus on the essential problem domain constructs to eliminate the complexities of non-essentials. Object-oriented (OO) modeling is a conceptual device that can be used to better understand a problem domain. It is analogous in this sense to general systems theory in its provision of a recipe to follow in defining and understanding a problem. Object-oriented analysis focuses first on the identification and enumeration of the objects that compose the system, rather than on system functionality. Constructed first, object models describe as fully as possible the objects, their attributes and behaviors, and the information they can exchange with their environments (Rumbaugh et al. 1991). A functional model complements the object model, defining interactions and associations among objects. These behaviors are defined by transformation rules, functions, and mappings, and may conform to constraints and follow various patterns of dependency. A dynamic model is the final complement, defining the sequencing and control of the problem domain. Object-oriented analysis involves the systematic construction of these three "orthogonal views" of a problem domain, as shown in Fig. 16.1. An object-oriented model includes an enumeration of its objects, the ways in which a system transforms its values, and an elaboration of the timing, sequencing and control of events.

There are many reasons to pursue the object-oriented approach. First, if a model is to form the foundation of experimental research, that foundation should be as stable as possible. The objects of most problem domains are much more stable than is their functionality. Indeed, most research focuses precisely on the effects of specified changes on a system's objects and operation. Object-oriented modeling establishes a solid foundation that provides a stable reference for subsequent use, reuse, and extension. Second, the modeling sequence is both rearranged and structured more explicitly than in relational modeling. Whereas most relational modeling focuses first on functionality, object-oriented modeling focuses first on

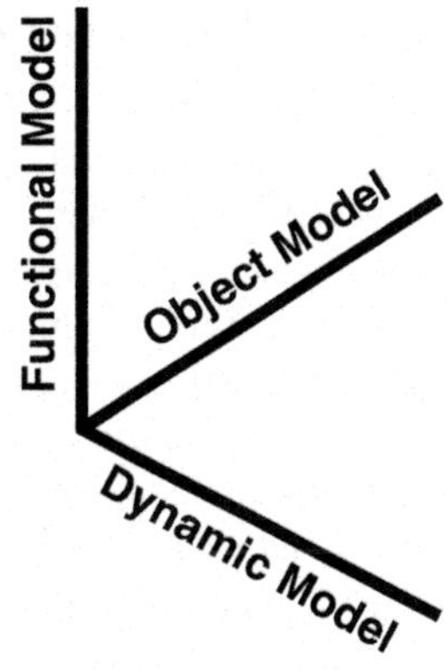

Fig. 16.1 An object-oriented model's orthogonal views

the model's objects. Because the recipes that we follow to build our understandings shape the processes and outcomes of inquiry, new recipes often lead to new questions, new hypotheses, and ultimately to a more comprehensive understanding of a problem domain.

A third reason for exploring object-oriented modeling is the potential to benefit from increased interaction. Scientists each have specialized areas of expertise. Adopting a common modeling approach and foundational reference model can enhance and facilitate communication of the essence of each application subdomain. Extensibility is a fourth and exceptionally strong reason for adopting object-oriented modeling. Object classes can be extended easily and independently without the need to modify interactions among class objects because of the encapsulated nature of class data and behavior.

Importantly for the present context, models can be developed incrementally. All problem domain modules need not be fully specified to productively develop subdomain modules. Teams of researchers can begin to collaborate much more effectively. A model of a production system, for example, might use a naïve representation of households until another researcher, with expertise in household consumption or residential choice behavior, develops a more comprehensive and realistic household module.

Finally, alternative behavioral propositions can be represented in class specifications. Suppose, for example, that a researcher wanted to isolate the systemic environmental impacts of introducing two alternative power-generating technologies. He or she could then design one new class for each technology, run the model simulation first with the existing technology class, and then once with each of the alternative technologies and compare the outcomes. This simulation approach parallels the "plug and play" design characteristics of modern personal computers, where parts with slightly different functionality (e.g., sound cards) can be interchanged freely. Because they have the same system interface, their inner workings can differ in important ways, yet still be compatible with the overall system.

16.3.1 The Case for Objects

Many model integration strategies have been less successful than they could have been, partly due to the failure of modelers to recognize the advantages of object-based modeling paradigms and more recently available supporting modeling platforms. Whereas most attempts at model integration link modules through aggregate and summary variables, module integration can be facilitated by *the explicit recognition of individual object integration* as a mechanism for linking modeling subdomains. As a simple example, consider that laborers who earn wages and salaries are the same individuals who shop, commute, migrate, choose residences that consume electricity and water, have children, etc. The cars they purchase are the ones they use in their journeys to work, and are the same ones that pollute the atmosphere. Laborers, therefore, constitute one logical class of objects in models of any of these activities. Thus, when modeling two of these problem subdomains together, maintaining the identity of individual laborers (among other objects) can be the integration linkage mechanism. With the exception of the related class of agent-based models (ABM), there are very few models that explicitly incorporate object identity.

A common modeling language can also promote cumulative and integrated model building. Mathematical formalization plays this role with some success, but mathematics is a low-level formalization, in the same sense that assembly language is a lower level programming language than is FORTRAN or Matlab®. Commonalities among subdomains, as a consequence, are not always readily apparent from their formal representations. Quite often, even subtle differences in modeling notation can be a barrier to effective cross-domain fertilization and integration. In the absence of a common modeling language, specialists in one subdomain often find it difficult to grasp quickly the essentials of a model in another.

The most frequently used objects of mainstream economic models are deterministic and stochastic equations, endogenous and exogenous variables, recursive and simultaneous blocks of equation systems, etc. In stark contrast, the object-oriented economic model comprises objects like households, firms, industries, and markets, that represent the entities of the economy more directly. The object-oriented model can be designed around objects along a continuum from individual agents to aggregates. Financial sectors or industries, for example, could either be modeled as aggregates or as individual banks or establishments, emphasizing the opportunities of object-oriented modeling for both micro level and macro levels. In an object-oriented program, a class of objects can represent anything from a typical agent to an entire interregional interindustrial system.

Fortunately, human and physical systems modelers can benefit from the experience of software engineers who have had to model increasingly complex computer-related systems that would quickly overwhelm any individual programmer. Computer and information sciences have made great strides in developing common workplaces and computer languages with effective diagrammatic toolkits that support a variety of conceptual representations, including object orientation. Most graphical user interfaces, e.g., are built with windows, panels, dialog boxes, text fields, dropdown lists and the like, which are modeled as objects with specified attributes and event-driven behaviors and that send and receive signals to and from other objects and algorithms. As a result of their efforts, computer modeling of complex systems via collaboration and teamwork is now commonplace.

16.3.2 Object-Oriented Modeling Fundamentals[1]

Object orientation is a systematic approach to modeling that can improve our conceptual understanding of research problem domains. Its modeling constructs, coupled with an intuitive graphical notation, provide an expressive set of conceptual descriptors that can enhance the model clarity. While object-oriented modeling shares much in common with a number of other approaches, such as Entity-Relationship (ER) modeling, ABM and simulation, and micro-simulation generally, the advantages of object-oriented modeling, per se, include its precise and easily understood terminology, its orthogonal object, functional and dynamic conceptual frames, graphical tools for depicting objects and associations, and its parallels with programming language terminology. Below, we review the fundamentals of object-oriented modeling, beginning with a more formal definition of objects.

Objects are the fundamental entities of the object-oriented model. They are abstractions of the essential aspects of a problem domain and are easily distinguished from one another in form and function. Objects are of various types, or classes, and are individual instances of the classes to which they belong. They are described by their properties: attributes and behaviors. An object's attributes are quantifiable characteristics that can take on data values. Its behaviors capture its functionality, and include the operations it can perform and the services it can provide, including self-contained operations and signals it can send and receive. Conducting a residential search, e.g., is a part of a household's functionality and is therefore one of several household object behaviors. Other behaviors can be much simpler, such as setting or reporting the value of an attribute to another object in response to an event.

Identity, classification, inheritance, aggregation, polymorphism, and encapsulation define the essence of an object-oriented model. ***Identity*** is established when an object is created (instantiated). Without identity, objects, classification, and encapsulation lack meaning. With identity, they can come into or go out of existence. Business establishments start up and shut down, can adopt and adapt managerial schemes, and can adopt new and abandon old technologies; individuals are born and die, and can change residences; and governments can implement, modify, or

[1]Parts of this section draw heavily on Jackson (1994, 1995). Seminal contributions and more complete descriptions can be found, inter alia, in Booch (1994), Rumbaugh et al. (1991), Coad and Yourdon (1991a), Coad and Yourdon (1991b), and Jackson (1995).

retract policies, all while maintaining their respective identities throughout their lifetimes. Because of object identity, all objects, as members or instances of classes, are distinct even if all of their attribute values and behaviors are identical. An object can change its attribute values, but still be identified as the same object.

Classification is an abstraction mechanism fundamental to human understanding. In object-oriented modeling, objects with identical properties belong to classes. A class is an invariant description of object structure. All establishment objects, for example, have "number of employees" as an attribute. The value of this attribute will differ from object instance to object instance, but all establishment objects will have this and other attributes in common. The act of classifying forces focus onto the essential, inherent aspects of the problem domain and its elements and provides a structured context within which modeling abstractions can be placed and ordered.

Inheritance refers to the class–subclass relationship. A subclass inherits the properties of, and is distinguished from, its super-class by new and distinctive properties. The inheritance mechanism is used to implement the *is a* (or *is a kind of*) relationship and serves to reduce repetition and complexity in model building. Subclasses at lower levels in a class hierarchy are derived from their antecedents, or superclasses. Inheritance allows different classes to share fundamental structure, which enhances the conceptual clarity of a model by reducing the number of distinct cases to be understood and analyzed. Inheritance also promotes model *extensibility*. Given a particular class hierarchy, extending it to model similar objects that have additional essential attributes or behaviors is straightforward.

A simple example of inheritance can be found in Járosi and Jackson's (2015) proof of concept technical document. They defined a household superclass (parent object) with a default Cobb-Douglas utility function, and from it derived a Stone-Geary type household subclass (child object). The child/parent analogy is apt, as children and subclasses inherit the attributes and behaviors of their parents and superclasses, respectively. Like children, subclass properties may be redefined and overwritten, and other properties (attributes and behaviors) can be added.

Objects are related through a variety of associations. ***Aggregation*** is a special type of association for which all objects of a given class are parts of a composite object. Actions taken on the composite can be automatically taken on the component parts. As an example, where no information is available, an industry might be modeled as a single entity, but where data are available and intra-industry variation is important, individual establishment objects might compose an industry aggregate. When the industry receives a signal to satisfy accumulated demand, its establishments receive the signal to provide their contributions to the industry response. Whereas generalization and inheritance describe the relationships among an object's associated classes and superclasses, aggregation relates objects of two distinct classes, one of which *is a part of* the other.

With ***polymorphism***, an operation of the same name can behave differently on objects of different classes, and an identically named attribute of two classes may be represented by different data structures. Operations of different classes can share the same semantics, but be implemented in a fashion appropriate to each. As an operation, for example, multiplication has a clear meaning, but its *implementation* differs with the nature of the operands. We can apply polymorphism to such concepts as industrial plant vs. human *aging*, service vs. manufacturing *production*, and wetland vs. cropland *conversion*. As a more concrete example, in traditional computable general equilibrium (CGE) models, it can be difficult to replace a one kind of production function by another, or to have industry specific functional forms. Even a small change in a single equation can cause unexpected, unintended, and even undetected consequences for the whole equation system. This happens because traditional modeling effectively forces researchers to think relationally rather than in terms of objects and behaviors. The one-two punch of encapsulation and polymorphism combines to underscore the advantages of the object-oriented approach.

Encapsulation refers to the process of hiding the internal details of object properties and behavioral implementations from view and tightly binding (or coupling) attributes and behaviors to objects. It reduces unnecessary interdependencies among objects in a problem domain and localizes any system changes. Through encapsulation, objects become virtually self-contained entities. They can be used confidently in one or many modules (and ultimately, models) in which they play an essential role. As long as the interface for an object is not diminished, it can be used, reused, modified and extended without fear of altering either the data values of other objects or the ability of other objects to access object data or trigger object behavior. Should a household object from a production model be integrated into a housing stock model, for example, it would be appropriate to add to it attributes such as square footage, but without altering other roles played by the household object in integrated problem domains. Likewise, should an industry switch technologies, only properties *within* that object need to be altered.

Class and inheritance relationships are consistent with the way in which humans organize information to understand better the world around them. Object identity provides a mechanism for linking different subdomains to capture interdependencies that surpass our ability to express analytically. Encapsulation ensures the integrity of data and behavior of objects, modules, and models, and protects against unintended consequences that are more likely to occur in classical structural programming approaches. Object models and associated class hierarchies are extensible. Encapsulation and extensibility should facilitate the cumulative science enterprise.

16.4 Object-Oriented OS^2 in Action

Systems models are ideal candidates for object-oriented open source development. They often comprise multiple subsystems, and subsystems also may comprise additional subsystems. The subsystems comprising each level can be simple additive collections or they can be interacting. Figure 16.2 conveys this idea graphically, where the larger system, represented by the gray circle, comprises three relatively independent subsystems, and three heavily interacting subsystems. Three of the first level subsystems are further composed of second level subsystems, and three

Fig. 16.2 Models as systems of systems

of these have third level subsystems. Each of these systems might correspond to distinct problem domains, and the larger system might span multiple disciplines.

As we progress in the knowledge building enterprise, each subsystem might well represent a problem domain that would encompass the entire knowledge base of a domain specialist. Likewise, a specialist in a system at any of these levels might well be required to make substantive improvements to a model of that system. Subsystem changes and their impact on the model of the whole system, however, can sometimes only be fully understood in the context of a larger and more comprehensive system model. Historically, modelers who wished to work on subsystems of larger comprehensive systems would have two options. The first is to become familiar enough with the encompassing system to develop a model that could be used as a kind of "backbone" that would provide at least a skeletal framework of salient system behaviors. They would then demonstrate the backbone model behavior with and without subsystem modifications to gain an understanding of partial effects. The second option is to identify a backbone model that is already in use, then attempt to gain access to it from the model's owners, and if successful, attempt to integrate their behavioral modeling improvements into the borrowed framework.

The first option has the disadvantage of requiring subsystem experts to devote time, energy, and intellectual capital to activities that lie outside of their primary fields of expertise. If there are multiple scientists working on the same problem domain, this clearly results in duplication of effort, since each must work outside their areas of expertise on backbone development, when, if there were an open source backbone available, none of them would need to redirect their efforts, and the time saved could instead be focused on researchers' own specialties. Perhaps less obvious is that if multiple experts develop subsystem modeling alternatives *along with* their own backbone models, then the difference in overall system behaviors will be a function not only of differences in subsystems, but also of the system backbones they have developed. This renders subsystem model comparisons difficult if not impossible, and further, it makes replication unlikely or even impossible.

The second option has its own disadvantages. First, it can be difficult to gain access to backbone models, either because such models are proprietary (either commercial or public laboratories where intellectual property is closely guarded), or because such models are so extensive that thousands of lines of code support the system models and transferring the models is difficult due to place or modeler dependency. The second drawback becomes apparent when the subsystem domain specialist is faced with the often daunting task of identifying specific mechanisms for integrating the new subsystem behavior within the larger modeling framework, and doing so without unintended consequences that often result when models are not developed with the kinds of modularity that supports extensions and enhancements. And third, models extended in this way remain closed to public view. Replicability under this option is also difficult if not impossible.

Object-oriented OS^2 modeling paves the way. Those with appropriate expertise can focus on developing the backbone. The wisdom of the crowd ensures that the salient backbone features are present and that each new backbone enhancement has endured the scrutiny of numerous others with similar expertise. Object-oriented OS^2 modeling can accommodate competing perceptions of appropriate system representations by providing an interface from which users can customize model features (e.g., endogenous vs. exogenous government sectors, various model closure assumptions, etc.). Such customizations can be documented in metadata configuration files, enabling replicability and simplified comparisons of outcomes from competing models. Because of the encapsulation and modularity of object orientation, modules with differing behavior can be substituted easily one for another in "plug-and-play" fashion, further facilitating model comparisons. Object-oriented OS^2 provides a foundation for ceteris paribus modeling.

In the remainder of this section, we present a model we are developing to serve as an exemplar for object-oriented OS^2 regional modeling. We review our problem-specific motivation, provide a description of the general class of models to which the exemplar belongs, and compare our model development and implementation approach to other modeling paradigms.

16.4.1 Motivation: Technology, Economy, and Environment

Environmental and socio-economic consequences of technological transitions are beginning to dominate scientific and policy discussions. Deepening our understanding of human and physical systems and their complex interactions has been a federal-level goal since the formation of the Committee on Human Dimensions of Global Change in 1989 by the National Research Council and other supporting agencies, and a great many related federal agency programs and initiatives have emerged since. Examples include the U.S. Department of Agriculture National Institute of Food and Agriculture program that targets improved economic, environmental, and social conditions, and National Science Foundation programs such as the Science, Engineering, and Education for Sustainability initiative aimed at informing "the societal actions needed for environmental and economic sustainability and human well-being", and the Environment, Society, and the Economy initiative to "encourage productive interdisciplinary collaborations between the geosciences and the social, behavioral, and economic sciences." Likewise, a recent Congressional Research Service report (Carter 2013) on the Water-Energy Nexus highlights the interdependencies among energy and water systems and calls for a more integrated approach to the challenges of confronting related issues that impact human welfare so forcefully.

Instead of comprehensive systems integration research, however, all too often what we see are models that, despite often achieving some level of integration, are developed and used only for specific problems and problem domains without the benefits of reuse and extension that would lead to *cumulative science and effective knowledge building*. Far too many scientific explorations begin with modelers reestablishing their own variations of modeling foundations that others already have formulated, on which their own conceptual and theoretical extensions and advances will be built. The commonalities among models that result from such individual research efforts are low, and model comparability and interoperability become excruciatingly difficult or simply impossible. What should be a steady march in a community-wide *cumulative knowledge-building enterprise* instead becomes an atomistic process where countless hours and substantial resources are wasted in foundation-building activities that duplicate the efforts of others. As a consequence, knowledge accumulates much more slowly than it otherwise could and should.

Because increasing specialization is now more common than expanding breadth of knowledge across domains, it is unlikely that individual researchers will be able to achieve these science integration goals on their own, so changing the current modus operandi is likely only by shifting to a more cooperative and collaborative knowledge-building environment that forms a scientific milieu in which researchers build on, incorporate, and benefit mutually from others' expertise through participation in a collaborative innovation network. Our vision of the future centers around OS^2 knowledge-building enterprises, with object-orientation as the foundation for organizing and managing the development of modeling applications across a range of problem domains. We now describe the Object-oriented Analysis

Fig. 16.3 Interlocking hierarchical systems

and Simulation of Industrial Systems (OASIS) model, which will be our foray into this kind of development in the economic and environmental systems modeling context. We envision a team of researchers working in a *community-wide knowledge building enterprise by developing the underlying OS^2 modeling framework that will provide a common modeling foundation for future integrated systems research.*

For an increasing number of research problem domains, subnational regions are the appropriate analytical units. That this is true for economic systems is evidenced by regionally focused programs of the U.S. Economic Development Administration (http://eda.gov/oie/ris/), and the CRS report on the energy-water nexus referenced above provides similar evidence for environmental, resource, and water issues. Of course, processes at the regional level often feed back, shape, and influence their national counterparts, just as regional economies compose their aggregate national counterparts, as in Fig. 16.3. Environmental systems and processes can operate locally, but not in isolation from the global. Energy, environment, and even health policy models are often developed without the benefits of integration with easily accessible and reproducible economic models, while those who do recognize the need to link other systems to regional and national economy very often resort to proprietary, commercial sources.

OASIS will model the U.S. and its regions, providing current and forecast input in the form of macroeconomic, household, and industry-level trends and constraints that establish the context for national economic systems models, nationally driven regional models, and integrating mechanisms for interregional and regional-to-national integration and feedbacks. The modeling platform will be open source and evolutionary, systematically embedding behaviors and characteristics of the backbone model that are deemed by the broader research community to be essential and stable, and weeding out those aspects that can be replaced by better representations. Its implementation will enable researchers to select from among system features that have yet to earn consensus approval, and from those that have been sanctioned by the user community but that might represent alternative behavioral assumptions. Indeed, an eventual suite of alternative modeling variations with explicitly identifiable commonalities and differences will promote direct and replicable model comparisons and contrasts.

A class of models that is particularly well suited to object-oriented modeling is known as space-time economic (STE) models. STE models can be calibrated and parameterized to represent the existing structure of an economy, and to forecast, incorporate, and respond to changes in that structure. In the process, temporal changes in prices, interest and wage rates, output, employment, income and the like are determined, carrying clear implications for socio-economic impacts across different groups in the economy. Barker (2004) has provided an excellent discussion of the relative strengths of the STE framework in the context of modeling the transition to sustainability. Unlike existing relational models, OASIS will be engineered from scratch as an object-oriented STE model. Its initial character will be influenced by existing STE models, but its implementation and eventual form will reflect not only the adaptability and flexibility of object orientation, but also the benefits of conceptual refinements by the initial project team and ultimately the broader research community.

Essential elements of the initial OASIS model will parallel many of the most common dynamic hybrid macroeconomic interindustry models developed and reported in the literature.[2] While model implementations differ, an idealized STE

---

[2]Some who have developed and used relational STE models include Dick Conway, who has used these models productively for decades in Washington State, Hawaii, and elsewhere; Geoffrey Hewings with models of Chicago, St. Louis, and the U.S. Midwest states region; Randall Jackson with models of Ohio, and the U.S., José Manuel Rueda-Cantuche and Kurt Kratena for the EU-27, Sergio Rey for various California regions; Clopper Almon, Douglas Meade and others at the University of Maryland with the INFORUM model of the U.S. and many other countries; and Guy West, who has applied interindustry econometric models to policy issues in Australia and its regions (for a small selection of related literature, see Conway 1990; Donaghy et al. 2007; Kim

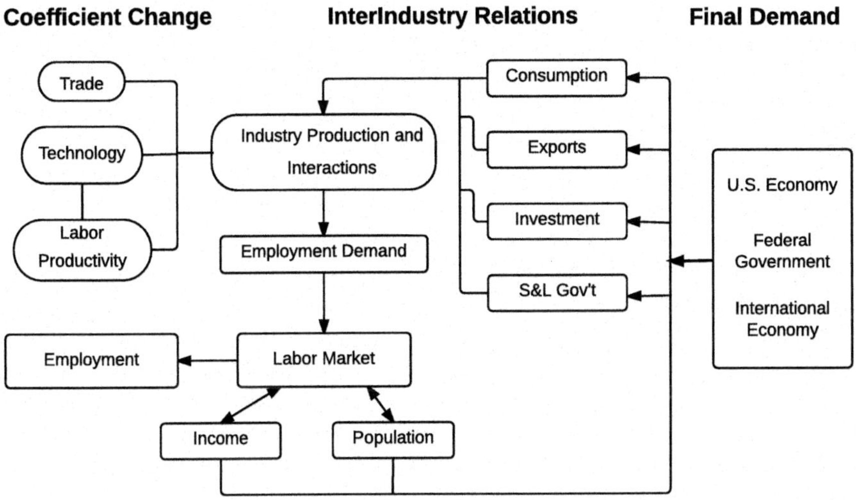

Fig. 16.4 Idealized structure of STE models

model structure is shown in Fig. 16.4. These models most commonly include econometrically specified forecasts of key economy-wide variables such as interest rates, unemployment rates, final demand activities, and population. Some regional models rely on exogenous national forecasts, while others generate national forecasts endogenously. Coupled, linked, or fully integrated with these economic drivers are industrial system relationships that tie economy-wide forecasts to industry-specific activity, and to households and household consumption activities through payments to labor. Payments to governments by industry and returns to capital are also tracked by industry, and labor and non-labor income can feed back to savings, investment, and additional consumption behavior. Models developed for different purposes have focused on specific aspects of system behavior, so while there is much in common across these models, there can be substantial differences. This allows for results that illuminate different system behaviors, but it also results in great difficulty in comparing the outcomes of different models. The OASIS backbone will facilitate the isolation of impacts of specific model behaviors by providing a common foundation on which behavioral extensions will be built.

Because of their position at the nexus of economy and environment, industries and their technologies will be represented explicitly as a primary class, providing a mechanism for linking systems. Technology plays a central and potentially unifying role in virtually all of the most critical issues that give rise to the need for integrated systems modeling. Human–environmental exchange takes place primarily through the operation of various technologies, be they transportation,

---

et al. 2016; Israilevich et al. 1996, 1997; Kratena et al. 2013; Rey 1997, 1998, 2000; Rey and Jackson 1999; West 1991; West and Jackson 1998, 2014).

agriculture, manufacturing, consumption, or power generation, and many of the most important such exchanges reside in the technologies used by industries in economic systems. Industrial processes use inputs from one another and from the environment, and their production activities alter air, water, and land characteristics. Hence, models that promise to integrate human and physical systems virtually all rely on mechanisms that provide meaningful representations of the economy, industry, technology, and environmental relationships.

Early OASIS subsystem enhancements will focus on industry and household objects. Industries are key to the modeling system because they dominate uses of the technologies that can be tied to both social and physical systems. Households are also key to system integration because of their critical role in driving economic activity via expressed demands, because they are the central providers of labor and are explicitly linked to industrial activity, and because differential demographic characteristics of households are dynamic and have been shown to have highly significant impact on consumption, housing, health, and environment (see e.g., the chapter by Hewings in this volume, and Kim et al. 2016). Developing alternative classes of households and industries will demonstrate key aspects of the object-oriented modeling approach and ways in which it speeds the knowledge building process.

The advantages of the object-oriented framework will be clear immediately. The OASIS model will have commodity supply- and demand-pool market objects that act as clearinghouses for commodities produced and demanded by industry and other economic entities. Indicative of the increased adaptability and extensibility of the object-oriented approach, consider the necessary actions to be taken when, as a simple example, a new industry is established in a region. In relational dynamic interindustry models, each industry's intermediate demand equation includes a term for demand from each and every other industry. Hence, adding one new industry to a traditional economic model with 200 industries necessitates determining and making corresponding changes to the existing 200 demand equations, and then adding the 201st equation—for the output module alone. Employment, income, and potentially other equations would have to be adjusted similarly. In the OASIS model, encapsulated behaviors and interfaces of industry objects will mean that adding a 201st industry will be a matter of object instantiation, since it is already a part the industry aggregation makes up the economic system. Default production behaviors production functions can optionally be replaced by alternative forms, e.g., allowing for economies of scale and input substitution, and each industry can have its own unique production functional form if and as desired.[3]

Another advantage derives from flexibility in terms introducing exchanges among industries and the environment. Water, resources, and emissions accounting can be added to or modified within the system on an industry by industry basis as new and improved data become available. As in other systems modeling frameworks

[3]A step further would allow for an industry to comprise collections of establishment level agents with more or less autonomous behaviors.

(commonly commercially based), environmental stores for accounting can be added to the OASIS model simply by creating those objects and modifying globally the respective industry class properties and object attributes. Additional system elements, such as environmental remediation processes, can be introduced as new classes and objects, with interfaces to environmental stores (as one approach). These simple examples demonstrate dramatically the advantages of encapsulation in object-oriented modeling frameworks.

16.4.2 STE Feasibility and Data Requirements

As a proof-of-concept exercise, we recently designed and implemented a CGE model of a small (3-sector) economy based on a hypothetical social accounting matrix (SAM). The model we developed recasts the conceptual basis of the SAM to model industries and households as objects, and the industrial system as an aggregation of industries. See Járosi and Jackson (2015) for details and accompanying computer code.

STE models are calibrated using a fairly extensive and wide-ranging base of supporting data. All of the data required for early versions of the OASIS model, however, are publicly available. Nearly all of the data are secondary data published by U.S. government agencies, and there is a variety of sources that make these data series available electronically. In addition to government agency websites, other groups compile and provide access to these data. Much of the data for an existing WVU hybrid econometric interindustry relational model, for example, are compiled and made available as a resource accompanying the freely and publicly available Fair econometric model.

16.4.3 Object Orientation vs. Other Modeling Approaches

Adopting the object-oriented approach in no way supplants established theory. On the contrary, object-oriented modeling provides a consistent foundation on which established theory can build. Even in cases where no simulation model might ever be implemented, the conceptual process of placing existing models within a single integrated framework (1) forces the exploration of relationships among problem domains that currently are unspecified, (2) potentially identifies inconsistencies among models, and (3) identifies directions for profitably extending existing model specifications.

16.4.3.1 Early Systems Microsimulation Modeling

Although there is a natural similarity between the object-oriented approach outlined and the microsimulation approaches of the early and mid-1960s, object-oriented modeling has much greater potential for success, and for many reasons. First, neither the hardware capacity nor the software tools were available then to model social science simulation aggressively. Today, there are graphical tools for designing software that not only assist us at the stage of conceptual design but in some cases can even automatically generate skeletal code in selected computer programming languages. Object-oriented programming languages now allow the simple expression of constructs that once required intensive and meticulous project oversight and programming efforts. An object-oriented conceptual model is a very short step from programming language code.

16.4.3.2 Modern MicroSimulation

There also is a separate body of literature founded on microsimulation methods. Caldwell (1983), Clarke and Wilson (1986), Clarke and Holm (1987), and Amrhein and MacKinnon (1988), for example, have used micro-simulation approaches in early urban and regional labor market and planning models, while Birkin and Clarke (2011) provide an overview and prospective of spatial microsimulation methods and applications. While the experiences and results of microsimulation efforts can help to identify critical model formulation and evaluation issues, microsimulation and object orientation are fundamentally different conceptually and operationally.

16.4.3.3 Agent-Based Modeling

Agents in ABM share a conceptual heritage with objects in object-oriented models. Although there are some strong commonalities, agents are generally autonomous entities that often require no external control mechanisms to initiate or govern their behaviors. Odell (2002, p. 42) explains that among their fundamental distinguishing attributes, agents are capable of watching "out for their own set of internal responsibilities," and "when and how an agent acts is determined by the agent." In contrast, he continues, "objects are considered passive because their methods are invoked only when some external entity sends them a message." Control in an object-oriented model is thus more centralized, which makes representation of a system of interrelated systems a much more tractable problem. Ultimately, of course, objects can comprise agents, and certain object behaviors might eventually take on characteristics of agents in ABM. There are other differences in terms of scope and computational requirements that lead us to prefer object orientation for our higher-level organizing structure.

16.4.3.4 Computable General Equilibrium (CGE) Modeling

CGE modeling is a well-established framework for impacts assessment research. It is founded squarely on neoclassical economics and produces outcomes from economic and policy shocks that correspond to values from restored equilibria in product, factor, and capital markets, optimizing with respect to firm and household behaviors. What distinguishes object-oriented models from CGE models is the focus on individual objects rather than relations. Object-oriented modeling allows us to specify as many different classes of elements in multiple systems as deemed appropriate and to track the behavior and status of individual elements within these classes—including, e.g., how household structures change and how the size composition of industries evolves. Although Barker (2004) and Scrieciu (2007) have cautioned against the use of CGE as a single integrated framework for sustainability impact assessment, behaviors similar to classical CGE models, including household utility maximization and firm profit maximization, or cost minimization could be incorporated into future versions of OASIS by modifying class behaviors. However, mechanisms available for linking a CGE model to transportation networks, land uses, and physical systems are much more limited, constrained, and opaque than they will be in the OASIS model. The focus on object identity provides options for specific mechanisms for subsystem model linkage and extensions. CGE modeling requires a relatively high level of economics training and computer programming skills to be used effectively, which could in turn limit the size of the community innovation network were CGE models to form the basis of an OASIS-like effort. Nevertheless, parallel object-oriented OS^2 CGE modeling could be pursued by researchers so inclined.

16.4.3.5 Inforum InterDyme

Of all of the STE models we have identified, the Inforum InterDyme system may be conceptually the closest to the modeling strategy proposed here. The INFORUM group has been among the most continually active and innovative in the U.S. Its InterDyme software is a package of programs for building interindustry dynamic macroeconomic models, developed by INFORUM and written in C++. Online documentation (http://www.inforum.umd.edu/papers/inforum/software/dyme.pdf) and personal correspondence with Inforum personnel suggests that the object-based character of their model lies primarily in algorithmic aspects like matrix, vector, equation, and time series objects, so the object-oriented conceptualizations in Inforum are fundamentally different from those of the proposed OASIS model. The Inforum models are viable econometric interindustry modeling options for certain analysts with strong and diverse programming and modeling skills, but our vision for OASIS is that of a much more easily accessible and user-friendly platform for a wide range of analysts.

16.4.4 Synergies and Flexibility

The long-run vision for OASIS is that of a flexible modeling foundation with a range of modeling options. We envision a graphical user interface for stable model versions that will present modeling default and alternative options to users in menu-like fashion. Industrial production function alternatives, household behavior options, model closure rules, and other modeling choices consistent with researchers' individual conceptual preferences will be selectable, and model metadata describing in detail the model characteristics and assumptions will be generated with each model simulation run. Depending on user selections, the model implemented might be closely aligned with CGE-type optimization models and features, or one with more linear input-output like behaviors, or a hybrid model wherein better known object behaviors are modeled with more sophistication, while less well-understood objects' behaviors are modeled more simply. Irrespective of model configuration, simulation and impacts forecasting research will be replicable and will form the basis for direct comparison of alternative futures with differences directly attributable to explicitly identifiable model differences.

16.5 Challenges and Opportunities

Shifting from a traditional to a new knowledge building paradigm will not be without its challenges. The first challenge will be communicating the benefits to science of the new paradigm well enough to attract a critical mass of researchers willing to invest their time and effort into building the initial modeling infrastructures—the system backbones—for various problem domains. The transition will begin with the development of backbones for easily identifiable systems of systems models, which will be vitally important platforms for demonstrating the advantages of working in a new way, including ease of model extension and use and speed of scientific advancement.

A second challenge will be overcoming objections from vested interests. Those with commercialized models may at first feel threatened by encroachment of “”free” alternatives. However, many individual consultants and even large companies provide licensed and supported versions of software that originally developed—and in many cases continues to develop—in open source communities. As just one example, RedHat® is a highly successful commercial distributor of the Linux operating system, which continues to be developed and available as a free and open source operating system. Other consultants will be in demand for their expertise in application and use of OS^2 modeling systems.

A third challenge will be arguments that stem from what we call modeling *religions*. Within regional science and economic impacts modeling, for example, there are those who belong to the *CGE church*, those who belong to the *STE church*, those that belong to the *church of input-output and social accounting*, the *church of cost-benefit analysis*, and so on. There will be cases where some of these might co-exist peacefully as alternative options within the same system of systems modeling project, but there will also be as much room as individuals choose to take for developing multiple projects. Ideally, there also will be subsystems that can be integrated with multiple projects. With the adoption of a consistent object-oriented approach and the appropriate attention to encapsulation and consistently defined object interfaces, domain experts can develop subsystems as modules for adoption and use in any cognate project. Class libraries grouped by problem domain will develop to support multiple application development goals.

The last challenge we address here concerns implications for the publication process, which is a foundation for merit determinations in several environments, and certainly for promotion and tenure decisions in academia. To be sure, journals like the *Journal of Statistical Software* satisfy the need for developers of R code, and we expect these and additional outlets to fill such needs. It will be possible to associate the progenitor of new object-oriented classes to be identified as such in the metadata that accompanies object-oriented libraries. Domain experts also will be able to publish analytical results that compare outcomes of baseline simulations to those that incorporate their new model behaviors. Further, they will be able to devote much more time than every before to the areas of their own expertise because they will be freed from having to develop their own super-system backbones to focus more directly on their own problem domains. The results they publish will be replicable and immediately open to evaluation—and hence, validation—by the larger user community. And once open to the user community, they will also be immediately available as the basis for further development, refinement, and enhancement.

16.6 Summary

The future of modeling in regional research, and indeed the majority of integrated human and physical modeling, will be one of networked individuals contributing to problem domains in which they share common interests, and advancing more specific knowledge in which their particular expertise lies. We believe that this future will take the form of an object-oriented OS^2 modeling paradigm that will accelerate the knowledge-building enterprise and deepen our understanding of the complex interactions among human and physical systems. Open science is an inclusive environment, open to participation by users and developers from all groups without reference to age, creed, or color. Therefore, it will include and serve underrepresented populations. It has the potential to contribute to deeper understanding and to inform policy across a wide array of human and physical problem domains, and because these domains can be integrated, it can do so in ways that identify unanticipated ecological impacts of changes in one system on others previously assumed to be largely independent.

The structure and operation of object-oriented OS^2 models like OASIS will move beyond initial formulations to embody the best conceptual developments of the participating community. This kind of modeling will dramatically reduce the need for researchers to duplicate foundational modeling backbones and data bases for integrated systems simulations, allowing scarce research resources to be directed instead to specific advances in knowledge and understanding. It will facilitate replication and comparative analysis and will clarify and make explanations for alternative futures from different simulations more transparent.

Object-oriented OS^2 will provide a common foundation for extensions to research across numerous problem domains and will allow valuable resources otherwise devoted to recreating and reinventing such foundations to be used much more effectively. It will significantly enhance the ability of regional modelers to generate reproducible research. It will enhance infrastructure for research and education, and it will accelerate knowledge creation. It will support policy analysis by providing comprehensive integrated models that are fully open and well documented and that reflect the state of the science. Object-oriented OS^2 will establish a modeling support infrastructure to accelerate scientific advancement in integrative systems modeling research, enhancing the productivity of individual researchers and building a cumulative body of knowledge more rapidly than is possible under today's more fragmented approaches.

Our OASIS project and the paradigm it represents will radically transform the way regional modeling and integrative science are conducted in many areas of social, behavioral, and even physical sciences. The results will be distinguished not only by the collective wisdom of the modeling community, but also by careful attention to the mechanisms that support replication and reproducibility. With the advantage of twenty first century technology, object-oriented OS^2 will deepen our understanding and radically accelerate the pace of knowledge building in coming decades. We see this as a fundamentally new knowledge building paradigm that will dominate future integrated systems research.

References

Amrhein CG, MacKinnon RD (1988) A micro-simulation model of a spatial labor market. Ann Assoc Am Geogr 78(1):112–131. doi:10.1111/j.1467-8306.1988.tb00194.x. http://dx.doi.org/10.1111/j.1467-8306.1988.tb00194.x

Barker T (2004) The transition to sustainability: a comparison of general equilibrium and space–time–economics approaches. Working paper 62, Tyndall Centre

Birkin M, Clarke M (2011) Spatial microsimulation models: a review and a glimpse into the future. In: Stillwell J, Clarke M (eds) Population dynamics and projection methods. Understanding population trends and processes, chap 9. Springer, Berlin, pp 193–208. doi:10.1007/978-90-481-8930-4_9. http://dx.doi.org/10.1007/978-90-481-8930-4_9

Booch G (1994) Object-oriented analysis and design with applications, 2nd edn. Benjamin-Cummings, Redwood City, CA

Caldwell SB (1983) Modeling demographic-economic interactions: micro, macro and linked micro/macro strategies. Socio Econ Plan Sci 17(5–6):365–372

Carter N (2013) Energy-water nexus: the energy sector's water use. Technical report, Congressional Research Service. https://www.fas.org/sgp/crs/misc/R43199.pdf

Clarke M, Holm E (1987) Microsimulation methods in spatial analysis and planning. Geogr Ann Ser B 69(2):145–164

Clarke M, Wilson AG (1986) A framework for dynamic comprehensive urban models: the integration of accounting and micro-simulation approaches. Sistemi Urbani 2(3):145–177

Coad P, Yourdon E (1991a) Object-oriented analysis. Prentice-Hall, Englewood Cliffs, NJ

Coad P, Yourdon E (1991b) Object-oriented design, vol 92. Prentice-Hall, Englewood Cliffs, NJ

Conway RS (1990) The Washington projection and simulation model: a regional interindustry econometric model. Int Reg Sci Rev 13(1–2):141–165

Donaghy KP, Balta-Ozkan N, Hewings GJ (2007) Modeling unexpected events in temporally disaggregated econometric input–output models of regional economies. Econ Syst Res 19(2):125–145

Gloor P (2002) Collaborative knowledge networks. eJETA Electron J E-Bus Technol Appl 1(3):1–11

Gloor PA (2006) Swarm creativity: competitive advantage through collaborative innovation networks. Oxford University Press, Oxford

Gloor PA, Heckmann C, Makedon F (2004) Ethical issues in collaborative innovation networks. Retrieved from http://wwwicknorg/documents/COIN4Ethicomppdf (date of download: 0512 2011)

Israilevich PR, Hewings GJ, Schindler GR, Mahidhara R (1996) The choice of an input-output table embedded in regional econometric input-output models. Pap Reg Sci 75(2):103–119

Israilevich PR, Hewings GJ, Sonis M, Schindler GR (1997) Forecasting structural change with a regional econometric input-output model. J Reg Sci 37(4):565–590

Jackson R (1994) Object-oriented modeling in regional science: an advocacy view. J Reg Sci 73(4):347–367

Jackson RW (1995) Directions in regional science. Int Reg Sci Rev 18(2):159–164

Járosi P, Jackson RW (2015) Object-oriented interindustry systems: proof of concept. Working papers technical document 2015-0. Regional Research Institute, West Virginia University. https://ideas.repec.org/p/rri/wpaper/2015td03.html

Kim K, Hewings GJD, Kratena K (2016) Household disaggregation and forecasting in a regional econometric input–output model. Lett Spat Resour Sci 9(1):73–91

Kratena K, Streicher G, Temurshoev U, Amores AF, Arto I, Mongelli I, Neuwahl F, Rueda-Cantuche JM, Andreoni V (2013) FIDELIO 1: fully interregional dynamic econometric long-term input-output model for the EU27. Publications Office, Luxembourg

Odell J (2002) Objects and agents compared. J Object Technol 1(1):41–53

Rey SJ (1997) Integrating regional econometric and input-output models: an evaluation of embedding strategies. Environ Plann A 29(6):1057–1072

Rey SJ (1998) The performance of alternative integration strategies for combining regional econometric and input-output models. Int Reg Sci Rev 21(1):1–35

Rey SJ (2000) Integrated regional econometric+ input-output modeling: issues and opportunities. Pap Reg Sci 79(3):271–292

Rey SJ (2014) Open regional science. Ann Reg Sci 52(3):825–837

Rey SJ, Jackson RW (1999) Labor-productivity changes in regional econometric + input-output models. Environ Plann A 31(9):1583–1599. doi:10.1068/a311583

Rumbaugh J, Blaha M, Premerlani W, Eddy F, Lorensen WE et al (1991) Object-oriented modeling and design. Prentice-Hall, Englewood Cliffs, NJ

Scrieciu SS (2007) The inherent dangers of using computable general equilibrium models as a single integrated modelling framework for sustainability impact assessment. A critical note on Böhringer and Löschel (2006). Ecol Econ 60(4):678–684

Wallach H (2016) Computational social sciences: towards a collaborative future. In: Alvarez R (ed) Computational social science: discovery and prediction. Cambridge University Press, Cambridge, pp 307–316

West GR (1991) A Queensland input-output econometric model: an overview. Aust Econ Pap 30(57):221–240

West GR, Jackson RW (1998) Input-Output+ econometric and econometric+ input-output: model differences or different models? J Reg Anal Policy 28(1):33–48

West GR, Jackson RW (2014) Simulating impacts on regional economies: a modeling alternative. In: Schaeffer P, Kouassi E (eds) Econometric methods for analyzing economic development, chap 9. IGI Global, Hershey, PA, pp 132–152

Randall Jackson is professor, Department of Geology and Geography, West Virginia University (WVU), and Director of the Regional Research Institute. His primary research interests are regional industrial systems modeling; energy, environmental, and economic systems interactions; and regional economic development. He is an adjunct professor in WVU's Department of Economics and Division of Resource Management, and in Geography at The Ohio State University (OSU). Previous faculty positions were at OSU and Northern Illinois University. Dr. Jackson earned the Ph.D. in geography and regional science from the University of Illinois at Urbana-Champaign in 1983.

Sergio Rey is professor, School of Geographical Sciences and Urban Planning, Arizona State University (ASU). His research interests focus on the development, implementation, and application of advanced methods of spatial and space-time data analysis. His substantive foci include regional inequality, convergence and growth dynamics as well as neighborhood change, segregation dynamics, spatial criminology and industrial networks. Previous faculty positions were at the Department of Geography, San Diego State University and a visiting professor at the Department of Economics, University of Queensland. Dr. Rey earned the Ph.D. in geography from the University of California Santa Babara in 1994.

Péter Járosi is research assistant professor, Regional Research Institute, West Virginia University (WVU). His primary research interests are regional industrial systems modeling; spatial computable general equilibrium models; corporate and public finance modeling; and regional economic development. Previous faculty positions were at Faculty of Business and Economics, University of Pécs, Hungary; MTA-PTE Innovation and Economic Growth Research Group of Hungarian Academy of Sciences; and College of Finance and Accountancy, Budapest Business School. Dr. Járosi earned his Ph.D. in regional science from the Doctoral School of Regional Policy and Economics, University of Pécs in 2011.

Chapter 17
Looking at John Snow's Cholera Map from the Twenty First Century: A Practical Primer on Reproducibility and Open Science

Daniel Arribas-Bel, Thomas de Graaff, and Sergio J. Rey

17.1 Introduction

In the fall of 2015 Ann Case and Economics Nobel Prize winner Agnus Deaton published a very influential paper in the Proceedings of the National Academy of Sciences (Case and Deaton 2015) concerning the increasing and alarmingly high mortality rates of white Americans aged 45–54. As possible reasons for this phenomenon, they suggested the devastating effects of suicide, alcohol and drug abuse. This article caused quite a great deal of upheaval, and political analysts and columnists even linked this with the electoral unrest amongst the white middle class. However, a comment of an anonymous blogger caused Andrew Gelman to rethink and recalculate the results of Case and Deaton. Namely, what if a shift *within* the age cohort of 45–54 would have happened now with more people being closer to 54 than to 45? Indeed, it turns out that, when correcting for age shifts within cohorts, the results of Case and Deaton are severely less pronounced (although the mortality rates of the white middle aged in the US still stand out compared to other countries).

The example above signifies that, even for Nobel Laureates, there is always a need to be able to reproduce and rethink scientific analyses, especially when the results are this influential. Mistakes can be made, and anecdotes like the above

D. Arribas-Bel (✉)
Department of Geography & Planning, University of Liverpool, Liverpool, UK
e-mail: D.Arribas-Bel@liverpool.ac.uk

T. de Graaff
Department of Spatial Economics, Vrije Universiteit Amsterdam, Amsterdam, The Netherlands
e-mail: t.de.graaff@vu.nl

S.J. Rey
School of Geographical Sciences and Urban Planning, Arizona State University, Tempe, AZ, USA
e-mail: srey@asu.edu

R. Jackson, P. Schaeffer (eds.), *Regional Research Frontiers - Vol. 2*,
Advances in Spatial Science, DOI 10.1007/978-3-319-50590-9_17

are abundant across all sciences. The scientific process is traditionally designed to correct itself, although this adjustment can be quite sluggish. To facilitate this self-correcting process and to minimize the number of errors within the data preparation, data analysis and results presentation phase, we argue that a proper workflow is needed: namely, one that facilitates reproducibility and Open Science.

In general, the need for more emphasis on research reproducibility and Open Science is increasingly recognised by universities, government institutions and even the public at large. Strangely, however, virtually no training is provided on workflow design and choice of appropriate tools. Students and researchers receive no guidance as to why or how they should adopt habits that favor Open Science principles in their research activity.[1] This applies as well to regional science where, given the emphasis on spatial data, maps and quantitative approaches, the need for a reproducible workflow is probably even more challenging than in most other social sciences. This chapter, therefore, focuses on the concept of workflows, reproducibility and Open Science, and how to apply them in a very practical sense. Moreover, it illustrates these concepts by providing a completely reproducible environment and hands-on example.

The next section deals with the concept of workflow, reproducibility and Open Science, introduces some specific workflows and tackles the question of why these approaches are relevant. In the third section, we give an example of a completely open and reproducible analysis of John Snow's famous cholera map from the nineteenth century. Although a proper workflow does not revolve around one single tool, but instead consists of a *coherent* set of tools and methodologies, we have chosen to use for this purpose the programming languages R and Python in combination with the Jupyter Notebook environment, because of its relative ease of use, accessibility and flexibility. The chapter concludes with a discussion of the advantages and the (perceived) disadvantages of our approach.

17.2 Workflow, Reproducibility and Open Science in Regional Science

The Business Dictionary (BusinessDictionary 2016) states that a workflow is a

> progression of steps (tasks, events, interactions) that comprise a work process, involve two or more persons, and create or add value to the organization's activities.

So a workflow in science is a set of steps (such as data gathering, data manipulation, analyses, presenting results), usually taken by multiple researchers, which leads to an outcome (the research findings). Reproducibility requires that the materials used to make a finding are available and that they are sufficient for

[1]See for notable exceptions Healy (2011) in the social sciences, and Rey (2014) and Arribas-Bel and de Graaff (2015) in regional science.

an independent researcher (including the original researcher herself) to recreate the finding. Open Science requires that all researchers have free and easy access to all materials used to make such a finding. Unfortunately, making your research open and reproducible often requires additional effort, and one may wonder whether it is worth it. Indeed, adopting a workflow directed at reproducibility and openness can often be costly in terms of time. However, there are significant gains to be made.

First, and the most obvious of all, the research becomes reproducible. This brings great benefits to the scientific community at large. Sharing code for estimations, figures and data management leads to a faster dispersion of knowledge. Secondly, it leads to larger transparency, and thus a higher probability of early error detection. Thirdly, research becomes more modular and portable, so that it is easier to cooperate with colleagues at a distance and to continue with parts of the research where others have left it. Fourthly, one of the most salient advantages of a reproducible workflow is that, in the long term, it makes the scientist more efficient. However, this will show up at the end of the research cycle, when somebody—an editor, a supervisor, a referee, a colleague, your own future self—may ask to redo (parts of the) analysis with slight modifications. In this context, having an automated process that prepares your data, analyses them and presents the final results is of great help. An additional benefit of a reproducible workflow is self-sanity. The effort put to explain to others what steps were taken and how they were approached provides an additional degree of confidence over the traditional case-scenario where documentation is scarce and unclear. Finally, reproducibility and especially openness increases the visibility of the research. Most notably, when code for a complex estimation is available alongside a paper, others will not only be more convinced of the results, but they also will be more likely to use it and give it proper credit.

Often, complete reproducibility in regional science is hard to achieve. Proprietary data, qualitative methods such as interviews and case studies and sampling issues in surveys often prohibit others from perfectly mimicking a study's results. However, by choosing appropriate tools, one can strive to work as reproducibly as possible. Making available coding books for surveys and interviews, protocols for case studies and data management code for proprietary data often significantly helps others to understand how the results have been obtained.

Recent years have seen a remarkable increase in tools and attention to reproducibility and openness. Unfortunately, most of these tools come from the realm of computer science and have not yet permeated into other domains, such as regional science. In general, there is not a particular set of tools that we advocate. However, there are some *types* of tools that in general are unavoidable when striving for an open and reproducible workflow, including:

- Data analysis and programming applications. For quantitative data analyses, one needs tools for data management and statistical analysis, such as the two most popular data science tools at the moment, `R` and `Python`.
- Typesetting applications. These are used to convey the text and results, whether on paper (typically using the `pdf` format) or on screen (using the `html`

language). Typically, `LaTeX` is often used for scientific purposes, especially because it is scriptable and produces high quality results. Nowadays, however, `Markdown` seems to be growing in popularity, mostly because of its very accessible and easy to learn syntax.

- Reference managers. These typically are combined with typesetting applications and form a database of references and a system to handle citations and reference lists. `BibTex`, `Mendeley`, and `Endnote` are popular applications.
- Version control systems. These enable the archiving of older file versions, while only one copy is ever in use (this avoids the usual awkward naming conventions for files, such as `FinalDraftVersion3.3.doc.final.really.docx`). In combination with central repositories, these version control systems act as well as backup applications. `Dropbox` is an example of a version control system, just as is the popular open source version control system `Git`.
- Literate programming environments. These are typically applications able to *weave* code, text and output together. At the moment, there are not many *general* literature programming environments. The most popular are probably the `knitr` package for `R`[2] and the `Jupyter` notebook for a multi-language environment. Moreover, these environments are able to write output to different formats (usually, `html`, `Markdown`, `LaTeX/pdf`, and the Open Office `.odt` format).

In general, tools for reproducible research need to be preferably open source and particularly *scriptable*. The lack of the latter makes it very difficult for other applications to communicate and "cooperate" with the tools used.

17.3 John Snow's Cholera Map

To demonstrate some of the ideas discussed above, we use a classic dataset in the history of spatial analysis: the cholera map by Dr. John Snow. His story is well known and better described elsewhere (Hempel 2006). Thanks to his mapping exercise of the location of cholera deaths in nineteenth century London, he was able to prove that the disease is in fact transmitted through contaminated water (associated to a specific pump), as opposed to the conventional thinking of the day, which stated that transmission occurred through the air. In this section, we will support Snow's view with the help of Exploratory Spatial Data Analysis (ESDA) tools. In the process, we will show how a reproducible and open workflow can in fact be applied by including the code required to produce the results presented.[3] In fact, the entire content, as well as its revision history, have been tracked using the Git version control software and can be downloaded from

---

[2]See for further information how to use R to make your research as reproducible as possible Gandrud (2013) and Stodden et al. (2014).

[3]Part of this section is based upon Lab 6 of Arribas-Bel (2016), available at http://darribas.org/gds15.

https://bitbucket.org/darribas/reproducible_john_snow. Equally, the code required to carry out the analysis is closely integrated in the paper and will be shown inline. A reproducible notebook version of this document, available from the online resource, allows the reader to not only see the code but to interactively execute it without decoupling it from the rest of the content in this chapter.

17.3.1 *Point Pattern Exploration*

This analysis will be performed using a combination of both the Python and R programming languages. In addition to both being free and open-source, they have become essential components of the data scientist's toolkit and are also enjoying growing adoption within academia and scientific environments. Thanks to the Jupyter Notebook (Perez 2015), both can be included alongside each other and the best of both worlds can be leveraged. We start with a visual map exploration by using data stored in the R package `HistData`. We then use this data for an analysis in the Python language. To do this, we need to import the Python interface to R.

```
import rpy2.robjects.conversion

import rpy2 as r
import rpy2.robjects

import rpy2.interactive as r
import rpy2.interactive.packages
```

The data for the original John Snow analysis is available in R as part of the package `HistData`, which we need to import together with the `ggplot2` package to create figures and maps.

```
r.packages.importr('HistData')
r.packages.importr('ggplot2')
```

In order to have a more streamlined analysis, we define a basic `ggplot` map using the data from `HistData` that we will call on later:

```
%%R

Snow_plot <- ggplot(Snow.deaths, aes(x = x, y=y)) +
        geom_point(data=Snow.deaths, aes(x=x, y=y),
                col="red", pch=19, cex=1.5) +
        geom_point(data=Snow.pumps, aes(x=x, y=y),
                col="black", pch=17, cex=4) +
        geom_text(data=Snow.pumps,
                aes(label = label, x = x, y = y+0.5))+
        xlim(6, 19.5) + ylim(4, 18.5) +
        geom_path(data=Snow.streets,
```

```
            aes(x=x,y=y,group=street), col="gray40") +
    ggtitle("Pumps and cholera deaths\n
                      in 19th century London")+
    theme(panel.background = element_rect(fill = "gray85"),
    plot.background = element_rect(fill = "gray85"),
    panel.grid.major = element_blank(),
    panel.grid.minor = element_blank(),
    axis.line=element_blank(),
    axis.text.x=element_blank(),
    axis.text.y=element_blank(),
    axis.ticks=element_blank(),
    axis.title.x=element_blank(),
    axis.title.y=element_blank(),
    plot.title = element_text(size = rel(2), face="bold"))
```

At this point, we can access the data:

```
%R head(Snow.deaths$x)
```

which produces the following results:

```
array([13.58801, 9.878124, 14.65398, 15.22057, 13.16265,
13.80617])
```

And move the coordinates from R to Python:

```
X = %R Snow.deaths$x
Y = %R Y=Snow.deaths$y
```

A first visual approximation to the distribution of cholera deaths can then be easily produced:

```
%R plot(X,Y)
```

which gives Fig. 17.1.

A more detailed map can also be produced by calling on the map we defined earlier:

```
%%R Snow_plot
```

which gives the spatial context of the coordinates as in Fig. 17.2.

We can start moving beyond simple visualization and into a more in-depth analysis by adding a kernel density estimate as follows:

```
%%R
## overlay bivariate kernel density contours of deaths
Snow_plot + geom_density_2d()
```

and overlaying it on top of our death locations map as in Fig. 17.3.

This already allows us to get a better insight into Snow's hypothesis of a contaminated pump (the one in Broad Street in particular). To further support this view, we will use some of the most common components of the ESDA toolbox.

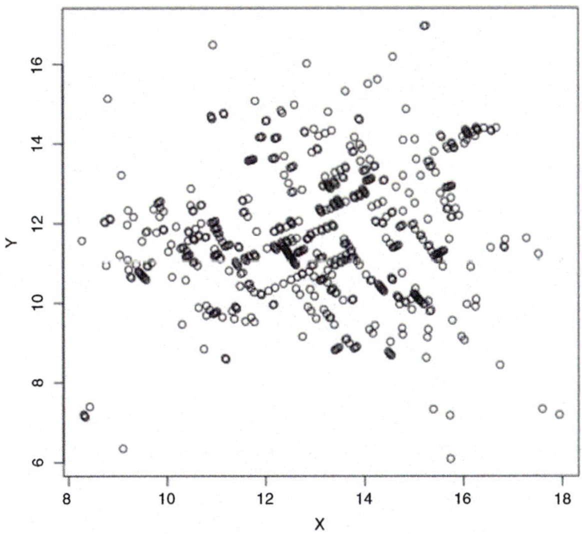

Fig. 17.1 *X* and *Y* coordinates of cholera deaths

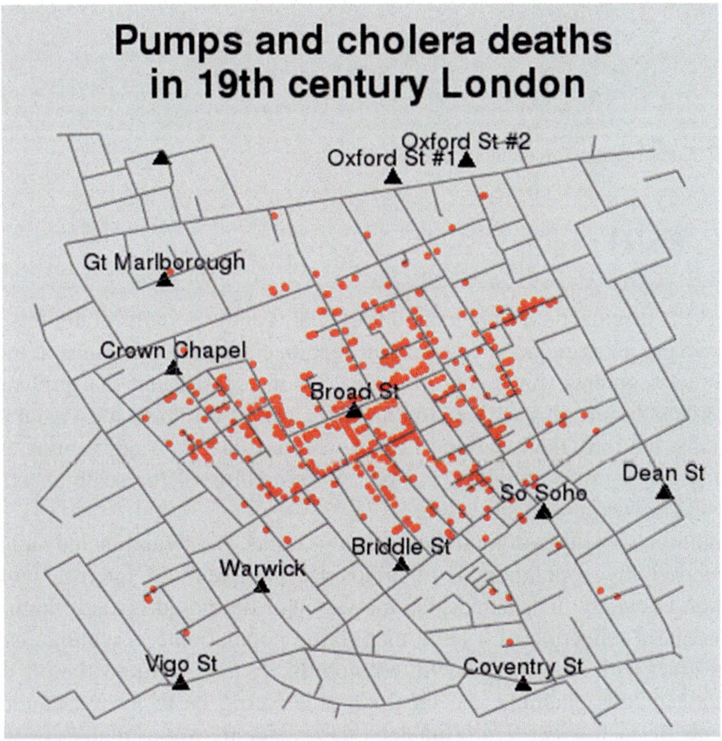

Fig. 17.2 Spatial point of map of cholera deaths

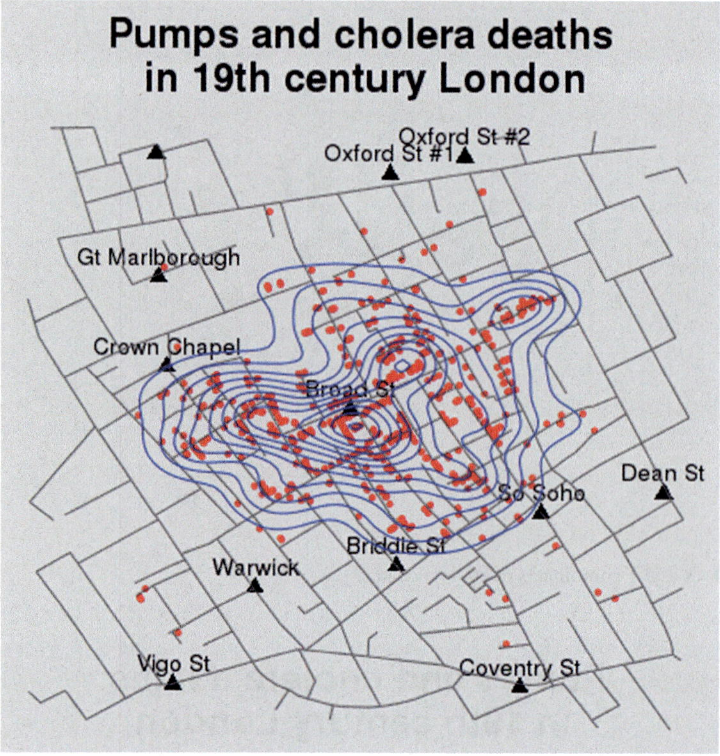

Fig. 17.3 Kernel estimation of cholera deaths

17.3.2 ESDA

Although the original data were locations of deaths at the point level, for this section we will access an aggregated version that reports cholera death counts at the street level. Street segments (lines, topologically) are the spatial unit that probably best characterizes the process we looking at; since we do not have individual data on house units, but only the location of those who passed away, aggregating at a unit like the street segment provides a good approximation of the scale at which the disease was occurring and spreading.

In addition, since the original data are raw counts, we should include a measure of the underlying population. If all maps are the events of interest, unless the population is evenly distributed, the analysis will be biased because high counts could just be a reflection of a large underlying population (everything else being equal, a street with more people will be more likely to have more cholera deaths). In the case of this example, the ideal variable would be to have a count of the inhabitants of each street. Unfortunately, these data are not available, so we need to find an approximation. This will inevitably imply making assumptions and potentially introducing a certain degree of measurement error. For the sake of this example, we will assume that, within the area of central London covered by the data, population was evenly spread across the street network. This means that the underlying population of one of our street segments is proportional to its length. Following this assumption, if we want to control for the underlying population of a street segment, a good approach could be to consider the number of cholera deaths per (100) metre(s)—a measure of density—rather than the raw count. The polygon file includes building blocks from the Ordnance Survey (OS data l' Crown copyright and database right, 2015).

This part of the analysis will be performed in Python, for which we need to import the libraries required:

```
%matplotlib inline
import seaborn as sns
import pandas as pd
import pysal as ps
import geopandas as gpd
import numpy as np
import matplotlib.pyplot as plt
```

17.3.2.1 Loading and Exploring the Data

Data in this case come from Robin Wilson.[4]

```
# Load point data
pumps = gpd.read_file('data/Pumps.shp')
# Load building blocks
blocks = gpd.read_file('data/polys.shp')
# Load street network
js = gpd.read_file('data/streets_js.shp')
```

To inspect the data and find out the structure as well as the variables included, we can use the head function:

```
print js.head().to_string()
```

with the following output

```
Deaths  Deaths_dens  geometry                                         segIdStr    seg_len
0  0      0.000000  LINESTRING (529521 180866, 529516 180862)  s0-1       6.403124
1  1      1.077897  LINESTRING (529521 180866, 529593 180925)  s0-2      92.773279
2  0      0.000000  LINESTRING (529521 180866, 529545 180836)  s0-3      38.418745
3  0      0.000000  LINESTRING (529516 180862, 529487 180835)  s1-25     39.623226
4  26    18.079549 LINESTRING (529516 180862, 529431 180978)  s1-27   143.808901
```

Before we move on to the analytical part, we can also create choropleth maps for line data. In the following code snippet, we build a choropleth using the Fisher-Jenks

[4]See: http://blog.rtwilson.com/john-snows-cholera-data-in-more-formats/.

classification for the density of cholera deaths in each street segment, and style it by adding a background color, building blocks and the location of the water pumps:

```
# Set up figure and axis
f, ax = plt.subplots(1, figsize=(9, 9))
# Plot building blocks
for poly in blocks['geometry']:
gpd.plotting.plot_multipolygon(ax, poly, facecolor='0.9')
# Quantile choropleth of deaths at the street level
js.plot(column='Deaths_dens', scheme='fisher_jenks',
        axes=ax, colormap='YlGn')
# Plot pumps
xys = np.array([(pt.x, pt.y) for pt in pumps.geometry])
ax.scatter(xys[:, 0], xys[:, 1], marker='^', color='k', s=50)
# Remove axis frame
ax.set_axis_off()
# Change background color of the figure
f.set_facecolor('0.75')
# Keep axes proportionate
plt.axis('equal')
# Title
f.suptitle('Cholera Deaths per 100m.', size=30)
# Draw
plt.show()
```

which produces Fig. 17.4.

17.3.2.2 Spatial Weights Matrix

A spatial weights matrix is the way geographical space is formally encoded into a numerical form so it is easy for a computer (or a statistical method) to understand. These matrices can be created based on several criteria: contiguity, distance, blocks, etc. Although usually spatial weights matrices are used with polygons or points, these ideas can also be applied with spatial networks made of line segments.

For this example, we will show how to build a simple contiguity matrix, which considers two observations as neighbors if they share one edge. For a street network as in our example, two street segments will be connected if they "touch" each other. Since lines only have one dimension, there is no room for the discussion between "queen" and "rook" criteria, but only one type of contiguity.

Building a contiguity matrix from a spatial network like the streets of London's Soho can be done with PySAL, but creating it is slightly different, technically. For this task, instead of the ps.queen_from_shapefile, we will use the network module of the library, which reads a line shapefile and creates a network representation of it. Once loaded, a contiguity matrix can be easily created using the contiguity weights attribute. To keep things aligned, we rename the IDs of the matrix to match those

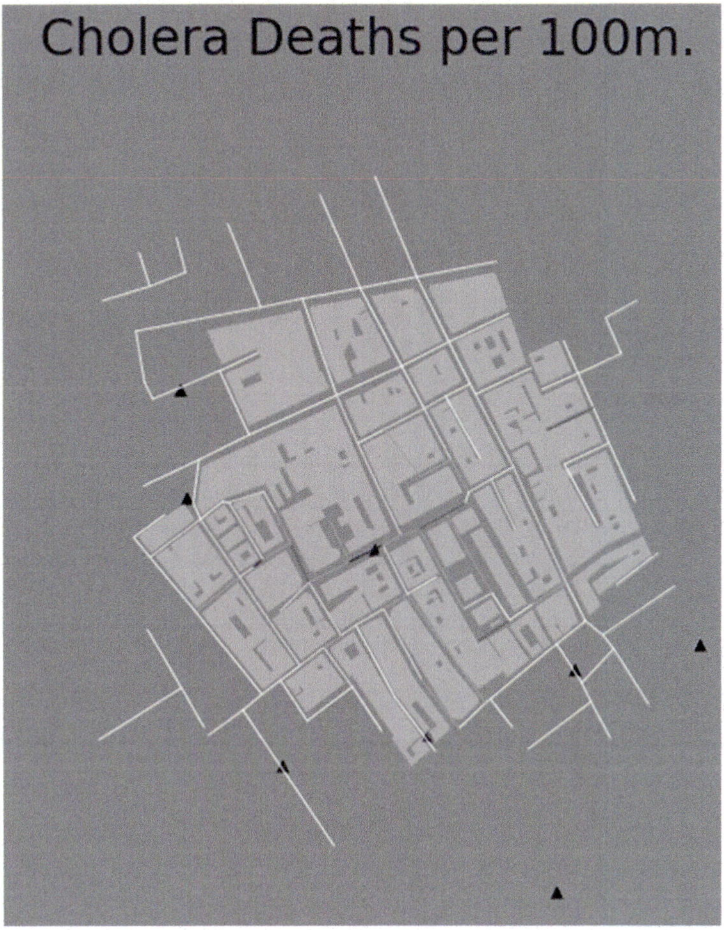

Fig. 17.4 Choropleth map of cholera deaths

in the table and, finally, we row-standardize the matrix, which is a standard ps.W object, like those we have been working with for the polygon and point cases:

```
# Load the network
ntw = ps.Network('data/streets_js.shp')
# Create the spatial weights matrix
w = ntw.contiguityweights(graph=False)
# Rename IDs to match those in the 'segIdStr' column
w.remap_ids(js['segIdStr'])
# Row standardize the matrix
w.transform = 'R'
```

Now, the w object we have just created comes from a line shapefile, but it is of the same type as if it came from a polygon or point topology. As such, we can inspect it in the same way. For example, we can check who is a neighbor of observation s0-1:

```
w['s0-1']
{u's0-2': 0.25, u's0-3': 0.25,
 u's1-25': 0.25, u's1-27': 0.25}
```

Note how, because we have row-standardized them, the weight given to each of the four neighbors is 0.25, which, all together, sum up to one.

17.3.2.3 Spatial Lag

Once we have the data and the spatial weights matrix ready, we can start by computing the spatial lag of the death density. Remember, the spatial lag is the product of the spatial weights matrix and a given variable and that, if W is row-standardized, the result amounts to the average value of the variable in the neighborhood of each observation. We can calculate the spatial lag for the variable Deaths_dens and store it directly in the main table with the following line of code:

```
js['w_Deaths_dens'] = ps.lag_spatial(w, js['Deaths_dens'])
```

Let us have a quick look at the resulting variable, as compared to the original one:

```
toprint = js[['segIdStr', 'Deaths_dens', 'w_Deaths_dens']].head()
# Note: next line is for printed version only. On interactive mode,
# you can simply execute 'toprint'
print toprint.to_string()
```

which yields:

```
  segIdStr  Deaths_dens  w_Deaths_dens
0     s0-1     0.000000       4.789361
1     s0-2     1.077897       0.000000
2     s0-3     0.000000       0.538948
3    s1-25     0.000000       6.026516
4    s1-27    18.079549       0.000000
```

The way to interpret the spatial lag (w_Deaths_dens) for the first observation is as follows: the street segment s0-2, which has a density of zero cholera deaths per 100 m, is surrounded by other streets which, on average, have 4.79 deaths per 100 m. For the purpose of illustration, we can check whether this is correct by querying the spatial weights matrix to find out the neighbors of s0-2:

```
w.neighbors['s0-1']
[u's0-2', u's0-3', u's1-25', u's1-27']
```

And then checking their values:

```
# Note that we first index the table on the index variable
neigh = js.set_index('segIdStr').loc[w.neighbors['s0-1'],
        'Deaths_dens']
neigh

segIdStr
s0-2      1.077897
s0-3      0.000000
s1-25     0.000000
s1-27    18.079549
Name: Deaths_dens, dtype: float64
```

And the average value, which we saw in the spatial lag is 4.79, can be calculated as follows:

```
neigh.mean()
4.7893612696592509
```

For some of the techniques we will be seeing below, it makes more sense to operate with the standardized version of a variable, rather than with the raw one. Standardizing means to subtract the average value and divide by the standard deviation each observation of the column. This can be done easily with a bit of basic algebra in Python:

```
js['Deaths_dens_std'] = (js['Deaths_dens'] -
        js['Deaths_dens'].mean())/js['Deaths_dens'].std()
```

Finally, to be able to explore the spatial patterns of the standardized values, sometimes called z values, we need to create its spatial lag:

```
js['w_Deaths_dens_std'] =
        ps.lag_spatial(w, js['Deaths_dens_std'])
```

17.3.2.4 Global Spatial Autocorrelation

Global spatial autocorrelation relates to the overall geographical pattern present in the data. Statistics designed to measure this trend thus characterize a map in terms of its degree of clustering and summarize it. This summary can be visual or numerical. In this section, we will walk through an example of each of them: the Moran Plot, and Moran's I statistic of spatial autocorrelation.

The Moran plot is a way of visualizing a spatial dataset to explore the nature and strength of spatial autocorrelation. It is essentially a traditional scatter plot in which the variable of interest is displayed against its spatial lag. To be able to interpret values as above or below the mean and their quantities in terms of standard deviations, the variable of interest is usually standardized by subtracting its mean and dividing it by its standard deviation.

Technically speaking, creating a Moran Plot is very similar to creating any other scatter plot in Python, provided we have standardized the variable and calculated its spatial lag beforehand:

```
# Setup the figure and axis
f, ax = plt.subplots(1, figsize=(9, 9))
# Plot values
sns.regplot(x='Deaths_dens_std', y='w_Deaths_dens_std',
        data=js)
# Add vertical and horizontal lines
plt.axvline(0, c='k', alpha=0.5)
plt.axhline(0, c='k', alpha=0.5)
# Display
plt.show()
```

which produces Fig. 17.5.

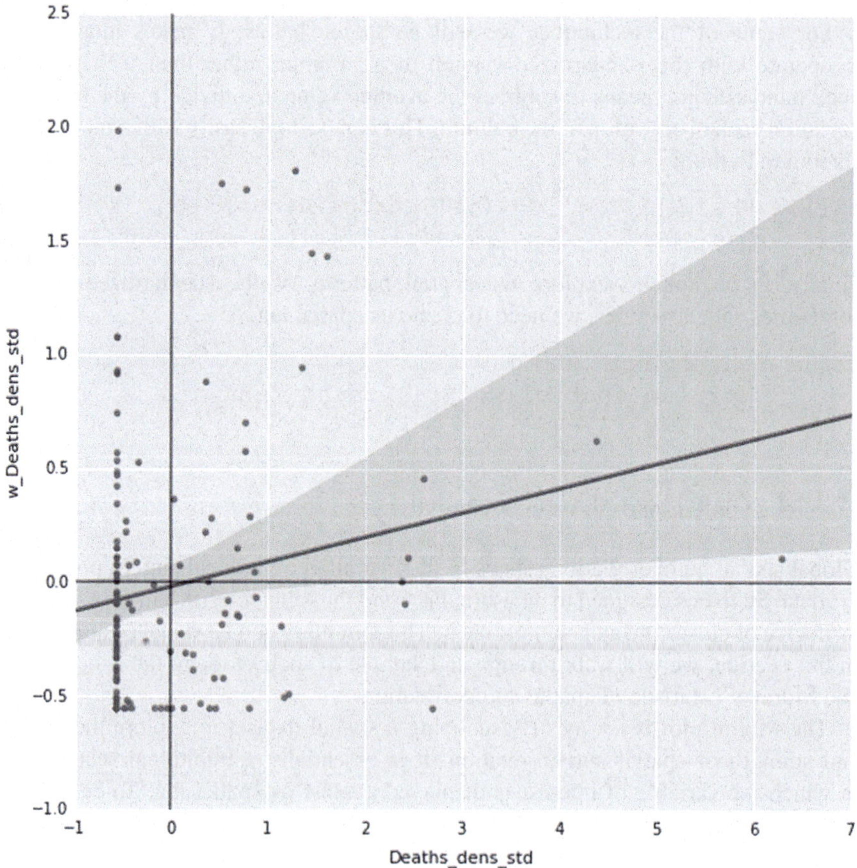

Fig. 17.5 Moran plot of cholera deaths

Figure 17.5 displays the relationship between `Deaths_dens_std` and its spatial lag which, because the *W* that was used is row-standardized, can be interpreted as the average standardized density of cholera deaths in the neighborhood of each observation. In order to guide the interpretation of the plot, a linear fit is also included in the post, together with confidence intervals. This line represents the best linear fit to the scatter plot or, in other words, what is the best way to represent the relationship between the two variables as a straight line. Because the line comes from a regression, we can also include a measure of the uncertainty about the fit in the form of confidence intervals (the shaded blue area around the line).

The plot displays a positive relationship between both variables. This is associated with the presence of positive spatial autocorrelation: similar values tend to be located close to each other. This means that the overall trend is for high values to be close to other high values, and for low values to be surrounded by other low values. This, however, does not mean that this is the only pattern in the dataset: there can of course be particular cases where high values are surrounded by low ones, and vice versa. But it means that, if we had to summarize the main pattern of the data in terms of how clustered similar values are, the best way would be to say they are positively correlated and, hence, clustered over space.

In the context of the example, the street segments in the dataset show positive spatial autocorrelation in the density of cholera deaths. This means that street segments with a high level of incidents per 100 m tend to be located adjacent to other street segments also with high number of deaths, and vice versa.

The Moran Plot is an excellent tool to explore the data and get a good sense of how many values are clustered over space. However, because it is a graphical device, it is sometimes hard to condense its insights into a more concise way. For these cases, a good approach is to come up with a statistical measure that summarizes the figure. This is exactly what Moran's *I* is meant to do.

Very much in the same way the mean summarizes a crucial element of the distribution of values in a non-spatial setting, so does Moran's *I* for a spatial dataset. Continuing the comparison, we can think of the mean as a single numerical value summarizing a histogram or a kernel density plot. Similarly, Moran's *I* captures much of the essence of the Moran Plot. In fact, there is an even closer connection between the two: the value of Moran's *I* corresponds with the slope of the linear fit overlayed on top of the Moran Plot.

In order to calculate Moran's *I* in our dataset, we can call a specific function in PySAL directly:

```
mi = ps.Moran(js['Deaths_dens'], w)
```

Note how we do not need to use the standardized version in this context as we will not represent it visually.

The method ps.Moran creates an object that contains much more information than the actual statistic. If we want to retrieve the value of the statistic, we can do it this way:

```
mi.I
0.10902663995497329
```

The other bit of information we will extract from Moran's I relates to statistical inference: how likely is it that the pattern we observe in the map and Moran's I is not generated by an entirely random process? If we considered the same variable but shuffled its locations randomly, would we obtain a map with similar characteristics?

The specific details of the mechanism to calculate this are beyond the scope of this paper, but it is important to know that a small enough p-value associated with the Moran's I of a map allows rejection of the hypothesis that the map is random. In other words, we can conclude that the map displays more spatial pattern that we would expect if the values had been randomly allocated to a particular location.

The most reliable p-value for Moran's I can be found in the attribute `p_sim`:

```
mi.p_sim
0.045999999999999999
```

That is just below 5% and, by standard terms, it would be considered statistically significant. Again, a full statistical explanation of what that really means and what its implications are is beyond the discussion in this context. But we can quickly elaborate on its intuition. What that 0.046 (or 4.6%) means is that, if we generated a large number of maps with the same values but randomly allocated over space, and calculated the Moran's I statistic for each of those maps, only 4.6% of them would display a larger (absolute) value than the one we obtain from the real data, and the other 95.4% of the random maps would receive a smaller (absolute) value of Moran's I. If we remember again that the value of Moran's I can also be interpreted as the slope of the Moran plot, what we have in this case is that the particular spatial arrangement of values over space we observe for the density of cholera deaths is more concentrated than if we were to randomly shuffle the death densities among the Soho streets, hence the statistical significance.

As a first step, the global autocorrelation analysis can teach us that observations do seem to be positively correlated over space. In terms of our initial goal to find evidence for John Snow's hypothesis that cholera was caused by water in a single contaminated pump, this view seems to align: if cholera was contaminated through the air, it should show a pattern over space—arguably a random one, since air is evenly spread over space—that is much less concentrated than if this was caused by an agent (water pump) that is located at a particular point in space.

17.3.2.5 Local Spatial Autocorrelation

Moran's I is a good tool to summarize a dataset into a single value that informs about its degree of clustering. However, it is not an appropriate measure to identify areas within the map where specific values are located. In other words, Moran's I can tell us whether values are clustered overall or not, but it will not inform us about where the clusters are. For that purpose, we need to use a local measure of spatial autocorrelation. Local measures consider each single observation in a dataset and operate on them, as opposed to on the overall data, as global measures do. Because of that, they are not good at summarizing a map, but they do provide further insight.

In this section, we will consider Local Indicators of Spatial Association (LISAs), a local counter part of global measures like Moran's I. At the core of these methods is a classification of the observations in a dataset into four groups derived from the Moran Plot: high values surrounded by high values (HH), low values nearby other low values (LL), high values among low values (HL), and vice versa (LH). Each of these groups are typically called "quadrants". An illustration of where each of these groups fall into the Moran Plot can be seen below:

```
# Setup the figure and axis
f, ax = plt.subplots(1, figsize=(9, 9))
# Plot values
sns.regplot(x='Deaths_dens_std', y='w_Deaths_dens_std', data=js)
# Add vertical and horizontal lines
plt.axvline(0, c='k', alpha=0.5)
plt.axhline(0, c='k', alpha=0.5)
ax.set_xlim(-2, 7)
ax.set_ylim(-2.5, 2.5)
plt.text(3, 1.5, "HH", fontsize=25)
plt.text(3, -1.5, "HL", fontsize=25)
plt.text(-1, 1.5, "LH", fontsize=25)
plt.text(-1, -1.5, "LL", fontsize=25)
# Display
plt.show()
```

which gives Fig. 17.6.

So far we have classified each observation in the dataset depending on its value and that of its neighbors. This is only halfway into identifying areas of unusual concentration of values. To know whether each of the locations is a statistically significant cluster of a given kind, we again need to compare it with what we would expect if the data were allocated in a completely random way. After all, by definition every observation will be of one kind of another based on the comparison above. However, what we are interested in is whether the strength with which the values are concentrated is unusually high.

This is exactly what LISAs are designed to do. As before, a more detailed description of their statistical underpinnings is beyond the scope in this context, but we will try to shed some light into the intuition of how they go about it. The core idea is to identify cases in which the comparison between the value of an observation and the average of its neighbors is either more similar (HH, LL) or dissimilar (HL, LH) than we would expect from pure chance. The mechanism to do this is similar to the one in the global Moran's I, but applied in this case to each observation, results in as many statistics as the original observations.

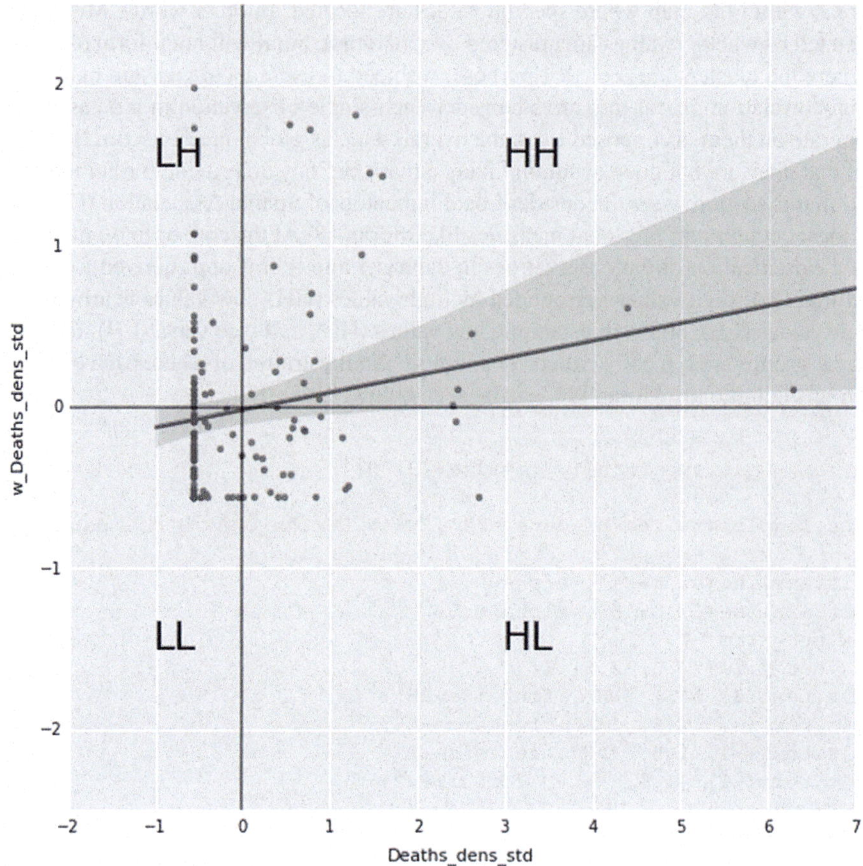

Fig. 17.6 Moran plot of cholera deaths with "quadrants"

LISAs are widely used in many fields to identify clusters of values in space. They are a very useful tool that can quickly return areas in which values are concentrated and provide suggestive evidence about the processes that might be at work. For that, they have a prime place in the exploratory toolbox. Examples of contexts where LISAs can be useful include: identification of spatial clusters of poverty in regions, detection of ethnic enclaves, delineation of areas of particularly high/low activity of any phenomenon, etc.

In Python, we can calculate LISAs in a very streamlined way thanks to PySAL:

```
lisa = ps.Moran_Local(js['Deaths_dens'].values, w)
```

All we need to pass is the variable of interest—density of deaths in this context—and the spatial weights that describes the neighborhood relations between the different observation that make up the dataset.

Because of their very nature, looking at the numerical result of LISAs is not always the most useful way to exploit all the information they can provide. Remember that we are calculating a statistic for every single observation in the data so, if we have many of them, it will be difficult to extract any meaningful pattern. Instead, what is typically done is to create a map, a cluster map as it is usually called, that extracts the significant observations (those that are highly unlikely to have come from pure chance) and plots them with a specific color depending on their quadrant category.

All of the needed pieces are contained inside the LISA object we have created above. But, to make the map making more straightforward, it is convenient to pull them out and insert them in the main data table, `js`:

```
# Break observations into significant or not
js['significant'] = lisa.p_sim < 0.05
# Store the quadrant they belong to
js['quadrant'] = lisa.q
```

Let us stop for second on these two steps. First, look at the significant column. Similarly as with global Moran's *I*, PySAL is automatically computing a *p*-value for each LISA. Because not every observation represents a statistically significant one, we want to identify those with a *p*-value small enough that to rule out the possibility of obtaining a similar situation from pure chance. Following a similar reasoning as with global Moran's *I*, we select 5% as the threshold for statistical significance. To identify these values, we create a variable, significant, that contains True if the *p*-value of the observation has satisfied the condition, and False otherwise. We can check this is the case:

```
js['significant'].head()
0     False
1     False
2     False
3     False
4      True
Name: significant, dtype: bool
```

And the first five *p*-values can be checked by:

```
lisa.p_sim[:5]
array([ 0.418,  0.085,  0.301,  0.467,  0.001])
```

Note how only the last one is smaller than 0.05, as the variable significant correctly identified.

The second column denotes the quadrant each observation belongs to. This one is easier as it comes built into the LISA object directly:

```
js['quadrant'].head()
0    3
1    3
```

```
2    3
3    3
4    4
Name: quadrant, dtype: int64
```

The correspondence between the numbers in the variable and the actual quadrants is as follows:

- 1: HH
- 2: LH
- 3: LL
- 4: HL

With these two elements, significant and quadrant, we can build a typical LISA cluster map.

```
# Setup the figure and axis
f, ax = plt.subplots(1, figsize=(9, 9))
# Plot building blocks
for poly in blocks['geometry']:
gpd.plotting.plot_multipolygon(ax, poly, facecolor='0.9')
# Plot baseline street network
for line in js['geometry']:
gpd.plotting.plot_multilinestring(ax, line, color='k')
# Plot HH clusters
hh = js.loc[(js['quadrant']==1) & (js['significant']==True),
        'geometry']
for line in hh:
gpd.plotting.plot_multilinestring(ax, line, color='red')
# Plot LL clusters
ll = js.loc[(js['quadrant']==3) & (js['significant']==True),
        'geometry']
for line in ll:
gpd.plotting.plot_multilinestring(ax, line, color='blue')
# Plot LH clusters
lh = js.loc[(js['quadrant']==2) & (js['significant']==True),
        'geometry']
for line in lh:
gpd.plotting.plot_multilinestring(ax, line, color='#83cef4')
# Plot HL clusters
hl = js.loc[(js['quadrant']==4) & (js['significant']==True),
        'geometry']
for line in hl:
gpd.plotting.plot_multilinestring(ax, line, color='#e59696')
#gpd.plotting.plot_multilinestring(ax, line, color='#e59696',
        linewidth=5)
# Plot pumps
xys = np.array([(pt.x, pt.y) for pt in pumps.geometry])
ax.scatter(xys[:, 0], xys[:, 1], marker='^', color='k', s=50)
# Style and draw
```

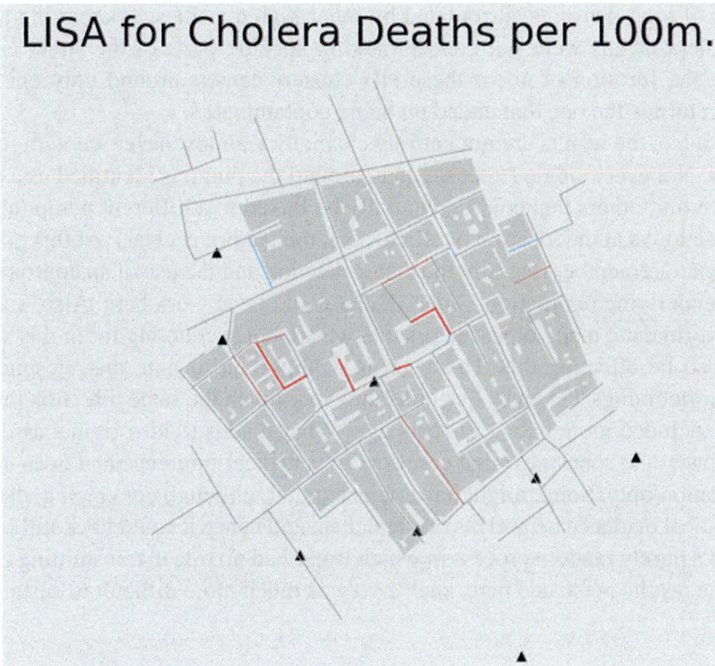

Fig. 17.7 LISA cluster map cholera deaths

```
f.suptitle('LISA for Cholera Deaths per 100m.', size=30)
f.set_facecolor('0.75')
ax.set_axis_off()
plt.axis('equal')
plt.show()
```

which yields Fig. 17.7.

Figure 17.7 displays the streets of the John Snow map of cholera and overlays on top of it the observations that have been identified by the LISA as clusters or spatial outliers. In bright red we find those street segments with an unusual concentration of high death density surrounded also by high death density. This corresponds with segments that are close to the contaminated pump, which is also displayed in the center of the map. In light red, we find the first type of spatial outliers. These are streets with high density but surrounded by low density. Finally, in light blue we find the other type of spatial outlier: streets with low densities surrounded by other streets with high density.

The substantive interpretation of a LISA map needs to relate its output to the original intention of the analyst who created the map. In this case, our original idea was to find support in the data for John Snow's thesis that cholera deaths were caused by a source that could be traced back to a contaminated water pump. The results seem to largely support this view. First, the LISA statistic identifies a few clusters of high densities surrounded by other high densities, discrediting the idea that cholera deaths were not concentrated in specific parts of the street network. Second, the location of all of these HH clusters centers around only one pump, which in turn is the one that ended up being contaminated.

Of course, the results are not entirely clean; they almost never are with real data analysis. Not every single street segment around the pump is identified as a cluster, while we find others that could potentially be linked to a different pump (although when one looks at the location of all clusters, the pattern is clear). At this point it is important to remember issues in the data collection and the use of an approximation for the underlying population. Some of that could be at work here. Also, since this is real world data, many other factors that we are not accounting for in this analysis could also be affecting this. However, it is important to note that, despite all of those shortcomings, the analysis points into very much the same direction that John Snow concluded more than 150 years ago. What it adds to his original assessment is the power and robustness that comes with statistical inference and does not with visualization only. Some might have objected that, although convincing, there was no statistical evidence behind his original map, and hence it could have still been the result of a purely random process in which water had no role in transmitting cholera. Upon the results presented here, such a view is much more difficult to sustain.

17.4 Concluding Remarks

This chapter deals with reproducibility and Open Science, specifically in the realm of regional science. The growing emphasis on geographically referenced data of increasing size and interest in quantitative approaches leads to an increasing need for training in workflow design and guidance in choosing appropriate tools. We argue that a proper workflow design has substantial benefits, including reproducibility (obviously) and efficiency. If it is possible to easily recreate the analysis and the resulting output in presentation or paper format, then slight changes induced by referees, supervisor or editors can be quickly processed. This is not only important in terms of time saving, but also in terms of accountability and transparency. In more practical terms, we illustrate the advocated approach by reproducing John Snow's famous cholera analysis from the nineteenth century, using a combination of R and Python code. The analysis includes contemporary spatial analytic methods, such as measuring global and local spatial autocorrelation measures.

In general, it is not so much the reproducible part but the openness part that some researchers find hard and counterintuitive to deal with. This is because the "publish or perish" ethos that dominates modern academic culture also rails against openness. Why open up all resources of your research so that others might benefit and scoop you in publishing first? A straightforward rebuttal to this would be: "Why publish then after all if you are hesitant to make all materials public?" And if you agree about this, why open up not only after the final phase when the paper has been accepted, but earlier in the research cycle? Some researchers are so extreme in this that they even share the writing of their research proposals with the outside world. Remember, with versioning control systems, such as Git, you can always prove, via timestamps, that you came up with the idea earlier then someone else.

Complete openness and thus complete reproducibility is often not feasible in the social sciences. Data could be proprietary or privacy-protected and expert interviews or case studies are notoriously hard to reproduce. And sometimes, you do in fact face cutthroat competition to get your research proposal rewarded or paper accepted. However, opening up your research, whether in an early, late or final phase definitely can reward you with large benefits. Mostly, because your research becomes more visible and is thus recognized earlier and credited. However, and most importantly, the scientific community most likely benefits the most as results, procedures, code and data are disseminated faster, more efficiently and with a much wider scope. As Rey (2009) has argued, free revealing of information can lead to increased private gains for the scientist as well as enhancing scientific knowledge production.

References

Arribas-Bel D (2016) Geographic data science'15. http://darribas.org/gds15

Arribas-Bel D, de Graaff T (2015) Woow-ii: workshop on open workflows. Region 2(2):1–2

BusinessDictionary (2016) Workflow [Online; accessed 15-June-2016]. http://www.businessdictionary.com/definition/workflow.html

Case A, Deaton A (2015) Rising morbidity and mortality in midlife among white non-hispanic americans in the 21st century. Proc Natl Acad Sci 112(49):15078–15083

Gandrud C (2013) Reproducible research with R and R studio. CRC, Boca Raton, FL

Healy K (2011) Choosing your workflow applications. Pol Methodologist 18(2):9–18

Hempel S (2006) The medical detective: John Snow and the mystery of cholera. Granta, London

Perez F (2015) Ipython: from interactive computing to computational narratives. In: 2015 AAAS Annual Meeting (12–16 February 2015)

Rey SJ (2009) Show me the code: spatial analysis and open source. J Geogr Syst 11:191–207

Rey SJ (2014) Open regional science. Ann Reg Sci 52(3):825–837

Stodden V, Leisch F, Peng RD (2014) Implementing reproducible research. CRC, Boca Raton, FL

Daniel Arribas-Bel is a Lecturer in Geographic Data Science at the University of Liverpool. He has held positions as Lecturer in Human Geography at the University of Birmingham, postdoctoral researcher at the Department of Spatial Economics at the VU University (Amsterdam), and postdoctoral researcher at the GeoDa Center for Geospatial Analysis and Computation at Arizona State University. Trained as an economist, Dani is interested in the spatial structure of cities and in the quantitative and computational methods required to leverage the power of the large amount of urban data increasingly becoming available. He is also part of the team of core developers of PySAL, the open-source library written in Python for spatial analysis.

Thomas de Graaff is assistant professor at the Department of Spatial Economics, Free University Amsterdam. His primary research interests are spatial interactions between households and firms; spatial econometrics; migration patterns; regional performance; and reproducibility of scientific research. Previous positions were at the Netherlands Bureau of Economic Policy Analysis (CPB) and the Netherland Environmental Assessment Agency (PBL). Dr. De Graaff earned the Ph.D. in economics from the Department of Spatial Economics at the Free University Amsterdam in 2002.

Sergio Rey is professor, School of Geographical Sciences and Urban Planning, Arizona State University (ASU). His research interests focus on the development, implementation, and application of advanced methods of spatial and space-time data analysis. His substantive foci include regional inequality, convergence and growth dynamics as well as neighborhood change, segregation dynamics, spatial criminology and industrial networks. Previous faculty positions were at the Department of Geography, San Diego State University and a visiting professor at the Department of Economics, University of Queensland. Dr. Rey earned the Ph.D. in geography from the University of California Santa Babara in 1994.

Printed by Printforce, the Netherlands